Hydrogen Energy

AF788697

Bahman Zohuri

Hydrogen Energy

Challenges and Solutions for a Cleaner Future

Springer

Bahman Zohuri
Galaxy Advanced Engineering Inc.,
University of New Mexico, Department
of Electrical and Computer Engineering
Albuquerque, NM, USA

ISBN 978-3-030-06665-9 ISBN 978-3-319-93461-7 (eBook)
https://doi.org/10.1007/978-3-319-93461-7

© Springer International Publishing AG, part of Springer Nature 2019
Softcover re-print of the Hardcover 1st edition 2019
This work is subject to copyright. All rights are reserved by the Publisher, whether the whole or part of the material is concerned, specifically the rights of translation, reprinting, reuse of illustrations, recitation, broadcasting, reproduction on microfilms or in any other physical way, and transmission or information storage and retrieval, electronic adaptation, computer software, or by similar or dissimilar methodology now known or hereafter developed.
The use of general descriptive names, registered names, trademarks, service marks, etc. in this publication does not imply, even in the absence of a specific statement, that such names are exempt from the relevant protective laws and regulations and therefore free for general use.
The publisher, the authors, and the editors are safe to assume that the advice and information in this book are believed to be true and accurate at the date of publication. Neither the publisher nor the authors or the editors give a warranty, express or implied, with respect to the material contained herein or for any errors or omissions that may have been made. The publisher remains neutral with regard to jurisdictional claims in published maps and institutional affiliations.

This Springer imprint is published by the registered company Springer Nature Switzerland AG
The registered company address is: Gewerbestrasse 11, 6330 Cham, Switzerland

This book is dedicated to my grandson, Darius

Preface

Hydrogen is the simplest and most abundant element on Earth—the hydrogen atom consists of only one proton and one electron. Hydrogen can store and deliver usable energy, but it doesn't typically exist by itself in nature and must be produced from compounds that contain it, which include diverse, domestic resources. Currently, most hydrogen is produced from fossil fuels, specifically natural gas. Electricity—from the grid or from renewable sources such as biomass, geothermal, solar, or wind—is also currently used to produce hydrogen. In the longer-term, solar energy and biomass can be used more directly to generate hydrogen as new technologies make alternative production methods cost competitive.

Hydrogen is high in energy, yet an engine that burns pure hydrogen produces almost no pollution. The National Aeronautics and Space Administration (NASA) has used liquid hydrogen since the 1970s to propel the space shuttle and other rockets into orbit. Hydrogen fuel cells power the shuttle's electrical systems, producing a clean by-product—pure water, which the crew drinks.

A fuel cell combines hydrogen and oxygen to produce electricity, heat, and water. Fuel cells are often compared to batteries as both convert the energy produced by a chemical reaction into usable electric power. However, the fuel cell will produce electricity as long as fuel (hydrogen) is supplied, never losing its charge.

In this book, I have tried to present in each chapter how hydrogen benefits our day-to-day lives and how this clean source of energy also plays a big role when it comes to being a new renewable source of energy as well as helping to enhance the decarbonization of our environment, given the fact the population is growing at a fast pace around the world.

Chapter 1 provides a basic understanding in the form of a fact sheet for the element hydrogen from both a chemical and physical properties point of view, including its characteristics and physical properties, uses, sources, and other data. Hydrogen is the lightest and most abundant element. A hydrogen atom consists of one proton and one electron; approximately 75% of the universe, but only a tiny fraction of the Earth, is comprised of hydrogen. Hydrogen is the oldest and cleanest element in the world and the first element on the periodic table. One of the most essential and powerful gases we use is hydrogen—in our cars, buses, space launches from the Cape, and so on.

Chapter 2 explains the current challenges faced by energy utilities in meeting the increasing demand for electricity as well as the need to immediately address environmental concerns such as climate change and decrease the pollution produced by transportation vehicles burning gasoline, for example. Hydrogen as a new—yet clean—source of energy can meet this demand. Due to population and economic growth, the global demand for energy is expected to increase by 50% over the next 25 years. This significant increase in demand along with the dwindling supply of fossil fuels has raised concerns over the security of the energy supply. In view of the increased energy demand and environmental pollution, different approaches such as distributed generation and demand-side management have been proposed and are widely being put into practice.

Chapter 3 describes the technical aspects of hydrogen-driven thermonuclear fusion energy from a high-level point of view to provide a general background for readers on this field of fusion physics both from the viewpoint of magnetic and inertial confinements of hydrogen elements in form of the its two isotopes of deuterium (D) and tritium (T). The thermonuclear fusion reactions in hydrogen isotopes are known to be a new source of energy in two widely different situations. At one extreme, relatively slow reactions in a very controlled, confined manner produces the energy emitted by the Sun and stars, whereas at the other extreme, rapid thermonuclear reactions are responsible for the strong thermonuclear power of the hydrogen bomb. Somewhere between these two extremes, it should be possible to bring about thermonuclear reactions under conditions that will allow the energy to be released from the hydrogen atoms at a controllable rate for electricity consumption and meet the need for a new, yet clean, source of energy.

Chapter 4 discusses what we have learned about the physics of cryogenics throughout history and its application in the storage of liquid hydrogen (LH_2). Cryogenics is the scientific field that addresses the production and effects of very low temperatures. The word originates from the Greek words *kryos* meaning "frost" and *genic* meaning "to produce." Using this definition, the term could be used to include all temperatures below the freezing point of water (0 °C).

Chapter 5 provides general information about hydrogen as the element driving the search for a renewable source of energy. It also shows that hydrogen can be found in many organic compounds as well as water—it is the most abundant element on Earth. However, it does not occur naturally as a gas; it is always combined with other elements, such as with oxygen to make water. Once separated from other elements, hydrogen can be burned as a fuel or converted into electricity. Fuel cells use the chemical energy of hydrogen or another fuel to cleanly and efficiently produce electricity—if hydrogen is the fuel, electricity, water, and heat are the only products. Fuel cells are unique in terms of the variety of their potential applications: they can provide power for systems as large as a utility power station and as small as a laptop computer.

Chapter 6 discusses nuclear hydrogen production plants, where hydrogen is an environmentally friendly energy carrier that, unlike electricity, can be stored in large quantities. It can be converted into electricity in fuel cells, with only heat and water as by-products. It is also compatible with combustion turbines and reciprocating engines to produce power with near-zero emission of pollutants. Therefore, hydrogen could play a major role in energy systems and serve all sectors of the economy, substituting for fossil fuels and helping mitigate global warming. The quest for better and cheaper production of this clean substance for consumption usage is an important task for engineers and scientists, who are working toward zero emissions and a decarbonized environment for the present and future generations.

Chapter 7 approaches the topic of large-scale production of hydrogen, so the demand by industry can be met for this, the simplest and most abundant element on Earth. The fast-paced growth of the need and demand for, and dependency on, raw materials such as hydrogen in today's industries is high on the list of political economy of most industrial countries around the globe. In particular, besides classical applications of hydrogen in industry, we have come to realize that it is a good source of renewable energy during on- and off-peak demand for electricity imposed on the grid by fast-growing industrial countries and their populations. However, the question of where the hydrogen comes from and how we can produce it remains. The "sustainable" routes are still too expensive. Steam reforming of hydrocarbons is considered to be the most feasible route today.

Chapter 8 considers the innovative approaches regarding hydrogen storage processes and technologies. Hydrogen storage is a significant challenge for the development and viability of hydrogen-powered vehicles. On-board hydrogen storage in the range of approximately 5–13 kg is required to enable a driving range of greater than 300 miles for the full platform of light-duty automotive vehicles using fuel cell power plants. In addition to production and distribution, costs are associated with hydrogen storage. Little public information is available on the cost of bulk gas storage, meaning a significant error margin exists for assumptions made on storage costs in this chapter. Hydrogen storage is a key enabling technology for the advancement of hydrogen and fuel cell technologies in applications such as stationary power, portable power, and transportation. Hydrogen has the highest energy per mass of any fuel; however, its low ambient temperature density results in a low energy per unit volume, therefore requiring the development of advanced storage methods that have the potential for higher energy density.

I hope the reader will enjoy this book as quick overview of this essential element of the Earth and its benefits to us as human beings.

Albuquerque, NM, USA
2016

Bahman Zohuri

Acknowledgment

I am indebted to the many people who aided, encouraged, and supported me beyond my expectations. Some are not around to see the results of their encouragement in the production of this book, yet I hope they know of my deepest appreciations. I especially want to thank my friends, to whom I am deeply indebted, who have continuously given their support without hesitation. These friends always kept me going in the right direction.

Above all, I offer very special thanks to my late mother and father, and to my children, in particular my son Sasha and grandson Darius. They have provided constant interest and encouragement, without which this book would not have been written. Their patience with my many absences from home and long hours in front of the computer to prepare the manuscript are especially appreciated.

Contents

About the Author

Bahman Zohuri currently works for Galaxy Advanced Engineering, Inc., a consulting firm that he started in 1991 when he left both the semiconductor and defense industries after many years working as a chief scientist. After graduating from the University of Illinois in the field of physics, applied mathematics, he then went to the University of New Mexico, where he studied nuclear engineering and mechanical engineering. He later joined Westinghouse Electric Corporation, where he performed thermal hydraulic analysis and studied natural circulation in an inherent shutdown, heat removal system (ISHRS) in the core of a liquid metal fast breeder reactor (LMFBR) as a secondary fully inherent shutdown system for secondary loop heat exchange. These designs were used in nuclear safety and reliability engineering for a self-actuated shutdown system. He designed a mercury heat pipe and electromagnetic pumps for large pool concepts of a LMFBR for heat rejection purposes for this reactor around 1978, when he received a patent for it. He was subsequently transferred to the defense division of Westinghouse, where he oversaw dynamic analysis and methods of launching and controlling MX missiles from canisters. The results were applied to MX launch seal performance and muzzle blast phenomena analysis (i.e., missile vibration and hydrodynamic shock formation). Dr. Zohuri was also involved in analytical calculations and computations in the study of non-linear ion waves in rarefying plasma, the results of which were applied to the propagation of so-called soliton waves and the resulting charge collector traces in the rarefaction characterization of the corona of laser-irradiated target pellets. As part of his graduate research work at Argonne National Laboratory, he performed computations and programming of multi-exchange integrals in surface physics and solid-state physics. He earned various patents in areas such as diffusion processes and diffusion furnace design while working as a senior process engineer at various semiconductor companies, such as Intel Corp., Varian Medical Systems, and National Semiconductor Corporation. He later joined Lockheed Martin Missile and Aerospace Corporation as Senior Chief Scientist and oversaw research and development (R&D) and the study of the vulnerability, survivability, and both radiation and laser hardening of different components of the Strategic Defense Initiative (SDI), known as Star Wars.

This included payloads (i.e., infrared sensors) for the Defense Support Program, the Boost Surveillance and Tracking System, and Space Surveillance and Tracking Satellite against laser and nuclear threats. While at Lockheed Martin, he also performed analyses of laser beam characteristics and nuclear radiation interactions with materials, transient radiation effects in electronics, electromagnetic pulses, system-generated electromagnetic pulses, single-event upset, blast, thermo-mechanical, hardness assurance, maintenance, and device technology.

Dr. Zohuri spent several years as a consultant at Galaxy Advanced Engineering serving Sandia National Laboratories, where he supported the development of operational hazard assessments for the Air Force Safety Center in collaboration with other researchers and third parties. Ultimately, the results were included in Air Force instructions issued specifically for directed energy weapons' operational safety. He completed the first version of a comprehensive library of detailed laser tools for airborne lasers, advanced tactical lasers, tactical high-energy lasers, and mobile/tactical high-energy lasers, for example.

He also oversaw SDI computer programs, in connection with Battle Management C^3I (command, control, communications, and intelligence) and artificial intelligence, and autonomous systems. He is the author of several publications and holds several patents, such as for a laser-activated radioactive decay and results of a through-bulkhead initiator. He has published the following works: *Heat Pipe Design and Technology: A Practical Approach* (CRC Press); *Dimensional Analysis and Self-Similarity Methods for Engineering and Scientists* (Springer); and *High Energy Laser (HEL): Tomorrow's Weapon in Directed Energy Weapons, Volume I* (Trafford Publishing Company). He has published other books with Springer Publishing Company: *Directed Energy Weapons: Physics of High Energy Lasers (HEL)*; *Thermodynamics in Nuclear Power Plant Systems*; and *Thermal–Hydraulic Analysis of Nuclear Reactors*.

Chapter 1
The Chemical Element Hydrogen

This chapter provides a basic understanding in the form of a fact sheet for the element hydrogen—the oldest and cleanest element in the world—including its characteristics and physical properties, uses, sources, and other data. Hydrogen is the lightest and most abundant element in the universe; approximately 75% of the universe, but only a tiny fraction of the Earth, is comprised of hydrogen. A hydrogen atom consists of one proton and one electron and it is the first element on the periodic table. One of the most powerful gases we use is hydrogen—in our cars, buses, space launches from the Cape, and so on.

1.1 Introduction

Hydrogen is a chemical element with the symbol H and atomic number 1. With a standard atomic weight of 1.008, hydrogen is the lightest element on the periodic table. Its monatomic form (H) is the most abundant chemical substance in the universe, constituting roughly 75% of all baryonic mass. Non-remnant stars are mainly composed of hydrogen in the plasma state. The most common isotope of hydrogen, called protium (a name that is rarely used; symbol ^1H), has one proton and no neutrons (see Figs. 1.1 and 1.2).

By definition in physical cosmology science, "recombination" refers to the epoch (i.e., chronology of the universe, as shown in Fig. 1.3) in which charged electrons and protons first became bound to form electrically neutral hydrogen atoms [1]. Recombination occurred about 378,000 years after the Big Bang (at a redshift of $z = 1100$).

The word "recombination" is misleading, since the Big Bang theory doesn't posit that protons and electrons had been combined before; however, the term exists for historical reasons as it was coined before the Big Bang hypothesis became the primary theory of the creation of the universe (see Fig. 1.4).

© Springer International Publishing AG, part of Springer Nature 2019

B. Zohuri, *Hydrogen Energy*, https://doi.org/10.1007/978-3-319-93461-7_1

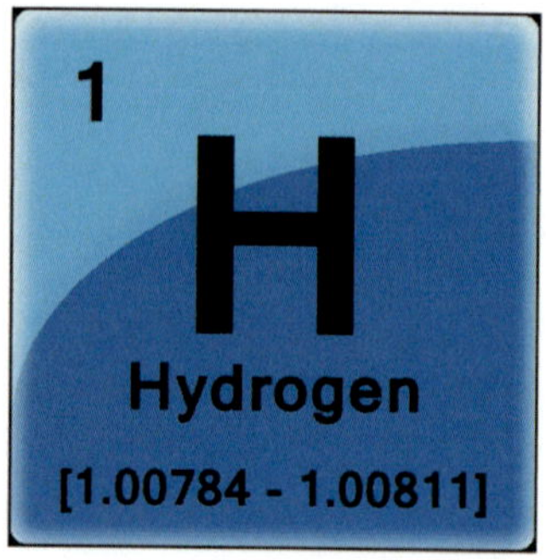

Fig. 1.1 Hydrogen atom in the chemical periodic table

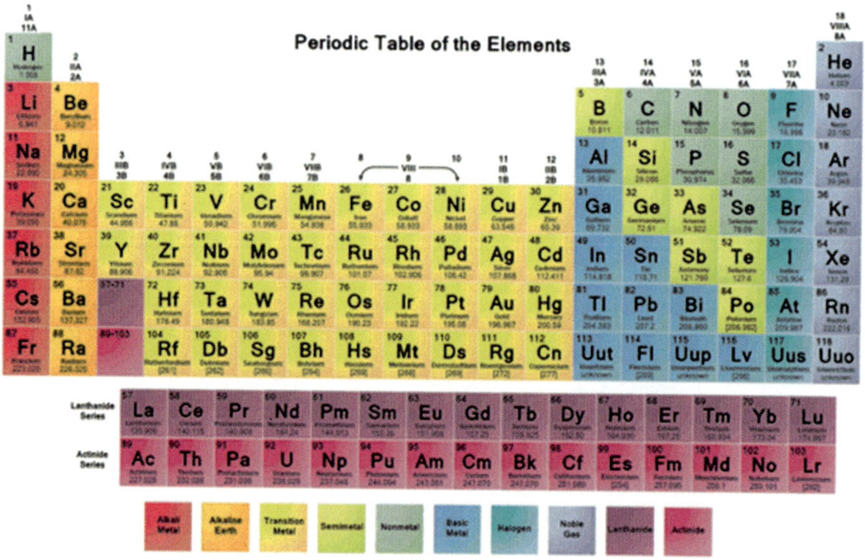

Fig. 1.2 The periodic table

Immediately after the Big Bang, the universe was a hot, dense plasma of photons, electrons, and quarks: the Quark epoch. At 10^{-6} s, the universe had expanded and cooled sufficiently to allow the formation of protons: the Hadron epoch. This plasma was effectively opaque to electromagnetic radiation due to Thomson scattering by free electrons, as the mean free path each photon could travel before encountering an electron was very short. This is the current state of the interior of the Sun. As the universe expanded, it also cooled. Eventually, this cooling of the universe occurred to the point that the formation of neutral hydrogen was energetically favored, and the fraction of free electrons and protons as compared with neutral hydrogen decreased to a few parts in 10,000 [1].

Recombination involves electrons binding to protons (hydrogen nuclei) to form neutral hydrogen atoms. Because direct recombination to the ground state (lowest energy) of hydrogen is very inefficient, these hydrogen atoms generally form with

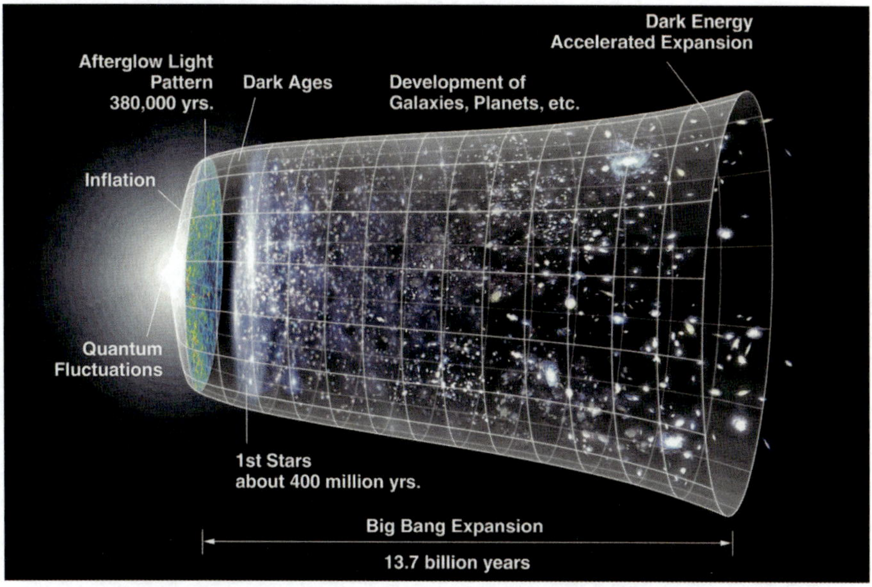

Fig. 1.3 Evolution of the (observable part) of the universe from the Big Bang (left) to the present

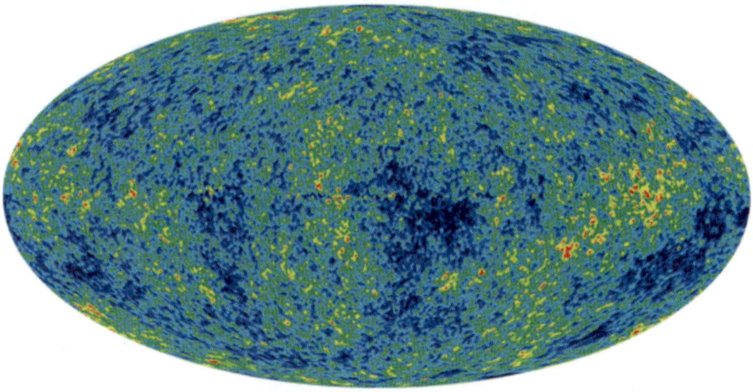

Fig. 1.4 The primary theory of the creation of the universe

the electrons in a high-energy state, and the electrons quickly transition to their low-energy state by emitting photons. Two main pathways exist: either from the 2p state by emitting a Lyman-α photon—these photons will almost always be reabsorbed by another hydrogen atom in its ground state—or from the 2s state by emitting two photons, which is very slow [1].

This production of photons is known as decoupling, which leads to recombination sometimes being called photon decoupling, but recombination and photon decoupling are distinct events. Once photons decoupled from matter, they traveled freely through the universe without interacting with matter and constitute what is

Table 1.1 Physical cosmology

Early universe
Expansion: future
Components: structure
Subject history

Table 1.2 Essential hydrogen facts

Element name	Hydrogen
Element symbol	H
Element number	1
Element category	Non-metal
Atomic weight	1.0079
Electron configuration	1S

observed today as cosmic microwave background radiation (in that sense, the cosmic background radiation is infrared black-body radiation emitted when the universe was at a temperature of some 4000 K, redshifted by a factor of 1100 from the visible spectrum to the microwave spectrum) [1].

At standard temperature and pressure, hydrogen is a colorless, odorless, tasteless, non-toxic, non-metallic, highly combustible diatomic gas with the molecular formula H_2. Since hydrogen readily forms covalent compounds with most non-metallic elements, most of the hydrogen on Earth exists in molecular forms such as water or organic compounds. Hydrogen plays a particularly important role in acid–base reactions because most of these involve the exchange of protons between soluble molecules. In ionic compounds, hydrogen can take the form of a negative charge (i.e., anion) when it is known as a hydride, or as a positively charged (i.e., cation) species denoted by the symbol H+. The hydrogen cation is written as though composed of a bare proton, but, in reality, hydrogen cations in ionic compounds are always more complex. As the only neutral atom for which the Schrödinger equation can be solved analytically, study of the energetics and bonding of the hydrogen atom has played a key role in the development of quantum mechanics.

The physical cosmology, as illustrated in Fig. 1.4, is shown in Table 1.1.

Table 1.2 describes the essential facts and chemical properties of hydrogen.

The element of hydrogen was discovered by Henry Cavendish in 1766, hydrogen having been produced for many years before it was recognized as a distinct element. The origin of the word hydrogen is from the Greek words *hydro*, meaning "water", and *gens*, meaning "forming." The element was given the name hydrogen by Antoine Lavoisier in 1783.

The physical properties of hydrogen are presented in Table 1.3 (for a vial containing ultrapure hydrogen gas). Hydrogen is a colorless gas that glows violet when it is ionized, as shown in Fig. 1.5.

Table 1.3 Physical and additional properties of hydrogen [2]

Phase at STP	Gas
Color	Colorless
Density	0.08988 g/L (0 °C, 101.325 kPa)
Melting point	14.01 K, −259.14 °C (−423.45 °F)
Boiling point	20.28 K, −252.87 °C
Triple point	13.8033 K (−259 °C), 7.042 kPa
Critical point	32.97 K, 1.293 Mpa
Heat of fusion	(H^2) 0.117 $kJ.mol^{-1}$
Heat of vaporization	(H^2) 0.904 $kJ.mol^{-1}$
Molar heat capacity	(H^2) 28.836 $J.mol^{-1}$ K^{-1}
Ground level	$2S_{1/2}$
Ionization potential	13.5984 eV
Specific heat	14.304 J/g K
Oxidation states	1, −1
Electronegativity	2.20 (Pauling scale)
Ionization energies	1st: 1312.0 $kJ.mol^{-1}$
Covalent radius	31 ± 5 pm
Van der Waals radius	120 pm
Crystal structure	Hexagonal
Magnetic ordering	Diamagnetic
Thermal conductivity	0.1805 $W.m^{-1}.K^{-1}$
Speed of sound (gas, 27 °C)	1310 m/s
CAS registry number	1333–74–0

CAS Chemical Abstracts Service, *STP* standard temperature and pressure

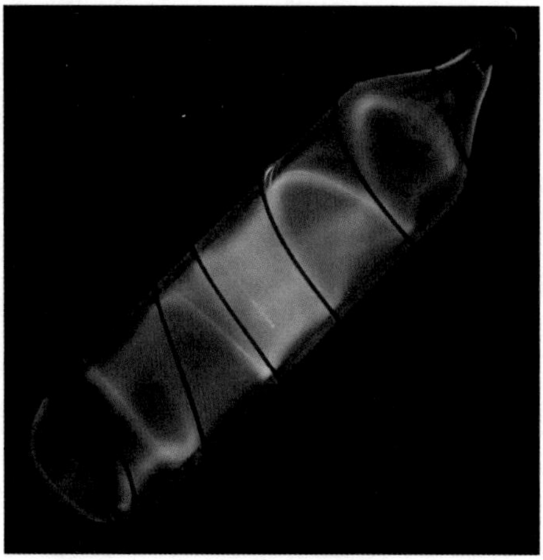

Fig. 1.5 Ionized hydrogen

1.2 The History of Hydrogen

Hydrogen is one of the most powerful gases we use—in our cars, buses, space launches from the Cape, and so on. Hydrogen has been found to be the best, oldest, and cleanest element, and is the first substance of the chemical periodic table (see Fig. 1.2).

Today, we are studying fuels for the future. But in science we must study the past in order to make today's world of hydrogen fuel and hydrogen fuel cells plausible [3].

Therefore, let us look at the history of this substance that is known as hydrogen.

1776 Hydrogen was first identified as a distinct element by British scientist Henry Cavendish after he produced hydrogen gas as a result of the reaction between zinc metal and hydrochloric acid. In a demonstration to the Royal Society of London, Cavendish applied a spark to hydrogen gas, yielding water. This discovery led to his later finding that water (H_2O) is made of hydrogen and oxygen.

1788 Building on the discoveries of Cavendish, French chemist Antoine Lavoisier gave hydrogen its name, which was derived from the Greek words *hydro* and *genes*, meaning "water" and "born of."

1800 English scientists William Nicholson and Sir Anthony Carlisle discovered that applying an electric current to water produced hydrogen and oxygen gases. This process was later termed "electrolysis."

1838 The fuel cell effect, combining hydrogen and oxygen gases to produce water and an electric current, was discovered by Swiss chemist Christian Friedrich Schoenbein.

1845 Sir William Grove, an English scientist and judge, demonstrated Schoenbein's discovery on a practical scale by creating a "gas battery." He earned the title "Father of the Fuel Cell" for his achievement.

1874 Jules Verne, an English author, prophetically examined the potential use of hydrogen as a fuel in his popular work of fiction entitled *The Mysterious Island*.

1889 Ludwig Mond and Charles Langer attempted to build the first fuel cell device using air and industrial coal gas. They named the device a fuel cell.

1920s A German engineer, Rudolf Erren, converted the internal combustion engines of trucks, buses, and submarines to use hydrogen or hydrogen mixtures. British scientist and Marxist writer J.B.S. Haldane introduced the concept of renewable hydrogen in his book *Daedalus or Science and the Future* by proposing that "there will be great power stations where during windy weather the surplus power will be used for the electrolytic decomposition of water into oxygen and hydrogen".

1937 After ten successful trans-Atlantic flights from Germany to the United States, the Hindenburg, a dirigible inflated with hydrogen gas, crashed upon landing in Lakewood, New Jersey, USA. The mystery of the crash was solved in 1997 when a study concluded that the explosion was not due to the hydrogen gas, but rather to a weather-related static electric discharge which ignited the airship's silver-colored canvas exterior covering, which had been treated with the key ingredients of solid rocket fuel.

1958 The United States formed the National Aeronautics and Space Administration (NASA). NASA's space program currently uses the most liquid hydrogen world-wide, primarily for rocket propulsion and as a fuel for fuel cells.

1959 Francis T. Bacon of Cambridge University in England built the first practical hydrogen–air fuel cell. The 5-kilowatt (kW) system powered a welding machine. He named his fuel cell design the "Bacon Cell." Later that year, Harry Karl Ihrig, an engineer for the Allis–Chalmers Manufacturing Company, demonstrated the first fuel cell vehicle: a 20-horsepower tractor. Hydrogen fuel cells, based on Bacon's design, have been used to generate onboard electricity, heat, and water for astronauts aboard the famous Apollo spacecraft and all subsequent space shuttle missions.

1970 Electrochemist John O'Mara Bockris coined the term "hydrogen economy" during a discussion at the General Motors (GM) Technical Center in Warren, Michigan, USA. He later published *Energy: The Solar-Hydrogen Alternative*, describing his envisioned hydrogen economy where cities in the United States could be supplied with energy derived from the Sun.

1972 The 1972 Gremlin, modified by the University of California at Los Angeles, was entered the 1972 Urban Vehicle Design Competition and won first prize for the lowest tailpipe emissions. Students converted the Gremlin's internal combustion engine to run on hydrogen supplied from an onboard tank.

1973 The OPEC (Organization of the Petroleum Exporting Countries) oil embargo and the resulting supply shock suggested that the era of cheap petroleum had ended, and that the world needed alternative fuels. The development of hydrogen fuel cells for conventional commercial applications began.

1974 The National Science Foundation transferred the Federal Hydrogen R&D Program to the United States Department of Energy. Professor T. Nejat Veziroglu of the University of Miami, Florida, organized The Hydrogen Economy Miami Energy (THEME) conference, the first international conference held to discuss hydrogen energy. Following the conference, the scientists and engineers who had attended it formed the International Association for Hydrogen Energy (IAHE).

1974 The International Energy Agency (IEA) was established in response to global oil market disruptions. IEA activities included the research and development of hydrogen energy technologies.

1988 The Soviet Union Tupolev Design Bureau successfully converted a 164-passenger TU-154 commercial jet to operate one of the jet's three engines on liquid hydrogen. The maiden flight lasted 21 min.

1989 The National Hydrogen Association (NHA) formed in the United States with ten members. Today, the NHA has nearly 100 members, including representatives from the automobile and aerospace industries, federal, state, and local governments, and energy providers. The International Organization for Standardization's Technical Committee for Hydrogen Technologies was also created.

1990 The world's first solar-powered hydrogen production plant at Solar-Wasserstoff-Bayern, a research and testing facility in southern Germany, became operational. The United States Congress passed the Spark M. Matsunaga Hydrogen, Research, Development and Demonstration Act (PL 101–566), which prescribed the formulation of a 5-year management and implementation plan for hydrogen research and development in the United States.

The Hydrogen Technical Advisory Panel (HTAP) was mandated by the Matsunaga Act to ensure consultation on and coordination of hydrogen research. Work on a methanol-fueled 10-kW proton exchange membrane (PEM) fuel cell began through a partnership including GM, Los Alamos National Laboratory, the Dow Chemical Company, and Canadian fuel cell developer, Ballard Power Systems.

1994 Daimler-Benz demonstrated its first New Electric CAR (NECAR-I) fuel cell vehicle at a press conference in Ulm, Germany.

1997 Retired NASA engineer Addison Bain challenged the belief that hydrogen caused the Hindenburg accident. The hydrogen, Bain demonstrated, did not cause the catastrophic fire but rather the combination of static electricity and highly flammable material on the skin of the airship. German car manufacturer Daimler-Benz and Ballard Power Systems announced a US$300 million research collaboration on hydrogen fuel cells for transportation.

1998 Iceland unveiled a plan to create the first hydrogen economy by 2030 with Daimler-Benz and Ballard Power Systems.

1999 The Royal Dutch/Shell Company committed to a hydrogen future by forming a hydrogen division. Europe's first hydrogen fueling stations were opened in the German cities of Hamburg and Munich.

A consortium of Icelandic institutions, headed by the financial group New Business Venture Fund, partnered with Royal Dutch/Shell Group, Daimler Chrysler (a merger of Daimler-Benz and Chrysler), and Norsk Hydro to form the Icelandic Hydrogen and Fuel Cell Company, Ltd. to further the hydrogen economy in Iceland.

2000 Ballard Power Systems presented the world's first production-ready PEM fuel cell for automotive applications at the Detroit Auto Show.

2003 President George W. Bush announced in his 2003 State of the Union Address a US$1.2 billion hydrogen fuel initiative to develop the technology for commercially viable hydrogen-powered fuel cells, such that "the first car driven by a child born today could be powered by fuel cells".

2004 United States Energy Secretary Spencer Abraham announced that over US $350 million would be devoted to hydrogen research and vehicle demonstration projects. This appropriation represented nearly one-third of President Bush's US $1.2 billion commitment to research in hydrogen and fuel cell technologies. The funding encompasses over 30 lead organizations and more than 100 partners selected through a competitive review process.

2004 The world's first fuel cell-powered submarine underwent deep-water trials (German navy).

2005 Twenty-three states in the United States have hydrogen initiatives in place.

Today–2050: Future Vision
In the future, water will replace fossil fuels as the primary resource for hydrogen. Hydrogen will be distributed via national networks of hydrogen transport pipelines and fueling stations. Hydrogen energy and fuel cell power will be clean, abundant, reliable, affordable, and an integral part of all sectors of the economy in all regions of the United States.

1.3 Summary

Thus, in summary, in the early 1500s the alchemist Paracelsus noted that the bubbles given off when iron filings were added to sulfuric acid were flammable. In 1671, Robert Boyle made the same observation. Neither followed up their discovery of hydrogen, and so Henry Cavendish gets the credit. In 1766 he collected these bubbles and showed that they were different from other gases. He later showed that when hydrogen burns it forms water, thereby ending the belief that water was an element. The gas was given its name hydrogen, meaning water-former, by Antoine Lavoisier.

In 1931, Harold Urey and his colleagues at Columbia University in the United States detected a second, rarer, form of hydrogen. This has twice the mass of normal hydrogen, and they named it deuterium.

1.4 Hydrogen Sources

In nature, hydrogen can be found in volcanic gases and some natural gases as a free element, as illustrated in Fig. 1.6, which shows the volcanic eruption of Stromboli Volcano (a small island in the Tyrrhenian Sea, off the cost of Sicily in Italy), the most recent major eruption of which took place as recently as April 13, 2009.

Hydrogen is created by decomposition of hydrocarbons via heat, the action of sodium hydroxide or potassium hydroxide on aluminum electrolysis of water, steam on heated carbon, or displacement from acids by metals.

As mentioned earlier in this chapter, hydrogen is the most abundant element in the universe (Fig. 1.7). The heavier elements were formed from hydrogen or other elements that were made from hydrogen. Although approximately 75% of the universe's elemental mass is hydrogen, the element is relatively rare on Earth.

Figure 1.7, an image taken by the Hubble Space Telescope, shows a region of ionized hydrogen in the Triangulum Galaxy.

Fig. 1.6 Volcanic eruption of Stromboli Volcano, Italy

Fig. 1.7 Ionized hydrogen within the Triangulum Galaxy

1.5 Hydrogen Isotopes

The three most stable isotypes of hydrogen occur naturally, and each have their own names (Fig. 1.8):

1. Protium (0 Neutron, Atomic Number $A = 1$);
2. Deuterium (1 Neutron, Atomic Number $A = 2$); and
3. Tritium (2 Neutrons, Atomic Number $A = 3$).

In fact, hydrogen is the only element with names for its common isotopes. Protium is the most abundant hydrogen isotope (see Fig. 1.9). H_4 to H_7 are extremely unstable isotopes that have been made in the laboratory but are not seen in nature. Both protium and deuterium are not radioactive. Tritium, however, decays into helium-3 through beta decay.

Protium is the most common isotope of the element hydrogen. It has one proton and one electron but no neutrons.

Fig. 1.8 The three most stable isotopes of hydrogen: protium ($A = 1$), deuterium ($A = 2$), and tritium ($A = 3$). A atomic number

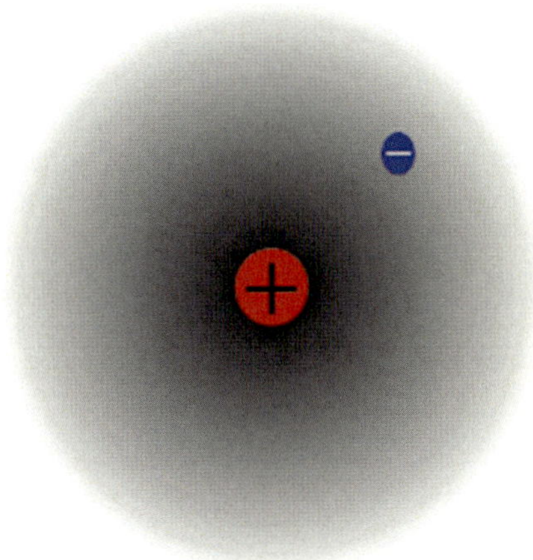

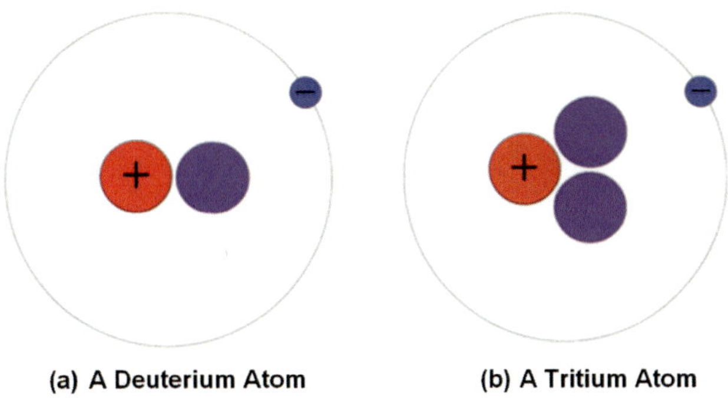

Fig. 1.9 Protium: hydrogen isotope

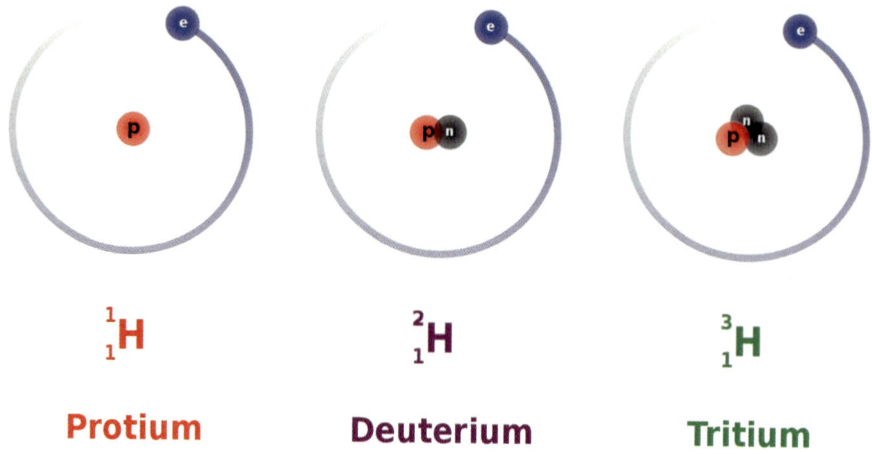

Fig. 1.10 Hydrogen isotopes of deuterium (**a**) and tritium (**b**)

Hydrogen is the only element whose isotopes have different names that are in common use today. The H_2 (or hydrogen-2) isotope is usually called deuterium (D, contains one proton, one neutron, and one electron; Fig. 1.10a), while the H_3 (or hydrogen-3) isotope is usually called tritium (T, contains one proton, two neutrons, and one electron; Fig. 1.10b). The symbols D and T (instead of H_2 and H_3) are sometimes used for deuterium and tritium, and it should be noted that the International Union of Pure and Applied Chemistry (IUPAC) states in the 2005 *Red Book* that the use of D and T is very common.

As mentioned earlier, H_2 (atomic mass 2.01410177811(12) u), the other stable hydrogen isotope, is known as deuterium and contains one proton and one neutron in its nucleus. The nucleus of deuterium is called a deuteron. Deuterium comprises 0.0026–0.0184% (by population, not by mass) of hydrogen samples on Earth, with the lower number tending to be found in samples of hydrogen gas and the higher enrichment (0.015% or 150 ppm) typical of ocean water. Deuterium on Earth has been enriched with respect to its initial concentration in the Big Bang and the outer solar system (about 27 ppm, by atom fraction) and its concentration in older parts of the Milky Way galaxy (about 23 ppm). Presumably, the differential concentration of deuterium in the inner solar system is due to the lower volatility of deuterium gas and compounds, enriching deuterium fractions in comets and planets exposed to significant heat from the Sun over billions of years of solar system evolution [4].

Deuterium is not radioactive and does not represent a significant toxicity hazard. Water enriched in molecules that include deuterium instead of protium is called heavy water. Deuterium and its compounds are used as a non-radioactive label in chemical experiments and in solvents for 1H-NMR (hydrogen-1 nuclear magnetic resonance) spectroscopy. Heavy water is used as a neutron moderator and coolant for nuclear reactors. Deuterium is also a potential fuel for commercial nuclear fusion [4].

H_3 (atomic mass 3.01604928199(23) u) is known as tritium and contains one proton and two neutrons in its nucleus. It is radioactive, decaying into helium-3

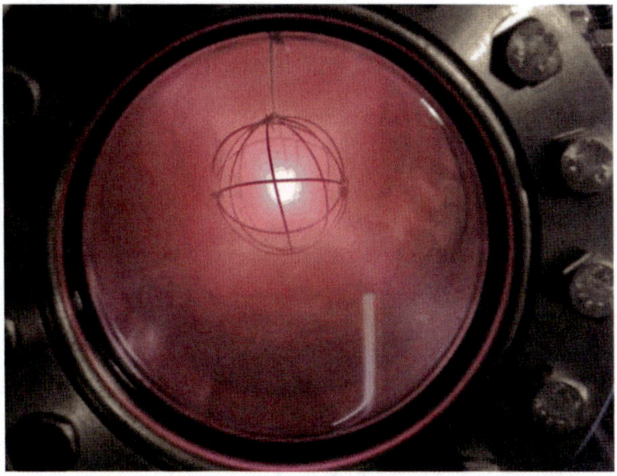

Fig. 1.11 Ionized deuterium in an Inertial Electrostatic Confinement (IEC) reactor

through beta decay with a half-life of 12.32 years [4]. Trace amounts of tritium occur naturally because of the interaction of cosmic rays with atmospheric gases. Tritium has also been released during nuclear weapons tests. It is used in thermonuclear fusion weapons, as a tracer in isotope geochemistry, and in specialized self-powered lighting devices.

The most common method of producing tritium is by bombarding a natural isotope of lithium, lithium-6, with neutrons in a nuclear reactor.

Tritium was once used routinely in chemical and biological labeling experiments as a radiolabel, which has become less common in recent times. D–T nuclear fusion uses tritium as its main reactant, along with deuterium, liberating energy through the loss of mass when the two nuclei collide and fuse at high temperatures.

Figure 1.11 demonstrates ionized deuterium in an Inertial Electrostatic Confinement (IEC); the characteristic pink or reddish glow displayed by ionized deuterium can be seen.

Note that IEC is a branch of fusion reaction research that uses an electric field to heat plasma to fusion conditions. Electric fields can work on charged particles (either ions or electrons), heating them to fusion conditions. This is typically done in a sphere, with material moving radially inward, but can also be done in a cylindrical or beam geometry. The electric field can be generated using a wire grid or a non-neutral plasma cloud [5].

Some other facts about hydrogen can be summarized here:

- Hydrogen is the lightest element; hydrogen gas is so light and diffusive that uncombined hydrogen can escape from the atmosphere.
- Hydrogen gas is a mixture of two molecular forms, ortho- and para-hydrogen, which differ by the spins of their electrons and nuclei. Normal hydrogen at room temperature consists of 25% para-hydrogen and 75% ortho-hydrogen. The ortho form cannot be prepared in the pure state. The two forms of hydrogen differ in energy, so their physical properties also differ.

- Hydrogen gas is extremely flammable.
- Hydrogen can take a negative charge (H^-) or a positive charge (H^+) in compounds.
- Hydrogen compounds are called hydrides.
- Ionized deuterium displays a characteristic reddish or pink glow.

1.6 Uses for Hydrogen

At first glance, hydrogen is the simplest element known to exist. A hydrogen atom has one proton and one electron. Hydrogen has the highest energy content of any common fuel by weight, but the lowest energy content by volume. It is the lightest element and a gas at normal temperature and pressure. Hydrogen is also the most abundant gas in the universe, and the source of all of the energy we receive from the Sun, and is also one of the most plentiful elements in the Earth's crust. Hydrogen as a gas (H_2), however, does not exist naturally on Earth; it is found only in compound form. Combined with oxygen, it forms water (H_2O). Combined with carbon, it forms organic compounds such as methane (CH^4), coal, and petroleum. It is found in all growing things—biomass.

Hydrogen is one of the most promising energy carriers for the future. It is a high efficiency, low polluting fuel that can be used for transportation, heating, and power generation in places where it is difficult to use electricity.

Hydrogen is a very important molecule with an enormous breadth and extent of application and use. It is currently being used in many industries, from chemical and refining to metallurgical, glass, and electronics. Hydrogen is primarily used as a reactant. But it is also being used as a fuel in space applications, as an "O_2 scavenger" in heat treating of metals, and for its low viscosity and density. Current uses of hydrogen in various industries throughout the world can be seen as, due to the increased use of heavier crude oils containing higher amounts of sulfur and nitrogen and to meet stringent emission standards, the need for hydrogen is experiencing very rapid growth in the petroleum refining industry. Hence, this application is discussed in more detail in the following sections and chapters.

1.6.1 Zeppelins and Airships

Historically, hydrogen was used during World War I (WWI) as a lifting substance for Zeppelins, with Germany leading such technology to transport passengers between Germany and the United States around the 1930s as well as these Zeppelins being used as means of weaponry war machine. The most famous passenger transport Zeppelin was called the Hindenburg—it eventually burned and crashed on May 6, 1937 at Lakehurst, New Jersey, USA (Fig. 1.12).

Fig. 1.12 Hindenburg disaster at Lakehurst, New Jersey, USA

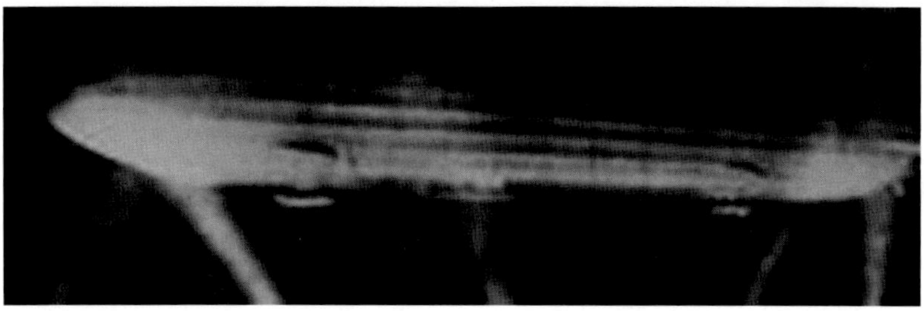

Fig. 1.13 Zeppelin raid over London in 1915

The idea of building these blimps originated with Count von Zeppelin, a retired German army officer, who created a flying weapon lighter than air, filled with hydrogen, and held together by a steel framework.

When WWI started in 1914, the German armed forces had several Zeppelins, each capable of travelling at about 85 mph and carrying up to 2 tons of bombs. With military deadlock on the Western Front, the Germans decided to use them against towns and cities in Britain. The first raid took place on the eastern coastal towns of Great Yarmouth and King's Lynn on January19, 1915, during which they were photographed over London (Fig. 1.13). Residents reported hearing an eerie throbbing sound above them, followed shortly afterwards by the sound of explosions in the streets. This was the first time that the full horror of aerial warfare was unleashed. Although people in England claimed that aerial attacks using Zeppelins held no military advantage over other aircraft of the era, it was all about instilling terror—and that is what these aerial bombardments did.

> The Zeppelins would come out of the dark - you could not see them, and it was totally random. You didn't know if you were running towards danger or away from its.

The aim of the Zeppelins was clear—the Germans hoped to destroy morale at home and force the British Government into abandoning the war in the trenches, but the sort of chaos and panic that the Germans had wanted was not created.

However, in 20 min a Zeppelin had dropped 3000 pounds of bombs, including 91 incendiaries, which had started 40 fires, gutted buildings, and left seven people dead. Not a single shot was fired in retaliation.

While Britain celebrated victories elsewhere, the Germans stepped things up with the so-called Super Zeppelins, but Britain had found the Zeppelin's Achilles heel—explosive bullets that could set alight the hydrogen that was floating them in the air. This would prove to be the Zeppelin's undoing.

Although the Zeppelin was embraced by both the Germans and the Allies during WWI, the Germans made far more extensive use of the rigid, hydrogen-filled airships. The concept of "strategic bombing"—targeted airstrikes on a particular location—didn't exist before the conflict. The advent of aerial warfare changed that, and also robbed the British of the protection afforded by the English Channel. The Zeppelin allowed Germany to bring the war to the English homeland.

During their brief, but deadly, dominance, the airships killed more than 500 people and injured more than 1000 all down eastern Britain. The last ever attempt to bomb Britain by a Zeppelin was over the Norfolk coast on August 5, 1918. Three years earlier, when a Zeppelin first appeared in the skies above Great Yarmouth, it was an invincible force, but now they were outclassed and dealt with swiftly. However, the Zeppelin had exposed the fact that those at home were now as vulnerable as those on the front line. The government became acutely aware they needed a comprehensive aerial defense system.

This led to the formation of the Royal Air Force (RAF) in 1918 and to the development of operations rooms such as the one at Duxford that proved so crucial in 1940 during the Battle of Britain and ultimately to victory in World War II (WWII).

The airships of the era were in some ways more capable than fixed-wing aircraft. They could fly higher, and farther, with greater payloads. But aerial raids were tricky. The large and slow airships flew at night and at high altitude to avoid being hit by artillery. That, however, made it hard for them to see their targets, and—given that this was the dawn of aerial bombardment—there was more than a little guesswork involved in knowing just how many bombs were needed to destroy whatever they were aiming at.

All these things aside, the airships were very successful for at least one thing: scaring the hell out of people. "They did more damage keeping people awake than actual physical damage," says Jeffery S. Underwood, a historian at the United States Air Force Museum at Wright-Patterson Air Force Base in Ohio. Zeppelins "scared the living daylights" out of the British. The Germans believed that bombing civilians would bring panic in the cities, leading the British government to collapse—or at least pull out of the fighting and leave the French on their own on the Western Front.

Zeppelins were also used for surveillance. Both sides used them to spot submarines, which were nearly invisible to ships but relatively easily seen from the air.

And airships were exceptionally useful for fleet maneuvers, carrying radios that could convey information to commanders on the ground. They also provided a measure of aerial protection for convoys. No less important was their tremendous cargo capacity. Zeppelins could carry men and munitions great distances, something that was not possible with the fixed-wing aircraft of the day.

Zeppelins remained popular after the war, and their development continued until the LZ 129 Hindenburg disaster in 1937. The crash, seen by millions of people in newspapers and newsreels, helped end the public's interest in traveling by airship. Today, they are used largely for promotional flights (the most famous airship is without doubt the Goodyear blimp, the latest iteration of which is *Wingfoot One*) and industrial purposes, though lighter-than-air airships have drawn renewed interest as military surveillance and communications platforms, as well as for passenger transport.

While WWI saw the rise of the submarine and the airplane as vital weapons of war, and even now they remain keystones of military might, the airship, on the other hand, became little more than a means of capturing fantastic aerial shots at the Super Bowl.

Airships "had a usefulness when employed correctly," says Underwood, who notes that although the technology has advanced, what we're doing today is "not a whole lot different from what they were doing" back then. Minus the scaring people part.

An airship was built by the Germans for the US Navy as a part of a post-WWI reparation agreement (Fig. 1.14). However, the USS Los Angeles airship ended up nearly vertical after its tail rose out of control while moored at the Naval Air Station at Lakehurst, New Jersey in 1927 (Fig. 1.15).

Commercially, most hydrogen is used to process fossil fuels and synthesize ammonia. Hydrogen is used in welding, hydrogenation of fats and oils, methanol production, hydrodealkylation, hydrocracking, and hydrodesulfurization. It is used to prepare rocket fuel, fill balloons, make fuel cells, make hydrochloric acid, and reduce metal ores.

Fig. 1.14 Airship over USS Los Angeles

Fig. 1.15 USS Los Angeles airship at the Naval Air Station, Lakehurst, New Jersey in 1927

Hydrogen is important in the proton–proton reaction and carbon–nitrogen cycle. Liquid hydrogen is used in cryogenics and superconductivity. Deuterium is used as a tracer and a moderator to slow neutrons. Tritium is used in the hydrogen (fusion) bomb. Tritium is also used in luminous paints and as a tracer.

The hydrogen fusion bomb, also known as a thermonuclear weapon, is a second-generation nuclear weapon design using a secondary nuclear fusion stage consisting of implosion tamper, fusion fuel, and spark plug which is bombarded by the energy released by the detonation of a primary fission bomb within, compressing the fuel material (tritium, deuterium, or lithium deuteride) and causing a fusion reaction. Some advanced designs use fast neutrons produced by this second stage to ignite a third fast fission or fusion stage. The fission bomb and fusion fuel are placed near each other in a special radiation-reflecting container called a radiation case that is designed to contain x-rays for as long as possible. The result is greatly increased explosive power when compared with single-stage fission weapons. The device is colloquially referred to as a "hydrogen bomb" or "H-bomb" because it employs the fusion of isotopes of hydrogen. Figure 1.16 shows the explosion of the test bomb "Mike" during Operation Ivy in the Marshall Islands, the first fusion bomb to be tested, on November 1, 1952.

Fig. 1.16 Explosion of the test bomb "Mike" during Operation Ivy

Hydrogen is one of the key starting materials used in the chemical industry. It is a fundamental building block for the manufacture of two of the most important chemical compounds made industrially: ammonia, and hence fertilizers, and methanol, used in the manufacture of many polymers. It is also used in the refining of oil, for example in reforming, one of the processes for obtaining high-grade petrol and in removing sulfur compounds from petroleum which would otherwise poison the catalytic converters fitted to cars. The uses of hydrogen are shown in Fig. 1.17.

By far the most important industrial chemical process for making hydrogen is through *steam reforming*, a method for producing hydrogen, carbon monoxide, or other useful products from hydrocarbon fuels such as natural gas. This is achieved in a processing device called a reformer which reacts steam at high temperature with the fossil fuel. The steam methane reformer is widely used in industry to make hydrogen. There is also interest in the development of much smaller units based on similar technology to produce hydrogen as a feedstock for fuel cells. Small-scale steam reforming units to supply fuel cells are currently the subject of research and development, typically involving the reforming of methanol, but other fuels are also being considered such as propane, gasoline, auto gas, diesel fuel, and ethanol.

In recent years, hydrogen (H_2) has gained much research emphasis as an energy carrier due to its environmental friendliness and wide range of energy applications. Approximately 95% of the hydrogen in the United States is generated via methane steam reforming, being utilized predominantly for petroleum refining and the

Fig. 1.17 Uses of hydrogen

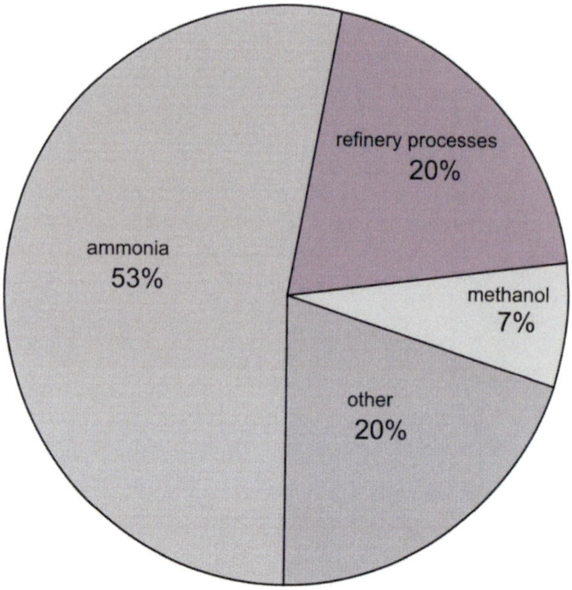

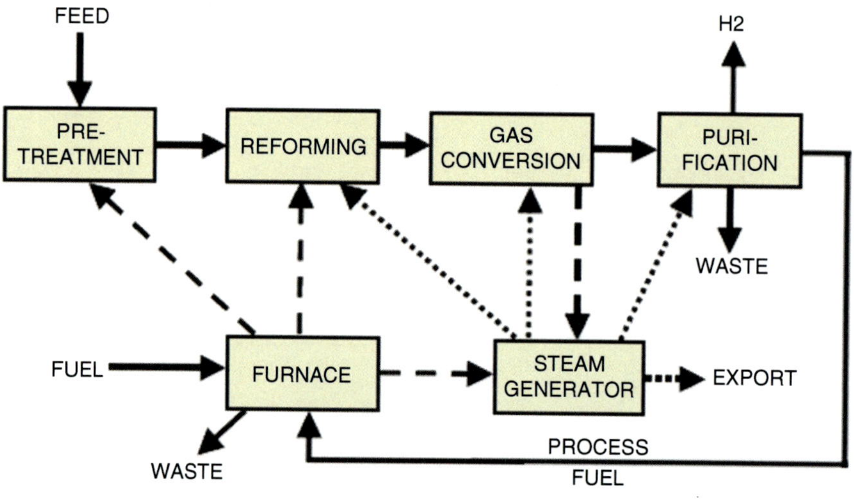

Fig. 1.18 Schematic representation of steam methane reforming

production of industrial commodities such as ammonia. In addition to the maturity of the technology, natural gas reforming is also the most economical of all hydrogen production pathways.

Methane steam reforming is a well-established process, as shown in Fig. 1.18. Steam and hydrocarbon enter the reactor as feedstock, and hydrogen and carbon dioxide are generated at the end of the process. The process is governed by the reactions is as follows:

$$
\begin{aligned}
&CH_4 + H_2O \leftrightarrow CO + 3H_2 && \Delta H_{295} = -206\ \text{kJ/mol} \\
&CO + H_2O \leftrightarrow CO_2 + H_2 && \Delta H_{298} = -41\ \text{kJ/mol} \\
&CH_4 + 2H_2O \leftrightarrow CO_2 + 4H_2 && \Delta H_{298} = 165\ \text{kJ/mol}
\end{aligned}
\tag{1.1}
$$

The steam-forming step, where methane reacts with water to produce carbon monoxide and hydrogen, is an endothermic process. Thus, the process is usually maintained at approximately 850 °C to obtain the desirable conversion. The second step is known as the water–gas shift reaction, where syngas reacts to recover hydrogen. Conventionally, the process is performed in multi-tubular fixed-bed reactors in the presence of a metal catalyst. However, the overall reaction is limited as both the steam reforming and the water–gas shift reactions are subjected to thermodynamic equilibrium. In recent years, research has been devoted to the use of catalytic membrane reactors in overcoming the equilibrium limitation.

Figure 1.18 provides a schematic representation of steam methane reforming.

The use of a steam reforming unit mitigates the problems of storage and distribution of hydrogen tanks for hydrogen vehicles. In addition, methane steam reforming operates at the highest efficiency compared with other current commercially available hydrogen production methods, such as partial oxidation of heavy oil and coal as well as coal gasification. Unfortunately, the production of hydrogen using steam reforming of natural gas does not eliminate greenhouse gas emissions. However, the carbon dioxide release is in fact lower for fuel cell vehicles powered by natural gas than those powered by gasoline.

The United States has an annual hydrogen production of approximately 9 million tons [9]. In order for hydrogen to become competitive in the energy market, the cost of production has to be lower than the other available alternatives. On top of the development of carbon dioxide capture and sequestration technology, another major challenge for methane steam reforming is to improve process efficiency and reduce production costs, particularly in keeping the price affordable in times of natural gas price fluctuations.

Part of hydrogen usage in industry is producing electricity from it. This is achieved by conveying the hydrogen gas to fuel cells where it combines with oxygen, the result being a chemical reaction that generates electricity and heat. The hydrogen gas can also just be burned to power vehicle engines. The by-products of this chemical reaction are water and carbon which are used to produce methane and coal.

1.7 Hydrogen Energy

In this section, I discuss the importance of hydrogen energy, what is hydrogen energy, and why it is important substance in our day-to-day lives, especially when we are working hard toward the goal of a clean future for our environment minus carbon dioxide and other dirty substances.

The universe consists of a mixture of a vast array of components, each of which has a vital role in the composition of the world. The most abundant components in the universe include hydrogen, nitrogen, and oxygen. Hydrogen is the most common component, making up 75% of the universe, and it plays a greater role in the sustainability of life. Apart from helping different living species to survive, hydrogen can be utilized to generate energy.

Hydrogen exists in almost all plant matter and also occurs naturally in water. Though the Sun has a greater percentage of hydrogen gas, the gas is so light that it virtually disappears from the surface of the Earth when conveyed by the Sun's rays. So, to effectively obtain hydrogen gas, it must be harnessed from water, natural gas, or biomass.

As mentioned at the beginning of this chapter, the hydrogen element is the most basic of all Earth's elements. As such, it is very abundant, but it does not really exist as a separate form of matter. Instead it is usually combined with other elements. To separate hydrogen gas from its companion substances takes a lot of work but it produces a powerful, nearly clean source of energy. As a gas, it can be used in fuel cells to power engines.

Hydrogen is a very important molecule with an enormous breadth and extent of application and use. It is currently being used in many industries, from chemical and refining to metallurgical, glass, and electronics. Hydrogen is primarily used as a reactant. But it is also being used as a fuel in space applications, as an "O_2 scavenger" in heat treating of metals, and for its low viscosity and density.

Hydrogen gas is extracted from water by a technique known as electrolysis, which involves running a high electric current through water to separate hydrogen and oxygen atoms. The electrolysis process is pretty expensive since it involves high energy expenditure, and the energy used to generate electricity in the electrolysis process is harnessed from fossil fuels such as oil, natural gas, or coal, although it can also be tapped from renewable energy sources such as solar, wind, and hydropower to ensure no greenhouse gas emissions. Obtaining volumes of oxygen by this method is still under research to establish a viable method of generating it domestically at a relatively low cost.

In summary, hydrogen is considered to be a secondary source of energy, commonly referred to as an energy carrier. Energy carriers are used to move, store, and deliver energy in a form that can be easily used. Electricity is the most well-known example of an energy carrier.

Hydrogen as an important energy carrier in the future has a number of advantages. For example, a large volume of hydrogen can be stored easily in a number of different ways. Hydrogen is also considered to be a high efficiency, low polluting fuel that can be used for transportation, heating, and power generation in places where it is difficult to use electricity. In some instances, it is cheaper to ship hydrogen by pipeline than sending electricity over long distances by wire.

1.7.1 Pros and Cons of Hydrogen Energy

In order to evaluate the advantages and disadvantages of hydrogen energy, in this section I list both and its pros and cons.

A. *Advantages of Hydrogen Energy*

1. *Renewable energy source and bountiful in supply*

   Hydrogen is a rich source of energy for many reasons, the main one being that it is bountiful in supply. While it may take a lot of resources to harness it, no other energy source is as infinite as hydrogen. That, essentially, means there is no possibility of it running out like other sources of energy.
2. *Practically a clean energy source*

   When hydrogen is burnt to produce fuel, the by-products are totally safe, which means they have no known side effects. Aeronautical companies actually use hydrogen as a source of drinking water; after the hydrogen is utilized, it is normally converted to drinking water for astronauts on spaceships or space stations.
3. *Hydrogen energy is non-toxic*

   This means that it does not cause any harm or destruction to human health. This aspect makes it preferred over other sources of fuel such as nuclear energy (i.e., fusion-driven energy) and natural gas, which are extremely hazardous or daunting to harness safely. It also allows hydrogen to be used in places where other forms of fuel may not be allowed.
4. *More efficient than other sources of energy*

   Hydrogen is a very efficient energy type since it has the ability to convey a lot of energy for every pound of fuel. This categorically means that an automobile that utilizes hydrogen energy will travel more miles than one with an equal amount of gasoline.
5. *Can be used to power space ships and future hydrogen-driven cars*

   The efficiency and power of hydrogen energy makes it an ideal fuel source for spaceships—its power is so high that it is able to quickly rocket spaceships to exploration missions. It is also the safest form of energy to perform such an energy-intensive task. Hydrogen energy is in fact three times more potent than gasoline and other fossil-based sources of fuel. This ideally means that you need less hydrogen to complete an enormous task.

   Currently, hydrogen is mainly used as a fuel in the NASA space program. Liquid hydrogen is used to propel space shuttles and other rockets, while hydrogen fuel cells power the electrical systems of the shuttle. The hydrogen fuel cell is also used to produce pure water for the shuttle crew.

   It also offers motive power for airplanes, boats, cars, and both portable and stationary fuel cell applications. The practical downside to using hydrogen in cars is that it is difficult to store in cryogenic or high-pressure tanks.
6. *Hydrogen fuel cells*

   Fuel cells directly convert the chemical energy in hydrogen to electricity, with pure water and heat as the only by-products. Hydrogen-powered fuel

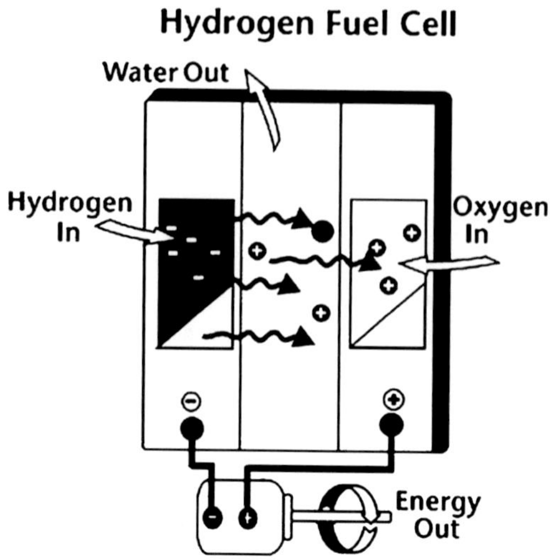

Fig. 1.19 Hydrogen fuel cell. (Courtesy of the US Department of Energy)

cells (see Fig. 1.19) are not only pollution-free, but a two- to three-fold increase in the efficiency can be experienced when compared with traditional combustion technologies.

Fuel cells can power almost any portable devices that normally use batteries. They can also power transportation such as vehicles, trucks, buses, and marine vessels, as well as provide auxiliary power to traditional transportation technologies. Hydrogen can play a particularly important role in the future by replacing the imported petroleum we currently use in our cars and trucks.

As noted earlier, these fuel cells are used to power the electrical systems of space shuttles during their mission in space.

B. *Disadvantages of Hydrogen Energy*

While hydrogen energy has a lot of admirable benefits, it is not really the outright preference for a clean and cheap energy source for most governments and companies. In a gaseous state, it is quite volatile. While this volatility gives it an edge over energy sources in terms of accomplishing numerous tasks, it equally renders it risky to use and work around. Some of the disadvantages of hydrogen energy include the following:

1. *Expensive source of energy*

Electrolysis and steam reforming, the two main processes of hydrogen extraction, are extremely expensive. This is the real reason it is not heavily used across the world. Today, hydrogen energy is chiefly used to power most hybrid vehicles. A lot of research and innovation is required to discover cheap and sustainable ways to harness this form of energy. Until then, hydrogen energy will remain exclusively for the rich.

2. *Storage complications*

   One of hydrogen's properties is that it has a low density. In fact, it is a lot less dense than gasoline. This means that it has to be compressed to a liquid state and stored the same way at lower temperatures to guarantee its effectiveness and efficiency as an energy source. This also explains why hydrogen must be stored and transported under high pressure at all times, which is why transportation and common use is far from feasible.
3. *Not the safest source of energy*

   The power of hydrogen should not be underestimated. Although gasoline is a little more dangerous than hydrogen, hydrogen is hugely flammable and frequently makes headlines for its potential dangers. Compared with gas, hydrogen lacks smell, which makes any leak detection almost impossible—sensors must be installed to detect leaks.
4. *Difficult to move around*

   It is a daunting task to transport hydrogen due to its lightness, whereas oil can be transported safely because it is mostly pushed through pipes and coal can conveniently be transported in dump trucks. Hydrogen also presents challenges when considering moving it in large quantities, which is why it is mostly only transported in small batches.
5. *Hydrogen energy cannot sustain the population*

   Despite the fact that hydrogen is bountiful in supply, the cost of harnessing it limits its extensive utilization. As is well-understood, it is quite challenging to disrupt the status quo. Energy from fossil fuels still rule the world. There is also no framework put in place to ensure cheap and sustainable hydrogen energy for the normal car owner in the future. Even if hydrogen were to become cheap right now, it would take years to become the most used source of energy since vehicles themselves and service stations would need to be customized to conform to hydrogen requirements. This would involve massive capital outlay.

   It is a fact that hydrogen energy is a renewable resource as it is abundantly available, and its impacts are hugely neglected; however, hydrogen companies will, in a real sense, need other forms of non-renewable energy such as fossil fuels (coal, natural gas, and oil) to separate it from oxygen. We may be able to minimize the over-reliance on fossils fuels when we embrace hydrogen energy, but it will be daunting to get rid of it from the system altogether.

Given these pros and cons, there are more advantages than disadvantages to hydrogen energy, although there is a need for advanced technology to produce this element and store it more efficiently.

In the future, hydrogen will join electricity as an important energy carrier since it can be made safely from renewable energy sources and is virtually non-polluting. It will also be used as a fuel for "zero-emission" vehicles, to heat homes and offices, produce electricity, and fuel aircraft.

Hydrogen has great potential as a way to reduce reliance on imported energy sources such as oil. However, before hydrogen can play a bigger energy role and

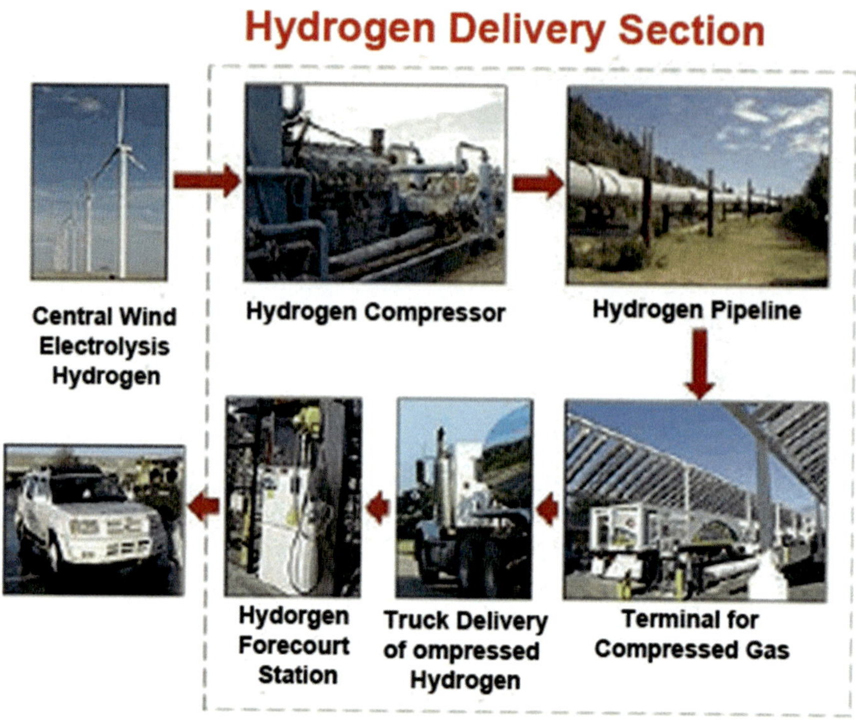

Fig. 1.20 Future hydrogen energy infrastructure. (Courtesy of the US Department of Energy)

become a widely used alternative to gasoline, many new facilities and systems must be built.

Figure 1.20 provides a demonstration of future hydrogen energy infrastructure in which the hydrogen is produced through a wind electrolysis system. The hydrogen is compressed up to pipeline pressure, and then fed into a transmission pipeline. The pipeline transports the hydrogen to a compressed gas terminal where it is loaded into compressed gas tube trailers. A truck delivers the tube trailers to a forecourt station where the hydrogen is further compressed, stored, and dispensed to fuel cell vehicles.

This technology has been implemented by Hydrogenics Corporation, a leading designer and manufacturer of hydrogen electrolyzers and fuel cell systems, who announced in 2016 that it has provided an integrated hydrogen generation system to Shell Hydrogen LLC. Hydrogenics provided an integrated hydrogen fueling station, based on its highly modular standard product line, including electrolyzer, compressor, storage, and dispensing systems. In order to meet the demanding space requirements of the fueling station, Hydrogenics implemented a canopy system where all the components are mounted on the roof of the station canopy, minimizing the footprint of the hydrogen station. In addition, Hydrogenics' electrolyzers produce

Fig. 1.21 Mercedes Benz B-Class with F-Cell

Fig. 1.22 Hydrogen fueling nozzle. (Courtesy of the US Office of Energy and Renewable Energy)

hydrogen from water, providing the most environmentally friendly approach to hydrogen generation.

In the future, hydrogen cars will burn hydrogen to produce energy, the only product of this burn being water residue, as follows:

$$2H_2(g) + O_2(g) \rightarrow 2H_2O(I) \quad \Delta H^{\ominus} = -286\,\mathrm{kJ\,mol^{-1}} \tag{1.2}$$

Technically, hydrogen is potentially an environmentally attractive fuel for the future. Figure 1.21 shows a Mercedes Benz B-Class F-Cell car, which uses a hydrogen substance as fuel, while Fig. 1.22 shows a hydrogen fueling nozzle [6].

One such fuel cell is the PEM cell, where PEM stands for proton exchange membrane or polymer electrolyte membrane.

All fuel cell electric vehicles (FCEVs) use pure hydrogen as a fuel, which is stored in pressure tanks (mainly at 700 atmospheres). A stream of pure hydrogen is delivered to the anode on one side of a membrane, which allows movement of cations but does not conduct either anions or electrons. One such material is a co-polymer of a sulfonated tetrafluoroethene (tetrafluoroethylene), incorporating perfluoroethenyl (perfluorovinyl) ether groups terminated with sulfonate groups onto a tetrafluoroethene backbone. The H+ ions (protons) are able to pass through the membrane as the proton on the $-SO_3H$ (sulfonic acid) groups "hop" from one acid site to another, thus protons are formed as follows:

$$H_2(g) \rightarrow 2H^+(aq) + 2e^- \tag{1.3}$$

The electrons pass through the external circuit from the anode to the cathode and so create an electrical potential. The protons permeate the membrane and react with oxygen at the cathode:

$$4H^+(aq) + O_2(g) + 4e^- \rightarrow 2H_2O(I) \tag{1.4}$$

Figure 1.23 illustrates the structure of a member cell. In this figure, both reactions are catalyzed by platinum. As platinum is extremely expensive, it is deposited as nanosized platinum particles onto carbon powder (Pt/C), which provides a large platinum surface area while the carbon allows for electrical connection between the

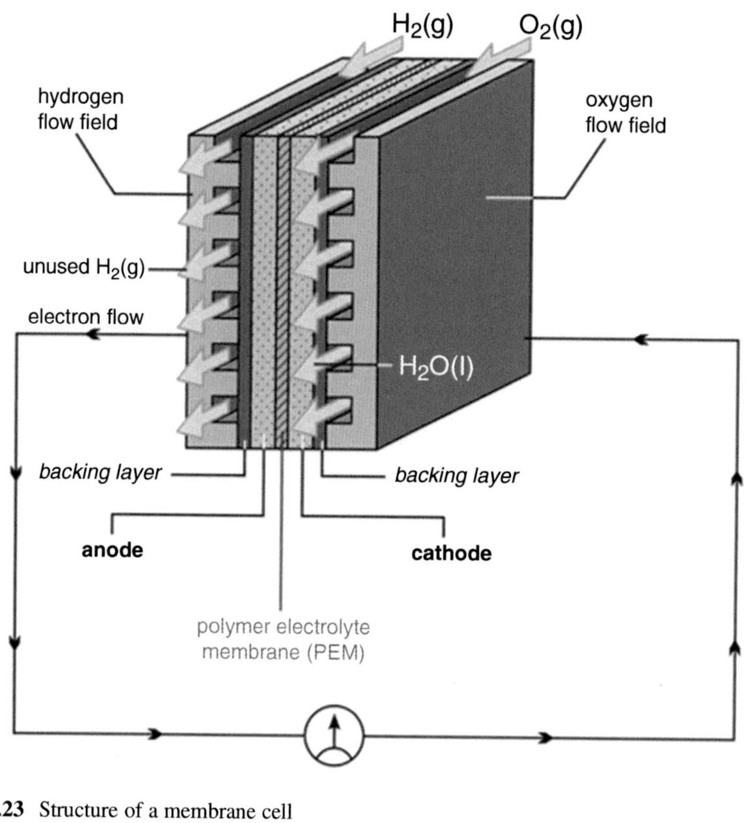

Fig. 1.23 Structure of a membrane cell

catalyst and the rest of the cell. Platinum is so effective because it has high catalytic activity and bonds to the hydrogen just strongly enough to facilitate electron transfer but not to inhibit the hydrogen from continuing to move around the cell [6].

The most effective ways of achieving the nanoscale platinum on carbon powder (Pt/C) use chemical solution deposition. The platinum particles are deposited onto carbon paper that is permeated with Polytetrafluoroethylene (PTFE) is a synthetic fluoropolymer of tetrafluoroethylene that has numerous applications. A second method of increasing the catalytic activity of platinum is to produce an alloy with other metals, such as nickel. It is thought that this reduces its tendency to bond to oxygen-containing ionic species, thereby increasing the number of available sites for oxygen adsorption and reduction.

The reactions take place at around 350 K and the process is therefore referred to as "cold combustion" in a low-temperature fuel cell [6].

Figure 1.24, several hundred cells are connected in series to form a so-called stack. The system potential thereby attained, amounting to 200 volts, is sufficient to power a vehicle.

Other, more recent membranes include polymers based on polybenzimidazole (PBI) (Fig. 1.25), which are stable at temperatures of about 500 K. At these higher temperatures, the membranes are more efficient and stronger, although there can be problems with their manufacture [6].

PBI itself is produced as a fiber that has a very high melting point. It is both highly resistant to heat and chemicals and has been used successfully to make a material

Fig. 1.24 Individual fuel cell. (Courtesy of Daimler AG)

Fig. 1.25 Organic chemical formation of polybenzimidazole [6]

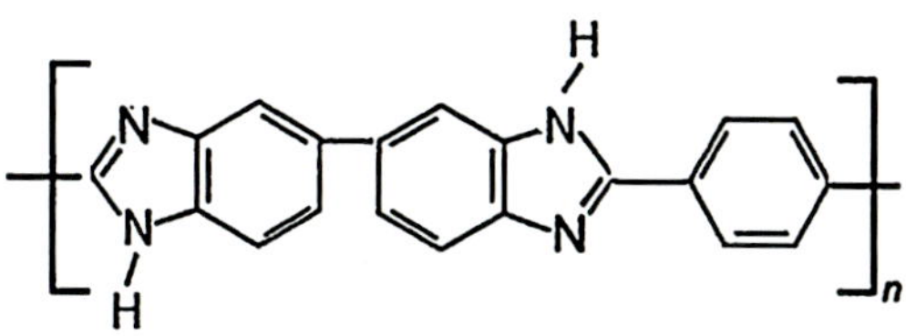

used by firemen, astronauts, and others exposed to danger from flames and other sources of high temperatures.

There is much research and development underway into how to use hydrogen as a fuel. In particular, progress has been made in the development of hydrogen-based fuel cells, but it is clear that the use of these cells to power cars that are to be used in conventional ways presents significant challenges associated with the creation of the hydrogen and its transport, storage, and distribution. Some of these difficulties are bypassed where cells are used to power transport vehicles within a business site that is large enough to produce, store, and distribute its own hydrogen to its own vehicles that work on the site and emit only water. Another approach would be to convert a liquid fuel into hydrogen, in situ in the car. For example, methanol can be converted into hydrogen and carbon dioxide by a reforming reaction similar to the process described earlier and used in a fuel cell, but currently no car manufacturer is working on such reformer systems for FCEVs. However, for stationary applications of fuel cells, reforming of hydrocarbons is reasonable, and projects are underway around the world to convert natural gas to hydrogen in mini-reformers.

1.8 Manufacture of Hydrogen

As mentioned in Sect. 1.5, by far the most important process for making hydrogen is the steam reforming process. The key parts of the process are the conversion of a carbon-containing material to a mixture of carbon monoxide and hydrogen followed by the conversion of carbon monoxide to carbon dioxide and the production of more hydrogen.

At present, the hydrocarbon used is generally methane or other light hydrocarbons obtained from natural gas or oil and coal. However, there is an increasing interest in using biomass, as outlined later in this section.

If hydrocarbons are used, the gas or vapor is mixed with a large excess of steam and passed through pipes containing nickel oxide, which is reduced to nickel during the reaction, and supported on alumina, in a furnace, which operates at high temperatures:

$$CH_4(g) + H_2O(g) \rightarrow 3H_2(g) + CO(g) \quad \Delta H^{\ominus} = +210\,\text{kJ mol}^{-1} \qquad (1.5)$$

The reaction is endothermic and accompanied by an increase in volume. It is thus favored by high temperatures and low partial pressures. The reaction is also favored by a high steam:hydrocarbon ratio. This increases the yield but increases operating (energy) costs. The high ratio also helps to reduce the amount of carbon deposited, which reduces the efficiency of the catalyst. The most effective way to reduce carbon deposition has been found to be impregnation of the catalyst with potassium carbonate.

In the second part of the process, the *shift reaction*, carbon monoxide is converted to carbon dioxide by reacting it with steam and so producing more hydrogen:

$$CO(g) + H_2O(g) \rightarrow H_2(g) + CO_2(g) \quad \Delta H^{\ominus} = -42\,kJ\,mol^{-1} \tag{1.6}$$

This reaction is significantly exothermic, and so high conversions to carbon dioxide and hydrogen are favored by low temperatures. This is difficult to control due to the heat created, and it has been common practice to separate the shift reaction into two stages: the bulk of the reaction being carried out at around 650 K over an iron catalyst, and the "polishing" reaction carried out around 450 K over a copper/zinc/alumina catalyst.

The carbon dioxide and any remaining carbon monoxide are then removed by passing the gases through a zeolite sieve. From time to time, the vessel containing the sieve is taken out of the gas stream and flushed with hydrogen to displace carbon dioxide and regenerate the sieve.

To obtain sustainable (i.e., zero emission of greenhouse gases) production of hydrogen, the carbon dioxide is captured and stored or used.

Thus, overall, one mole of methane and two moles of steam are theoretically converted into four moles of hydrogen, although this theoretical yield is not achieved as the reactions do not continue to completion:

$$CH_4(g) + 2H_2O(g) \rightarrow 4H_2(g) + CO_2(g) \tag{1.7}$$

Hydrogen is produced from a variety of feedstocks, mostly from natural gas, coal, or naphtha. The ways in which hydrogen is obtained from these feedstocks are dealt with separately.

In some countries (notably China), methane and other gases based on oil are in short supply and need to be imported. This has led to a significant change in the choice of fuel used in reformers in these countries. Instead of a hydrocarbon gas, coal is used as it is more readily available.

Research is being undertaken to see whether biomass instead of coal or oil can be used effectively to manufacture hydrogen. One of the key issues is to minimize the energy used in collecting the biomass and transporting it to the place of use, which may involve the use of non-renewable energy sources. Financial and environmental costs can be high, relative to the savings in switching from non-renewable resources to biomass.

There are two key points to consider:

(a) *How do we manufacture hydrogen in as "green" a way as possible?*

Hydrogen, as discussed earlier, is produced on a large scale from fossil fuels (natural gas, coal), thus also producing the environmentally damaging greenhouse gas, carbon dioxide.

The obvious route to produce hydrogen is by the reversing the process of burning hydrogen to produce water, that is, the electrolysis of water. The overall equation is as follows:

$$2H_2O(I) \rightarrow 2H_2(g) + O_2(g) \tag{1.8}$$

However, this requires electricity from power stations. If the power station uses fossil fuels, it defeats the purpose, namely to produce a fuel without the production of carbon dioxide. Other forms of generating power, such as nuclear, wind and geothermal, do not have this disadvantage, but these routes are not readily available in many countries.

However the hydrogen is produced, there are problems in distributing the gas in an economic but environmentally friendly way.

In the existing supply network, hydrogen is delivered under pressure to hydrogen refueling stations (HRS) by specially constructed tankers either as liquid hydrogen or as compressed gaseous hydrogen and then transferred into storage vessels. An alternative and cleaner method is via a pipeline and there are small existing pipeline networks, for example in Germany, the Netherlands, and California, which are used to deliver hydrogen straight from the point of manufacture.

Hydrogen refueling units are being developed which produce hydrogen in situ in the garage. By the end of 2016, it is expected that there will be several hundred such refueling sites in the United States, Germany, Japan, and the United Kingdom, which will be able to deliver hydrogen to a car at pressures between 350 and 700 atmospheres (see Figs. 1.20 and 1.22).

Figure 1.26 shows an electrolyzer constructed on two levels in a special 4 m compartment within a garage. It can produce 100 kWh worth of hydrogen fuel per

Fig. 1.26 Infrastructure electrolyzer. (Courtesy of the Linde Group)

day. The post on the upper level is the ionic compressor which enables the hydrogen to be delivered to a car at an appropriate pressure.

As noted earlier, there are practical difficulties in the distribution and storage of hydrogen. One approach would be to convert a liquid fuel into hydrogen, in situ in the car. For example, methanol has been used in experimental cars. The fuel was converted into hydrogen and carbon dioxide by a reforming reaction, similar to the process described earlier for the large-scale manufacture of hydrogen. This demands a very high level of engineering skills to produce conversion units which are light enough for a car but strong enough to withstand all the problems caused by continuous vibrations.

An enormous amount of research is also being undertaken on using sunlight as the energy source, one being via bio-photolysis. This involves the production of algae in water through photosynthesis, followed by bacterial decomposition of the algae to produce hydrogen. An important discovery was that by depriving the algae of sulfur, normal photosynthesis is inhibited and instead an enzyme is activated and hydrogen, not oxygen, is produced in light. Present research is concerned with making these processes more efficient.

(b) *How do we use the hydrogen efficiently in engines to produce energy?*

As stated earlier, we know that hydrogen is potentially an environmentally attractive fuel for the future.

Hydrogen could be burnt in an engine in a similar way to petrol, the gas burning in air to release energy. However, although there is an advantage over using a hydrocarbon as there is no carbon dioxide produced, harmful nitrogen oxides would still be formed from the reaction of nitrogen and oxygen in the hot engine.

However, in a fuel cell, hydrogen reacts with oxygen without burning. The energy released is used to generate electricity, which is used to drive an electric motor. This is a more environmentally friendly system and over the last decades much research has been undertaken by chemists and engineers to produce very efficient fuel cells where about half of the energy from the reaction between hydrogen and oxygen to produce water is released as electrical potential.

Although most hydrogen is produced from natural gas today, the Fuel Cell Technologies Office is exploring a variety of ways to produce hydrogen from renewable resources. Figure 1.27 shows a typical plant in which hydrogen is produced from renewable resources.

In summary, since hydrogen does not exist on Earth as a gas, it must be separated from other compounds. Two of the most common methods used for the production of hydrogen are electrolysis or water splitting and steam reforming.

Steam reforming is currently the least expensive method for producing hydrogen. It is used in industries to separate hydrogen atoms from carbon atoms in methane. Because methane is a fossil fuel, the process of steam reforming results in greenhouse gas emissions, which is linked to global warming.

Fig. 1.27 Hydrogen production plant

Electrolysis involves passing an electric current through water to separate water into its basic elements, hydrogen and oxygen. Hydrogen is then collected at the negatively charged cathode and oxygen at the positive anode. Hydrogen produced by electrolysis is extremely pure, and results in no emissions as electricity from renewable energy sources can be used. Unfortunately, electrolysis is currently a very expensive process.

There are also several experimental methods of producing hydrogen such as photo-electrolysis and biomass gasification. Scientists have also discovered that some algae and bacteria produce hydrogen under certain conditions, using sunlight as their energy source.

1.9 Why Is This Pathway Being Considered?

Today, reforming low-cost natural gas can provide hydrogen for FCEVs as well as other applications. In the long-term, the United States Department of Energy expects that hydrogen production from natural gas will be augmented with production from renewable, nuclear, coal (with carbon capture and storage), and other low-carbon, domestic energy resources [7].

Petroleum use and emissions are lower than for gasoline-powered internal combustion engine vehicles. The only product from an FCEV tailpipe is water vapor, and even with the upstream process of producing hydrogen from natural gas as well as delivering and storing it for use in FCEVs, the total greenhouse gas emissions are cut in half and petroleum is reduced by over 90% compared with today's gasoline vehicles [7].

References

1. https://en.wikipedia.org/wiki/Recombination_(cosmology)
2. https://www.thoughtco.com/hydrogen-facts-element-1-or-h-607917
3. US Department of Energy. https://www.eere.energy.gov/hydrogenandfuelcells
4. https://en.wikipedia.org/wiki/Isotopes_of_hydrogen
5. https://en.wikipedia.org/wiki/Inertial_electrostatic_confinement
6. http://www.essentialchemicalindustry.org/chemicals/hydrogen.html
7. https://www.energy.gov/eere/fuelcells/hydrogen-production-natural-gas-reforming

Chapter 2
Hydrogen-Powered Fuel Cell and Hybrid Automobiles of the Near Future

Currently, providers of energy utilities are presented with the challenges of increased energy demand and the need to immediately address environmental concerns such as climate change and decreasing the pollution produced by transportation vehicles burning gasoline, for example. Due to population and economic growth, the global demand for energy is expected to increase by 50% over the next 25 years. This significant demand increase along with the dwindling supply of fossil fuels has raised concerns over the security of the energy supply. In view of the increased energy demand and environmental pollution, different approaches such as distributed generation and demand-side management have been proposed and are widely being put into practice. However, optimal utilization of the existing energy infrastructure is an issue that also needs to be addressed properly to deal with the major challenges that energy utilities are facing. A hydrogen vehicle is a vehicle that uses hydrogen as its onboard fuel for motive power. Hydrogen vehicles include hydrogen-fueled space rockets as well as automobiles and other transportation vehicles. The power plants of such vehicles convert the chemical energy of hydrogen to mechanical energy either by burning hydrogen in an internal combustion engine or by reacting hydrogen with oxygen in a fuel cell to run electric motors. Widespread use of hydrogen to fuel transportation is a key element of a proposed hydrogen economy.

2.1 Introduction

In the past, energy infrastructure projects have typically been planned and operated independently, but the present trend is to move towards an *integrated view* because of the common technical and economic interactions between different types of energy infrastructure. Of specific interest is the integration of the energy demand for electrical power with the energy demand for transportation fuel. For effective and

© Springer International Publishing AG, part of Springer Nature 2019

B. Zohuri, *Hydrogen Energy*, https://doi.org/10.1007/978-3-319-93461-7_2

coordinated planning and operation, each energy system needs to be investigated in the context of an integrated system.

For example, congestion on a particular electrical transmission path can be relieved by shifting part of the energy flow to another energy network, such as one that uses hydrogen as an energy vector [1], as discussed in this chapter. However, complex problems arise in this integrated approach, given the size of the problem and the multiple economic and technical interactions between various subsystems [2]. Accordingly, appropriate models and tools need to be developed to investigate all these interactions, considering the different types of energy infrastructures and energy vectors involved in such integrated systems.

The transport sector is one of the largest and fastest growing contributors to energy demand, urban air pollution, and greenhouse gases: for example, in Canada, the transport sector represents almost 35% of the total energy demand and is the second highest source of greenhouse gas emissions [3, 4]. In view of these issues and the challenges associated with the supply of oil, the issue of alternative fuels, in particular hydrogen, for meeting the future energy demand of the transport sector has gained much attention. Thus, the ultimate goal is to find a zero-emission transportation fuel such as hydrogen that can be derived from a wide range of primary energy resources.

Hydrogen as an energy carrier can link or interface multiple energy resources for multiple end-uses; this has led to the development of the "hydrogen economy" concept [5–9], which originated in the early 1970s. If non-fossil energy sources for hydrogen production are utilized, hydrogen as a new energy carrier in the framework of integrated energy systems will significantly help relieve many environmental concerns. For example, no nitrogen oxide, sulfur oxide, and carbon dioxide emissions will be generated by hydrogen-powered vehicles in the urban airshed [10]; also, decentralized generation of electricity and heat using stationary fuel cells would yield no smog precursor emissions.

Hydrogen can also be considered as a new alternative for electricity storage. For electrically powered, zero-emission vehicles, batteries are limited in range, do not allow for rapid refueling, and are costly; onboard use of hydrogen allows for zero-emission vehicles with an extended range and rapid refueling. Furthermore, hydrogen production through the electrolysis process has approximately 80% efficiency, which means that most of the electrical energy can be stored and distributed as hydrogen [5, 11]. Also, the economics of production, storage, and utilization of electrolysis-based hydrogen become quite relevant in the context of competitive electricity markets, given the significant price differences between peak and low-price periods [2, 12]; this economic argument becomes more compelling if environmental pollution costs are considered [13, 14]. In addition, since traditional generation plants are most efficient when operating at rated load levels, and considering congestion problems in the electricity transmission system during the normal operation of a power grid, the use of hydrogen as an energy carrier to increase the efficiency and reliability of the grid is certainly attractive [1]. Furthermore, as an example, penetration of fuel cell vehicles (FCVs) into the transport sector in Ontario,

Canada introduces an energy storage capacity for Ontario's grid and this in turn facilitates the integration of intermittent energy resources such as wind and solar [2].

At the present level of technological development, there are still a variety of challenges regarding the production, distribution, storage, and use of hydrogen [15–18]. Therefore, it is necessary to develop appropriate tools and models to study a transition to a hydrogen economy to better understand the specific advantages and disadvantages, with the consideration of costs.

2.2 Fuel Cells

A fuel cell is similar to a battery as it produces electricity from chemical reactions. Most batteries are used up when the chemicals inside them run out; however, the chemicals in fuel cells do not get used up—they come from a separate fuel tank. So long as there is fuel in the tank, the fuel cell keeps running and producing electricity. Today, most fuel cells use hydrogen as the fuel in that tank.

The way a fuel cell works is that, in the fuel cell, hydrogen combines with oxygen from the air to make water. This is the same chemical reaction that happens when hydrogen burns. In fact, in a fuel cell, the reaction makes electricity. As described in Chap. 1, and also shown here in Fig. 2.1, the hydrogen element only consists of one proton (+) and one electron (−), with no neutrons, within nucleus of its atom. To make electricity, the proton and electron need to be split apart, and a fuel cell can perform this function.

Therefore, the way the hydrogen fuel cell operates is that each side of the fuel cell has a metal bar called an electrode, as depicted in Fig. 2.2.

As Fig. 2.2 shows, the electrodes are like the positive and negative poles of a battery and they form a loop, or circuits. The circuit carries an electrical current.

Fig. 2.1 Hydrogen atom

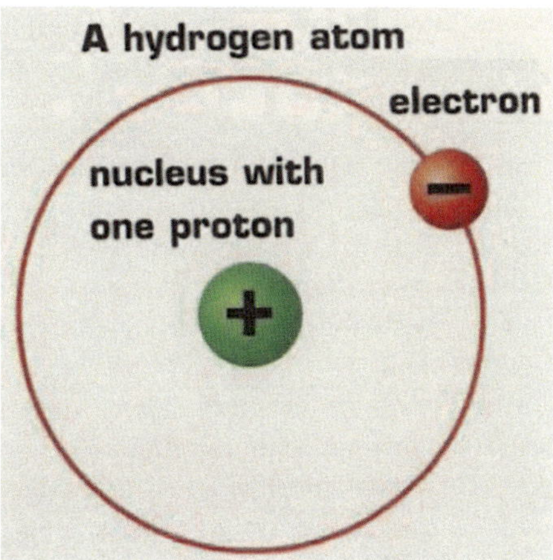

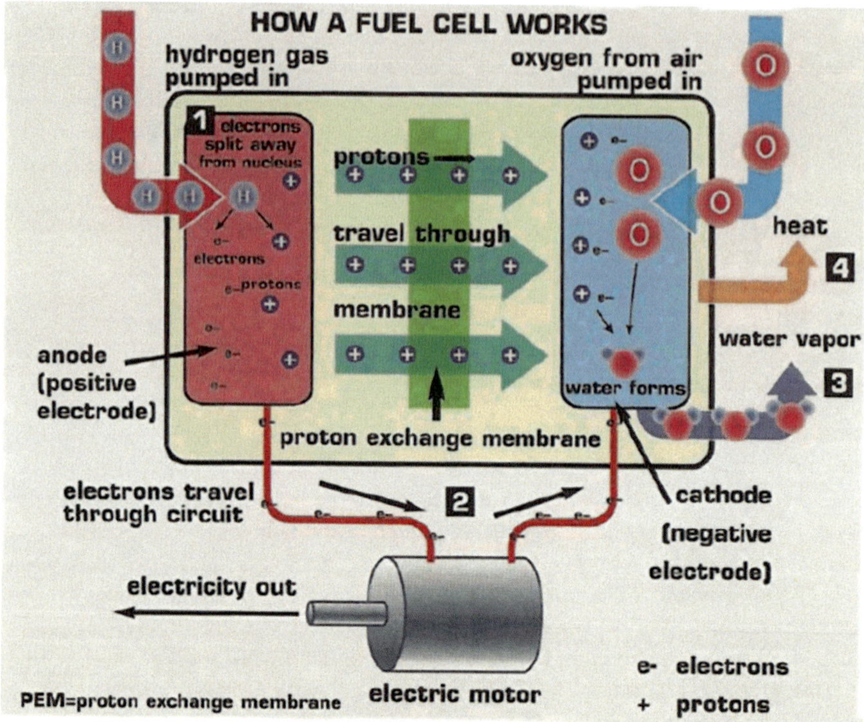

Fig. 2.2 Operation of a fuel cell. See text for explanation of numbering

Furthermore, in Fig. 2.2 we observe the following (see corresponding numbers on the figure):

1. Electron splits;
2. Electrons travel through the circuit to make electricity, then the protons travel to the cathode. At this stage, hydrogen reforms and unites with oxygen;
3. Water vaporizes; and
4. Heat is produced.

As the hydrogen splitting process takes place, the hydrogen comes in one side of the fuel cell and the electrons in the hydrogen are attracted to the positive electrode. This is called the anode, where the electrons split away from the protons. The electrons travel through the electrical circuit and on the way produce electricity; the electrons continue to go through the circuit, and, finally, end up on the other side of the fuel cell.

Meanwhile, on the other side of the cell, the protons pass through a sheet of plastic in the middle of the fuel cell. In this case, the plastic is called a proton exchange membrane (PEM). At this stage, the protons move toward the negative electrode, or cathode, on the other side of the PEM, and only the protons can pass through the PEM. Once through, the protons meet up with the electrons that came through the circuit. They then join to form hydrogen atoms again. Oxygen is pumped into the side of the fuel cell near the cathode, then the hydrogen atoms combine with the oxygen to make water—water has two hydrogen atoms and one oxygen atom as per its chemical formation and organic formula.

The flow of the electrons through the circuit in the fuel cell creates an electric current, making electricity. As the current flows, we can use the electricity to light a lightbulb or power an electric motor—we can use the electricity produced by fuel cells in many ways.

Fuel cells are an important technology for a potentially wide variety of applications, including micropower, auxiliary power, transportation power, stationary power for buildings and other distributed generation applications, and central power. These applications will be across a large number of industries worldwide.

Hydrogen fuel cells are very efficient, reliable, work well, and do not waste energy, while also being a decarbonized type of fuel that does not produce carbon dioxide or any other polluting chemicals.

Engines that burn hydrogen fuel or fossil fuels are not able to turn all of that fuel's energy into power. Some fuel is wasted in the burning process due to combustion; however, using a fuel cell is a different ball game. Fuel cells make electricity from the hydrogen instead of burning it as fuel (see Chap. 5, Sect. 5.3, "Hydrogen Fuel Cells").

Electric power made this way is an efficient form of energy because the fuel cells are more efficient than an ordinary engine. Although at the present time fuel cells cost a lot more to make, given today's technology available for their production, the total cost of ownership (TCO) will reduce as fuel cell production technology advances. The fuel cell in Fig. 2.3 is as powerful as a 2.0-L car engine but much smaller, as illustrated by comparing it to the size of a pen and clipboard.

Progress continues in fuel cell technology today and ongoing research into FCVs is being performed by most well-known car manufactures around the world.

Fig. 2.3 Comparison of the size of a fuel cell in respect to a pen and clipboard

Importantly, polymer electrolyte fuel cells, molten carbonate fuel cells (MCFC), and solid oxide fuel cells (SOFCs) have been demonstrated at a commercial size in power plants.

Phosphoric acid fuel cells (PAFCs) have entered the marketplace with more than 220 power plants delivered. Highlighting this commercial entry, the phosphoric acid power plant fleet has demonstrated more than 95% availability and several units have passed 40,000 h of operation. One unit has operated for over 49,000 h.

Generally speaking, there are two levels of fuel cell research:

(i) Study of the fundamental physical and electrochemical processes (reaction kinetics, ionic transport, capillary phenomena, etc.); and
(ii) Engineering design of cells, stacks, and fuel cell systems. The coupling between these two levels is provided by the physical modelling of fuel cells.

This relatively new discipline aims to understand the basic transport and kinetic phenomena in a real cell and stack environment. Physical modeling uses first-level data to pave the way for better cell, stack, and system designs [19].

Overall, there are many types of fuel cells currently in operation in a wide range of applications, including:

- MCFCs,
- SOFCs,
- PAFCs,
- Direct methanol fuel cells (DMFCs) and
- Low- and high-temperature PEM fuel cells.

Of particular note, on October 8, 2015 the inaugural National Hydrogen and Fuel Cell Day officially recognized and celebrated the benefits provided by fuel cell and hydrogen energy technologies.

2.3 Hydrogen Fuel Cell Applications

The next few years will see a massive emergence of hydrogen fuel cells as an alternative energy option in both transportation and domestic use. The long-range expectation is that hydrogen will be used as a fuel, produced from renewable energy, fossil, or nuclear sources, offering an environmentally acceptable and efficient source of power/energy.

In recent years, many scientists and engineers have moved into the field of hydrogen and fuel cells because it is both exciting and well-funded. The aim is to transform the way energy is delivered and used over the coming years, with major changes in technologies for production, distribution, and conversion, as a response to the political instability of many oil-producing countries, uncertainty about resources, and increasing concerns over environmental effects.

Examples of recent applications of hydrogen/fuel cells are in:

1. Solar energy
2. Space technology
3. Hydrogen/fuel cell cars
4. Hydrogen-driven industrial applications of the physics of cryogenics [20]
5. Renewable energy [21]
6. Direct uses of hydrogen.

There are so few that we can name them all. There are plenty of reference materials and text books on this topic and literature can also be found on the internet. Readers who are interested in any of these subjects should refer to these materials, although a few of these applications are mentioned in this book in the following chapters.

With a battery and gas storage system, the advantages that each technology provides can be achieved and this creates a basis for marking the transition to a battery-free hydrogen system in the future.

2.4 Near-Future Hydrogen-Driven Cars and Industry Milestones

In 2015, Toyota started selling the Mirai, its fuel cell vehicle, and Hyundai continued commercial sales of its fuel cell Tucson (USA) and ix35 (outside the USA) models. Honda unveiled its fuel cell electric vehicles (FCEVs) concept at several auto shows in 2015 and the Clarity became available in late 2016. Several other companies are working to develop hydrogen cars.

As already noted, on October 8, 2015 the inaugural National Hydrogen and Fuel Cell Day officially recognized and celebrated the benefits provided by fuel cell and hydrogen energy technologies. The date selected references the atomic weight of hydrogen: 1.008. As part of the day's events, the United States Senate passed Senate Resolution 2171 declaring October 8 to be National Hydrogen and Fuel Cell Day and the House of Representatives introduced House Resolution 4682 expressing their support. Fuel cell and hydrogen energy companies and their industry allies, friends, and champions celebrated National Hydrogen and Fuel Cell Day with a range of activities and announcements in the United States and abroad.

As of 2014, 95% of hydrogen is made from natural gas. It can be produced using renewable sources, but that is an expensive process [22]. Integrated wind-to-hydrogen (power-to-gas) plants, using electrolysis of water, are exploring technologies to deliver costs low enough, and quantities great enough, to compete with hydrogen production using natural gas [23]. The drawbacks of hydrogen use are high carbon emissions when produced from natural gas, capital cost burden, low energy content per unit volume, production and compression of hydrogen, and the large investment in infrastructure that would be required to fuel vehicles [24, 25].

In 2015, industrial gas company Linde reached a major milestone—more than 1 million hydrogen fills for the fuel cell-driven material handling equipment (MHE) at BMW's plant in Spartanburg, South Carolina, USA. BMW operates more than 350 fuel cell-powered forklifts to service the plant's production and logistics functions, making it the largest single-site fuel cell forklift fleet in the world.

Fuel cells promise clean and efficient energy conversion, low-to-zero emissions at the point of use (of particular benefit for urban environments), and flexibility in terms of the primary power source used to generate the hydrogen or hydrogen-rich fuel. Industry and government have partnered in North America, Europe, and Asia to develop proton exchange membrane fuel cells (PEFCs) for stationary, portable, and transportation applications [26, 27]. Industrial research and development for these uses has been ongoing for more than a decade for this class of fuel cells, with most development efforts going into transportation applications. This prolonged development effort has led to notable improvements in power density, energy conversion efficiency, and lifetime of fuel cells, and yet very few individuals own FCVs [28].

FCV demonstration programs continue, although not at as quickly as had been hoped. Fuel cells are closer to entering into specific niche markets such as co-generation, forklift traction power, backup power/industrial battery replacement, and consumer electronics, all of which have great promise but represent smaller markets than the automotive markets might become in the future. While investors have become more cautious about fuel cells, given the long period in which investments did not yield large growth as hoped, the current state of the technology continues to advance and has matured over time. Now is a good time to review fuel cell technology, its promise, and the challenges yet to be undertaken. Many opportunities still lie in the fields of electrochemistry, electrochemical engineering, mass transport, and materials science [28].

Figure 2.4 presents an estimated cost per kilowatt for fuel cell stacks from mass production and estimated cost point for PEFCs.

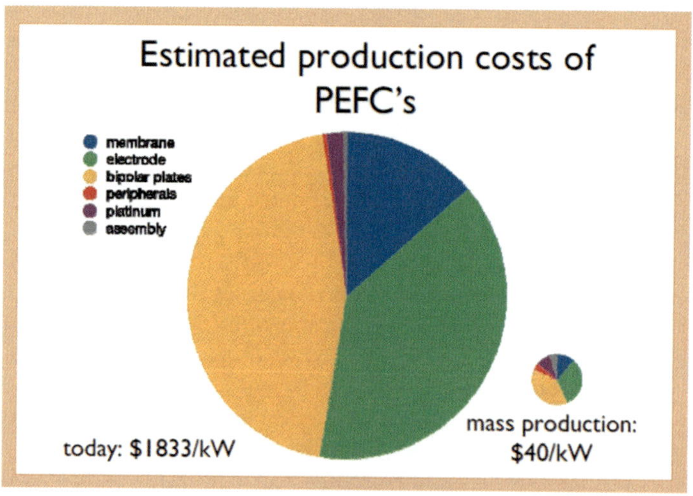

Fig. 2.4 Estimated costs per kW for fuel cell stacks and proton exchange membrane fuel cells [29]

The three main markets for fuel cell technology are stationary power, transportation, and portable power. Stationary power includes any application in which fuel cells are operated at a fixed location for primary power, backup power, or combined heat and power . Transportation applications include motive power for passenger cars, buses and other FCEVs, specialty vehicles, MHE, and auxiliary power units for off-road vehicles. Portable power applications include fuel cells that are not permanently installed or fuel cells in a portable device.

In support of the efforts by the California Air Resources Board (ARB) to incorporate automobile Original Equipment Manufacturers (OEM) input into its June 2015 AB-8 report, the California Fuel Cell Partnership (CaFCP) OEM Advisory Group members—American Honda, General Motors, Hyundai, Mercedes-Benz, Nissan, Toyota, and Volkswagen—developed a consensus list of recommended station priority locations for the next 19 hydrogen stations in California, focused on the Northern California, Southern California, and Central Valley regions.

2.5 European Hydrogen and Fuel Cell Projects

Europe's energy supply is characterized today by structural weaknesses and geopolitical, social, and environmental shortcomings, particularly regarding security of supply and climate change. Energy is a major determinant of economic growth and these deficiencies can have a direct impact on the European Union's (EU) growth, stability, and the well-being of its citizens. Energy supply security, mitigating climate change, and economic competitiveness are therefore the main drivers for energy research, within the context of sustainable development—a high-level EU objective [30].

Hydrogen and electricity together represent one of the most promising ways to realize sustainable energy, whilst fuel cells provide the most efficient device for converting hydrogen, and possibly other fuels, into electricity. Hydrogen and fuel cells open the way to integrated "open energy systems" that simultaneously address all of the major energy and environmental challenges and have the flexibility to adapt to the diverse and intermittent renewable energy sources that will be available in the Europe of 2030. Hydrogen can be produced from carbon-free or carbon-neutral energy sources or from fossil fuels with carbon dioxide capture and storage. Thus, the use of hydrogen could drastically reduce greenhouse gas emissions from the energy sector. Fuel cells are intrinsically clean and very efficient—up to double the efficiency of internal combustion engines—and capable of converting hydrogen and other fuels to electricity, heat, and power. They can also be sited close to the point of end use, allowing exploitation of the heat generated in the process [30].

There are several strategic areas of research currently being pursued for hydrogen:

- *Clean production*: Development and techno–socio–economic assessment of cost-effective pathways for hydrogen production from existing and novel processes;

- *Storage*: Exploration of innovative methods, including hybrid storage systems, which could lead to breakthrough solutions
- *Basic materials:* Functional materials for electrolyzers and fuel processors, and novel materials for hydrogen storage and hydrogen separation and purification;
- *Safety*: Pre-normative Research and Technological Development (RTD) required for the preparation of regulations and safety standards at an EU and global level; and
- *Preparing the transition to a hydrogen energy economy*: Supporting the consolidation of current EU efforts on hydrogen pathway analysis and road mapping.

EU-funded research in the area of fuel cell systems is aimed at reducing the cost and improving the performance, durability, and safety of fuel cell systems for stationary and transport applications to enable them to compete with conventional combustion technologies. This will include materials and process development, optimization, and simplification of fuel cell components and subsystems as well as modeling, testing, and characterization. The long-term goal is to achieve commercial viability for many applications by 2020 [30].

The main EU funding mechanism for research, technological development, and demonstration is the Framework Program (FP), which is mainly implemented through calls for proposals. Based on the treaty establishing the EU, the FP has to serve two main strategic objectives: strengthening the scientific and technological bases of industry and encouraging international competitiveness while promoting research activities in support of other EU policies [30].

2.6 Hydrogen Transportation Concepts by Geographic Region

In the United States, the state of California is leading the move towards having hydrogen refueling stations, and other countries such as Germany, Japan, the United Kingdom, and Canada are also looking into having infrastructure such as hydrogen refueling stations in place and to produce hydrogen for delivery from these stations.

For example, in Ontario, Canada, to determine the demand for hydrogen and required size of electrolytic hydrogen production plants (HPPs) in different zones of the province, it is first necessary to determine the number of cars in each zone during the 2008–2025 period, which in turn requires estimation of the zonal population levels during the planning period. Therefore, the populations of cities and towns of more than 10,000 inhabitants were used to find the population of each zone, considering the geographic location of the zones. The population of each zone was then proportionally scaled up such that the sum of the zonal populations would equal the 12,861,940 population estimated for Ontario on January 1, 2008 by Statistics Canada [31]. Ontario, as part of this effort, uses the annual base-load growth rate for each zone [32] to find the zonal population in the study period. The total projected population of Ontario in 2025 estimated in this way (15,663,374) is equal to that reported by the Ministry of Finance [33], confirming the adequacy of the assumptions used here.

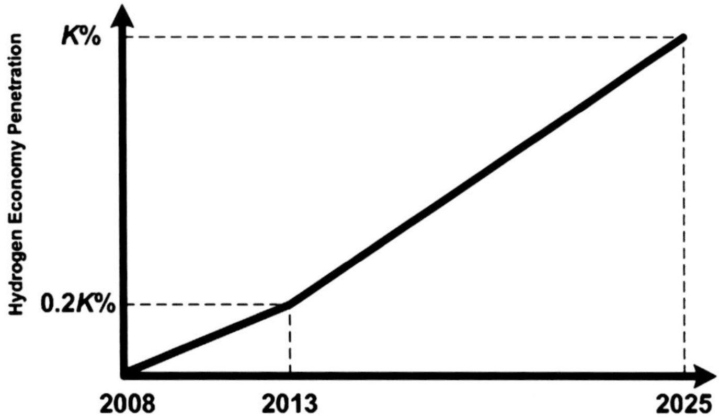

Fig. 2.5 Assumed hydrogen economy transition in Ontario, Canada [32]

However, as mentioned in Chap. 1 of this book, there are practical difficulties in the distribution and storage of hydrogen. One approach would be to convert a liquid fuel into hydrogen, in situ in the car. For example, methanol has been used in experimental cars. The fuel was converted into hydrogen and carbon dioxide by a reforming reaction, similar to the process described earlier for the large-scale manufacture of hydrogen. This demands a very high level of engineering skills to produce conversion units which are light enough for a car but strong enough to withstand all the problems caused by continuous vibrations.

Hajimiragha et al. [32] discuss the hydrogen economy in Ontario with the assumption, as depicted in Fig. 2.5, that there would be initial sluggish development up to 2013 due to the need to develop the required hydrogen-related infrastructure and the possible slow adoption of hydrogen-based vehicles, with a faster slope thereafter. This figure specifies the level of a hydrogen economy penetration in each year, that is, the percentage of the total cars on the road that need to be fueled by hydrogen, to meet the target value in 2025. Observe that 0.2 *K* % in this figure means that only 20% of the target value *K* can be realized by 2013. The target value *K* is increased in the model to find the maximum feasibility limits. This hydrogen economy transition is assumed to be similar in all ten zones of Ontario.

Based on 20,000 km annual mileage and 100 km/kg average fuel economy, the hydrogen demand of a light FCV is calculated to be approximately 0.55 kg/day. This number yields the daily hydrogen demand in each zone, considering the number of cars and the hydrogen economy penetration levels to be realized each year. For example, for a 1% hydrogen economy penetration, the total demand of hydrogen in Ontario in 2025 is found to be almost 47 ton/day, of which the Toronto, South West, and West zones would be the greatest consumers with almost 46%, 17%, and 9% share of the hydrogen demand, respectively [32].

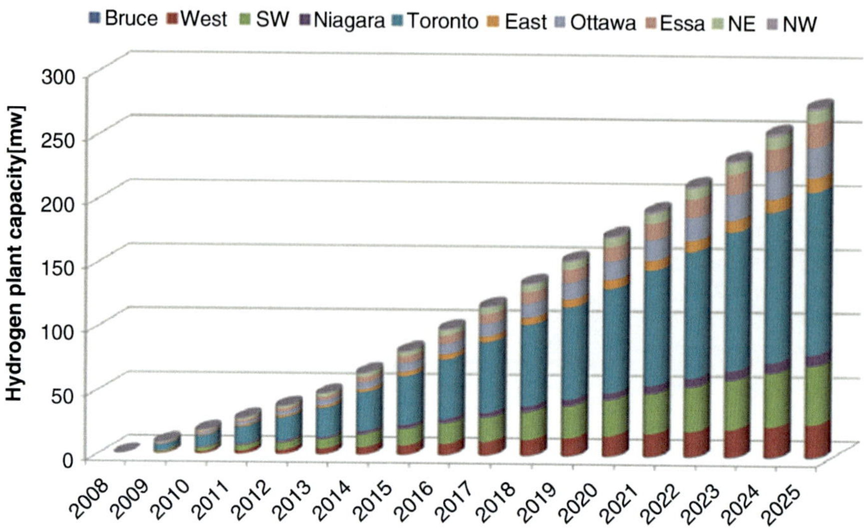

Fig. 2.6 Required zonal capacity of electrolytic hydrogen production plants for 1% hydrogen economy realization [32]

The zonal hydrogen demand during the planning period allows for determination of the required capacity of the HPPs. Thus, based on the higher heating value (HHV) of hydrogen, the power requirement of a hydrogen production unit composed of a rectifier, electrolyzer, and compressor for 24 kg/day hydrogen production is equal to 56.36 kW, assuming 70% efficiency. Based on 68 h of operation per week, including 8 and 14 h of operation during weekdays and weekends, respectively, the actual hydrogen production rate of this unit will be almost 9.7 kg/day. Therefore, the required size of the plant producing 1 ton/day hydrogen is equal to 5.8 MW based on 68 h of operation per week. Figure 2.6 illustrates the zonal capacity of HPPs which would need to be developed in the 2008–2025 period to establish a 1% hydrogen economy penetration across Ontario; observe that almost 275 MW of power is needed by 2025, with a minimum share in the Bruce zone with less than 600 kW and a maximum share in Toronto with more than 127 MW.

The ideal case for zonal transportation according to the hydrogen transportation concept, depicted in Fig. 2.7, is one in which both import and export possibilities exist for each zone [32].

However, it should be noted that the cost of hydrogen transportation based on compressed gas trucks is comprised of operation and investment cost components. The investment costs correspond to truck cabs and compressed gas tube trailers, and the operation costs include diesel costs and driver wages, as well as insurance, licensing, maintenance, and repair costs [32]. The corresponding hydrogen transportation cost parameters for 2008 as estimated in the study by Hajimiragha et al. [32] are shown in Table 2.1.

Hajimiragha et al. [32] analyzed their calculation based on a step function, as shown in Eq. 2.1, where they look at the total operation costs (TOCs) of hydrogen

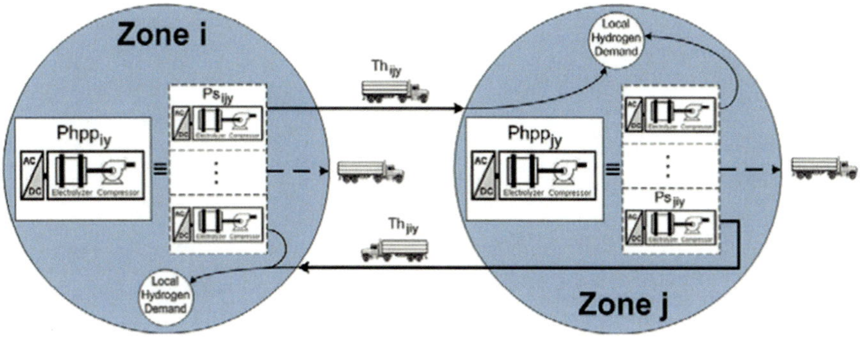

Fig. 2.7 Demonstration of the hydrogen transfer concept [32]

Table 2.1 Hydrogen transportation cost parameters [32]

Maximum capacity of included gas trucks (ton)	0.4
Cab capital cost (CAD)	100,000
Tube trailers capital cost (CAD)	240,000
Diesel cost (CAD/km)	0.4212
Insurance, licensing, maintenance, and repair costs (CAD/km)	0.4
Driver wage (CAD/km)	0.3
Cab lifetime (years)	5
Tube trailers lifetime (years)	20

CAD Canadian dollars

transfer between zones i and j, where Eq. 2.1 depends on amount of hydrogen transferred:

$$\mathrm{TOC}_{ijy} = m \cdot \mathrm{OC}_y \cdot d_{ij} \tag{2.1}$$

where m stands for the number of trucks, OC_y represents the operation costs of one compressed gas truck in Year y in Canadian dollars (CAD)/km, and d_{ij} is the approximate distance between zones i and j in km.

Based on the data provided in Table 2.2, the operation costs of a compressed gas truck amount to CAD$1.1212/km for 2008. Based on typical inflation rates in Ontario, it is assumed that these operation costs will increase at 2.5% per year for the following years up to 2025. Finally, assuming that truck cab prices will probably rise due to inflation, but tube trailer prices will likely decrease since the prices could currently be considered high, it was assumed that the total investment costs for both combined could be expected to remain approximately constant during the planning years.

In general, the number of hydrogen stations opening has grown worldwide, with 54 new hydrogen stations opened in 2015 [30]. The majority (48) are publicly accessible, with the others used either for fleet customers or for bus fueling. Japan led the way with 28 new stations, Europe had 19, and seven stations were opened in

Table 2.2 Minimum acceptable hydrogen prices for different operating timeframes

Operation hours during weekdays	8	8	8	8	8	8	8
Operation hours during weekends	12	14	16	18	20	22	24
Minimum hydrogen selling price (CAD/kg)	4.90	4.78	4.67	4.57	4.50	4.42	4.33
NPV (CAD)	1222.96	176.231	950.517	805.536	1656.26	309.526	1255.63
IRR (%)]	8	8	8	8	8	8	8
PI (–)	1.00347	1.00050	1.00268	1.00226	1.00464	1.00086	1.00349

CAD Canadian dollars; *NPV* Net present value; *IRR* Internal rate of return; *PI* Profitability index

the United States—six in California and one in Colorado at the National Renewable Energy Laboratory (NREL). Plans were also announced in 2015 for 104 more hydrogen stations: 35 in the United States, five in Japan, and 64 in Europe, with 34 of those to be located in Germany. See Table 27 of the *Fuel Cell Technologies Market Report 2015* by the US Department of Energy [34] for examples of hydrogen stations opened or announced during 2015 and Table 28 for California hydrogen station funding awards. Table 29 of the same report [34] provides examples of commercial hydrogen refueling stations that are available today.

2.7 Energy and Global Warming

In order to deal with the problem of global warming, we need to ask ourselves how much energy we need each day to go on with our day-to-day life. As human beings, like animals, we need to eat food to provide energy in our bodies for motion. But we need to use many other kinds of energy in our daily lives, such as energy to light our homes and buildings and to keep them warm or cool, and to cook our food. We also need energy to grow basic food and transport it to stores for the end user to purchase for meals. Energy is also used for transportation and to power all kinds of machinery that may impact our daily lives. Figure 2.8 illustrates energy consumption in the United States by form of usage.

Today, close to 85% of our energy needs are fulfilled by oil, coal, and natural gas. We burn fossil fuels as a means of producing energy and these fossil fuels are formed from plants and animals that lived a long time ago. The world's supply of fossil fuels will not last forever, however, especially when trying to meet the future energy needs of the next generation. Every year we use 1–2% more energy than the year before. By 2050, we will be using more than twice as much energy as produced in 2000 by burning fossil, oil, gas, and so on. Given the fast pace of population growth around the world, more industries are needed to build around them also, and thus energy needs grow accordingly.

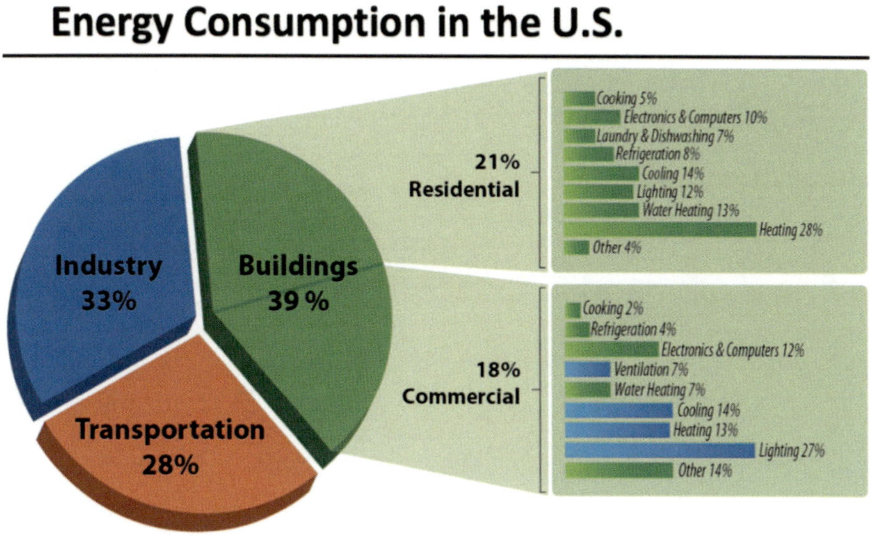

Fig. 2.8 Forms of energy usage in the United States. (Courtesy of the US Department of Energy)

A few countries, for example in the Middle East, have large fossil fuel reservoirs, but many countries are forced to buy the fossil fuels that they need to meet their energy needs. Some fossil fuel supplies might be used up sooner than expected and planned for. The faster we use up the supply of fossil fuels due to demand, the faster prices of crude oil and fossil fuels will rise, and as result some countries with low-end gross domestic product (GDP) may not be able to afford to buy enough oil-based fuel for their day-to-day consumption.

However, the end result of burning these fossil fuels to produce energy is the production of a tremendous amount of pollution globally that will enhance global warming growth and has adverse effects on our environment as well as on animals, particularly wild animals.

The global effects of burning fossil fuels, coal, and gas via global warming can be observed in the Earth's polar ice caps located at the North and South Poles, which are melting faster than usual. We can observe changes in the lives of people and animals living at the Poles. For example, polar bears (Fig. 2.9) and other animals need lots of ice in their Artic habitat to survive.

Global warming is also slowly turning parts of the world into desert due to acid rain. In some places, farmers are having trouble growing food crops and each year more hurricanes and other extreme types of weather occur, which is the result of this unusual growth of global warming. Forest fires are also becoming more common every year. The rate of global warming will keep rising for at least another 50 years, unless we start reducing this rate by turning to cleaner types of fuel as the source of energy for our growing needs.

Hydrogen is one near-term solution that we need to look at it. It is a wonderful fuel, which has the following advantages:

Fig. 2.9 Artic polar bears

Table 2.3 Pros and cons of hydrogen as a source of fuel

Positives	Problems
Renewable source of energy	Can be expensive to produce
Clean source of energy	Producing hydrogen fuel uses fossil fuel
Can be produced in large and small amounts	Harder to store than some other fuels
Can be made anywhere	Can be difficult to transport
Produces much more energy than fossil fuels when burned	Needs careful handling
Great possibilities for the future	Everyday use will require huge changes in vehicle design

- Hydrogen is the simplest element as a basic substance.
- Hydrogen is the first known element, and scientists believe it formed soon after the universe began.
- Hydrogen is the most abundant substance in the universe.
- Hydrogen powers the Sun and all the stars in Galaxy that we know of. Huge clouds of hydrogen gas also float in space.
- Hydrogen is part of many substances found on Earth, so long as we have an unlimited source of water within seven oceans and other places.
- Hydrogen has more energy per weight than any other fuel.

However, hydrogen fuels also have pros and cons at this time, as listed in Table 2.3.

To stop or slow global warming, we need to release fewer waste gases. One way to do this is to burn fewer fossil fuels and another is to use fuels that are renewable, where the source of energy does not run out. One kind of renewable energy that can replace fossil fuels is hydrogen gas, which is rich in energy and is an exciting fuel of the future.

2.8 Hydrogen for the Future

We need to make a lot of changes from what we are used to at the present time in order to switch to using hydrogen as a fuel source, either for transportation or as a new source of energy, as well as using it as a renewable energy element. It involves more than changing to fuel cell cars—we also need to evaluate how these cars and other vehicles will get the hydrogen they need? We will have to build new hydrogen fuel filling stations to match this demand. These hydrogen filling stations will also need to store the hydrogen fuel as liquid hydrogen in tanks safely. The hydrogen must be turned into a gas before being used as part of this innovative approach, as illustrated in Fig. 2.10.

However, while electric charging stations have continued to spring up across the United States, hydrogen fueling stations have not experienced the same rapid growth. This is an issue in California, where Assembly Bill 8 stipulates a goal of 100 functioning public hydrogen stations by 2020. The latest forecast predicts the state will come up short. According to a joint agency staff report from the California Energy Commission and the California ARB, AB-8, published in November 2017, the state was home to 31 hydrogen retail locations at that time. That number has risen to 35 today, according to the Department of Energy's Alternative Fuel Data Center. Another 34 are planned for the future, meaning California will still be roughly 30 retail locations short of its 100-station goal as we approach 2020.

All current and planned retail locations have used funds from the California Energy Commission's Alternative and Renewable Fuel and Vehicle Technology Program. In the report, the Energy Commission and California Air Resource Board (CARB) presented proposed funding changes to the program to accelerate station

Fig. 2.10 Typical infrastructure of hydrogen filling stations

development, including reducing the amount allocated to each station to increase the total number of grants it can make.

We also need facilities and factories to manufacture hydrogen from its unlimited source, which occupies two-thirds of the globe as water. In the near future, power plants that supply our homes with electricity need to switch from using generators that burn fossil fuels as part of green technology and decarbonization of environment as well as to prevent global warming—but this is a huge job.

We will not be able to use our current oil pipelines to transport hydrogen as hydrogen gas can escape through the tiniest hole in a container. Instead, we will need tankers and huge ships with large tanks and pipelines that are airtight for the transportation of this precious fuel.

Given all the demands currently placed on the green environment with existence of carbon dioxide, along with global population growth and the corresponding increased need for cars, we have no choice but to look into hydrogen-driven automobiles. We need to do what we can to prevent the global warming effect.

For example, as part the Green Car Effort, companies such as Mercedes-Benz have created a hybrid bus, which combines fuel cell and diesel electric technology together, as illustrated in Fig. 2.11. The Mercedes-Benz Citaro Fuel Cell Hybrid city bus was presented by Daimler at a conference for public transport authorities in Vienna, Austria in 2009 [35]. The bus is unique in that it employs Daimler's hydrogen-powered fuel cell along with their diesel electric hybrid powertrain called the Citaro G Blue Tec Hybrid. It runs with little to no noise and emits few pollutants, making it ideal for inner-city environments. At the time of the conference, the company expected to make 30 of these revolutionary buses and test them via the European mass transit system.

Fig. 2.11 Daimler Citaro hybrid bus

In 2011 Mercedes-Benz reported that these fuel cell buses had performed as expected and were continuing to operate on a daily basis. According to Mercedes-Benz, the Citaro buses demonstrate the viability of its "environmentally friendly fuel cell drive." At that time, the 36 fuel cell Citaros in operation had logged more than 2.2 million kilometers (1.3 million miles), an average of around 36,000 miles per Citaro.

2.8.1 A High Price

We must do all we can within our power now and in the future to slow global warming, and of course one way is to cut emissions, which are the wastes produced by the present fuels in our cars in the form of gasoline and oil or fossil fuels from our power plants.

However, the issues we encounter in the quest for a new source of energy that provides a clean way to consume it and has no impact on our environment are the TCO and return on investment (ROI).

Fuel cells for automobiles and power plants are a very expensive proposition, given the technology available today to produce them. It will cost a lot of money to allow the current vehicles and power plants to handle hydrogen as fuel. We cannot switch over to hydrogen power until is is cheap enough for everyday usage.

2.8.2 Looking Ahead

From a manufacturing point of view, it is always better and cheaper to mass produce an item or good. Thus, the more we use hydrogen for fuel or utilize any other applications driven by hydrogen, the faster the cost will drop, both from a production perspective and at sale points, such as fill stations.

At first, we might use hydrogen mixed with natural gas—such a mixture produces almost no pollution. Using this mix in any car engine or, for that matter, in any engines and power plants, will help reduce global warming to start with by migrating from traditional fuel to such a mixture.

The cheaper fuel cells get, the more they will be used by consumers. We will use fuel cells in trucks, boats, and even aircraft, although hydrogen is already used as part of rocket fuel for space exploration. Hospitals might use a fuel cell generator for backup electricity in case of a power failure, particularly in the emergency room during surgical operations.

Highrises and buildings that use fuel cells for power will gain two benefits from such a source of energy: the fuel cells will provide electricity, and these buildings will also be warmed by the heat produced by the fuel cells.

A lot of nations globally are involved in the mission to achieve the zero-emission goal. For example, Iceland plans to cut its carbon dioxide emissions to zero by the year 2050 and has already begun working toward changing the kinds of fuel it uses. Because of its location, Iceland uses plenty of geothermal energy as part of the natural heat from Earth stored underground as hot water and steam. Nearly every home in Iceland has a heating system that is heated by geothermal energy. Most of Iceland's vehicles still run on traditional fuel such as oil or a polymer aspect of it such as gasoline. But Iceland is now starting to use hydrogen-powered buses and is making its way toward hydrogen-driven cars for day-to-day transportation by its residents.

The hydrogen used by these buses, for the time being, is produced from water and electricity and the electricity is made without using fossil fuels—thus, the hydrogen used to power Iceland's vehicles is truly a clean energy. Iceland may well reach its goal of zero emissions in the near term.

Beside Mercedes-Benz and its hydrogen drive bus, the Honda Motor and Toyota companies are developing and testing cars powered by a fuel cell stack, as a group of fuel cells, as illustrated in Fig. 2.12.

The fuel cell stack will do two jobs simultaneously. It will:

1. Make enough hydrogen fuel to power a fuel-cell car; and
2. Produce enough extra heat to warm a house.

In the case of a car driven by a fuel cell, during the chemical reaction between hydrogen and oxygen, FCVs generate electricity to power a motor. Instead of gasoline they are fueled by hydrogen, an environmentally friendly energy source that can be produced from a variety of raw materials.

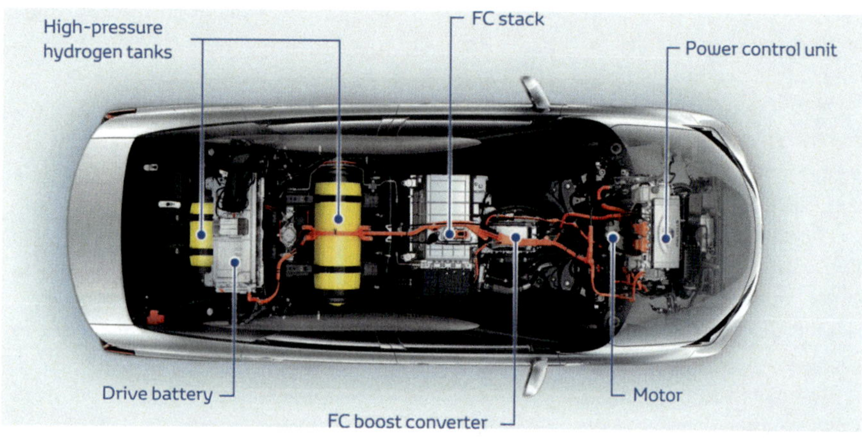

Fig. 2.12 Top view of a fuel cell hydrogen vehicle (FCHV)

2.8.3 The Dawn of Hydrogen as the Future of Fuel Cells

Fifty to 60 years from now, we should be able to make hydrogen cheaply enough from unlimited water sources around the globe to allow hydrogen to become our main energy source. All the hydrogen power we use will then produce only one kind of waste: water residue. The water can be used or released into the air as water vapor. In the air, it will form clouds and later fall as rain, which will run into rivers, lakes, and oceans and can be reused to make more hydrogen for fuel as a recycling process.

Hydrogen fuel cells will play an effective and important part in meeting our future energy needs in terms of electricity and renewable energy as well as being a step towards our goal of zero emissions to prevent carbon dioxide generation. The energy to make hydrogen will come from several sources, although we may need to use fossil fuels to make the first batch of hydrogen; however, wherever possible, we may use wind, solar, nuclear, or other clean sources of energy. Someday, we may even have "hydrogen farms," which will make hydrogen from algae.

As our energy needs grow, we will learn to make clean energy as well as using the energy in a more efficient way. Either way, hydrogen fuel will help us with such achievements.

2.9 Summary

To summarize, I quote United States Congressman Dan Lipinski (2006) (Democrat—Illinois): "Hydrogen can fill critical energy needs beyond transportation. Hydrogen can also be used to heat and generate electricity for homes. The future possibilities of this energy source are enormous".

References

1. G. Taljan, M. Fowler, C. Canizares, G. Verbic, Hydrogen storage for mixed wind--nuclear power plants in the context of a hydrogen economy. Int J Hydrog Energy **33**(17), 4463–4475 (2008)
2. A. Hajimiragha, M.W. Fowler, C.A. Cañizares. Hydrogen economy impact on optimal planning and operation of integrated energy systems. In: *Proceedings of the International Conference & Workshop on Micro-Cogeneration Technologies & Applications*, Ottawa, April 2008
3. Transport Canada. http://www.tc.gc.ca/environment/menu.htm#climatechange. Accessed 2008
4. R.S.M. Arico. Measuring the oil vulnerability of Canadian cities. Simon Fraser University; Spring 2007. http://ir.lib.sfu.ca/bitstream/1892/4211/1/etd2765.pdf. Accessed 2008
5. J. Rifkin, *The hydrogen economy: the creation of the worldwide energy web and the redistribution of power on earth* (Penguin Putnam, New York, 2002)
6. National Research Council and National Academy of Engineering, *The hydrogen economy: opportunities, costs, barriers and R&D needs* (The National Academies Press, Washington, DC, 2004)
7. M.L. Wald, Questions about a hydrogen economy. Sci. Am. **290**, 40–47 (2004)

8. A. Benthem, G. Kramera, R. Ramerb, An options approach to investment in a hydrogen infrastructure. Energy Policy **34**(17), 2949–2963 (2006)
9. S. Ramesohl, F. Merten, Energy system aspects of hydrogen as an alternative fuel in transport. Energy Policy **34**(11), 1251–1259 (2006)
10. K. Adamson, Hydrogen from renewable resources—the hundred-year commitment. Energy Policy **32**(10), 1231–1242 (2004)
11. W. McDowall, M. Eames, Forecasts, scenarios, visions, backcasts and roadmaps to the hydrogen economy: a review of the hydrogen futures literature. Energy Policy **34**(11), 1236–1250 (2006)
12. F.A. Felder, A. Hajos, Using restructured electricity markets in the hydro-gen transition: the PJM case. In: Proc. IEEE: Special Issue on the Hydrogen Economy **94**(10), 1864–1879 (2006)
13. S. Prince-Richard, M. Whale, N. Djilali, A techno-economic analysis of decentralized electrolytic hydrogen production for fuel cell vehicles. Int J Hydrog Energy **30**(11), 1159–1179 (2005)
14. Yohe G.W., Lasco R.D., Ahmad Q.K., Arnell N.W., Cohen S.J., Hope C., Janetos A.C., Perez R.T., 2007: Perspectives on climate change and sustainability. Climate Change 2007: Impacts, Adaptation and Vulnerability. Contribution of Working Group II to the Fourth Assessment Report of the Intergovernmental Panel on Climate Change, Parry M.L., Canziani O.F., Palutikof J.P., van der Linden P.J., Hanson C.E., Eds., Cambridge University Press, Cambridge, UK, 811–841
15. A.B. Lovins, Twenty hydrogen myths. Rocky Mountain Institute, June 2003. http://www.rmi.org/images/other/Energy/E03-05-20HydrogenMyths.pdf. Accessed 2008
16. R. Hammerschlaga, P. Mazzab, Questioning hydrogen. Energy Policy **33**(16), 2039–2043 (2005)
17. U. Bossel, Does a hydrogen economy make sense? In: Proc. IEEE: Special Issue on the Hydrogen Economy **94**(10), 1826–1837 (2006)
18. C.J. Andrews, Formulating and implementing public policy for new energy carriers. In: Proc. IEEE: Special Issue on the Hydrogen Economy **94**(10), 1852–1863 (2006)
19. A. Kulikovsky, *Analytical Modelling of Fuel Cells*, 1st edn. (Elsevier, USA, 2010)
20. B. Zohuri, *Physics of Cryogenics: An Ultralow Temperature Phenomenon*, 1st edn. (Elsevier, USA, 2017)
21. B. Zohuri, *Hybrid Energy Systems: Driving Reliable Renewable Sources of Energy Storage*, 1st edn. (Springer, Cham, 2018)
22. J. Romm. Tesla trumps Toyota: why hydrogen cars can't compete with pure electric cars. https://thinkprogress.org/tesla-trumps-toyota-why-hydrogen-cars-cant-compete-with-pure-electric-cars-326468e3dbc2/
23. Wind-to-Hydrogen Project. Hydrogen and Fuel Cells Research. Golden: National Renewable Energy Laboratory, U.S. Department of Energy. September 2009. Archived from the original on 26 August 2009.
24. B. Berman, Fuel cells at center stage. *New York Times* (2013)
25. A. Davies, Honda is working on hydrogen technology that will generate power inside your car. Business Insider. 2013 Nov 23
26. U.S. DOE, Multi-Year Research, Development and Demonstration Plan: planned program activities for 2005–2015. http://www1.eere.energy.gov/hydrogenandfuelcells/mypp/
27. European Hydrogen and Fuel Cell Technology Platform Implementation Plan – Status 2006, in European Hydrogen and Fuel Cell Technology Platform (HFP) (2007)
28. J.P. Meyers, Getting back into gear: fuel cell development after the hype. https://www.electrochem.org/dl/interface/wtr/wtr08/wtr08_p36-39.pdf
29. H. Tsuchiya, O. Kobayashi, Mass production cost of PEM fuel cell by learning curve. Int. J. Hydrog. Energy **29**, 985 (2004)
30. European Commission, Community Research, "European Hydrogen and Fuel Cell Program", Sixth Framework Program, EUR 21241. http://europa.eu.int/comm/research/rtdinfo_en.html
31. Statistics Canada. http://www.statcan.ca/Daily/English/080327/d080327d.htm Accessed 2008

32. A. Hajimiragha, M. Fowler, C. Cañizares, Hydrogen economy transition in Ontario-Canada considering the electricity grid constraints. Int J Hydrog Energy **34**, 5275–5293 (2009)
33. Ministry of Finance, Ontario, Canada. 2018. http://www.fin.gov.on.ca/english/economy/demographics/projections/2007/demog07t1.html
34. U.S. Department of Energy, Energy Efficiency and Renewable Energy "Fuel Cell Technologies Market Report 2015"
35. Mercedes hybrid bus combined fuel cell and diesel electric technology. Green Car Reports. https://www.greencarreports.com/news/1021808_mercedes-hybrid-bus-combines-fuel-cell-and-diesel-electric-technology

Chapter 3
Hydrogen Driving Thermonuclear Fusion Energy

The thermonuclear fusion reactions in hydrogen isotopes are known to be a new source of energy in two very different situations. At one extreme, relatively slow reactions in a very controlled, confined manner produces the energy emitted by the Sun and stars, whereas at the other, rapid thermonuclear reactions are responsible for the strong thermonuclear power of the hydrogen bomb. Somewhere between these two extremes, it should be possible to bring about thermonuclear reactions under conditions that will allow the energy to be released from hydrogen atoms at a controllable rate for electricity consumption and to meet the need for a new, yet clean, source of energy. We are mainly concerned with confinement of plasma at terrestrial temperatures (i.e., very hot plasmas); our primary interest is in applications to controlled fusion research in magnetic and inertial confinement reactors, such as the tokomak reactor machine, a magnetic confinement device and which involves a laser-driven pellet fusion chamber and micro-balloon glass, which contains the two isotopes of hydrogen to achieve inertial confinement. This chapter takes a very high-level and general approach to the subject of the two types of confinement of interest in the plasma physics of high temperatures for the purpose of thermonuclear fusion reactions that will take place between the two isotopes of hydrogen (H), namely deuterium (D) and tritium (T). The main purpose of this chapter is to provide the necessary general background for scientists and engineers who are planning to enter the field of research into another source of clean energy via controlled thermonuclear reactions [1–3].

3.1 Introduction

Our increased energy demands going forward has forced us to investigate alternative production of energy in a clean way, in addition to producing it via nuclear fission and fossil fuels. Controlled thermonuclear fusion reaction has been suggested as an alternative way of generating energy, either via magnetic confinement or inertial

© Springer International Publishing AG, part of Springer Nature 2019

B. Zohuri, *Hydrogen Energy*, https://doi.org/10.1007/978-3-319-93461-7_3

confinement of plasma to generate heat to produce steam and as result produce electricity to meet such increases in energy demand. Each of these approaches has their own technical and scientific challenges, which scientists need to overcome. This chapter discusses ways of confining plasma and the systems of the confinement, which are able to impose a controlled thermonuclear fusion reaction for this purpose.

Controlled thermonuclear reactions, in which there is a net gain of energy, are possible in principle, and thus could serve as a source of useful energy power—a clean one—from the two hydrogen (H) isotopes that in chemistry are known to us as deuterium (D) and tritium (T). The source of these raw materials would be water or perhaps lithium minerals, which are available all over the world.

The achievement of economical controlled thermonuclear power would have highly significant consequences, both economically and politically. For this compelling reason, and also because of the exceptionally challenging nature of the scientific and technical problems involved, the study of thermonuclear reactions in elements of hydrogen, which is being actively pursued in many countries around the world, will inevitably attract increasing interest.

3.2 Energy Released from Thermonuclear Fusion Reactions

Considering today's technology driving thermonuclear fusion for the release of clean energy, it is probable that the basic "fuel" material for high-temperature fusion reaction power will be deuterium (D) and tritium (T), the two isotopes of hydrogen (H), which are 1–6500 atoms of ordinary hydrogen in water for deuterium, while the percentage of tritium in natural hydrogen is $4 \times 10^{-15}\%$. However, recently some Japanese scientists have claimed that similar fusion reactions may take place in deuterium at low temperatures, with the potential to produce energy power that may be less costly than the earlier approach.

Deuterium was discovered in 1931. Since a neutron weighs just a bit more than a proton, deuterium is slightly more than twice as heavy as protium. Two atoms of deuterium (sometimes called heavy hydrogen) combined with one of oxygen is called heavy water. Oxygen is the same in both regular water and heavy water. It possesses the bulk of the mass. So heavy water is only about 10% heavier than drinking water. Chemically, heavy water resembles regular water, but there are important differences. Figure 3.1 provides an illustration of the deuterium isotope atom.

Percentage wise, the difference in mass between H-1 and H-2 (i.e., D) is much greater than the difference between the isotopes of other elements. Mass affects bond length, and bond length affects other properties. Because water is so important to life, heavy hydrogen is very useful in biological research.

Fig. 3.1 Deuterium isotope atom

Nuclear magnetic resonance (NMR) and infrared (IR) spectroscopic properties differ considerably between protium and deuterium. Thus, labeling organic compounds with one or more deuterium atoms can help in the study of organic reactions. In NMR, various deuterated solvents prove most useful in structure elucidation.

Hydrogen without a neutron is protium. Hydrogen with one neutron is deuterium. Hydrogen with two neutrons is tritium. Tritium alone of the three is less than completely stable—it is radioactive. The ratios of these three forms of hydrogen in the vicinity of the Earth is approximately:

99.98% H-1 (protium) 0.016% H-2 (deuterium) <0.01% H-3 (tritium)

Tritium is an uncommon product of the nuclear fission of uranium-235, plutonium-239, and uranium-233, with production of about one atom for each 10,000 fissions. The release or recovery of tritium needs to be considered in the operation of nuclear reactors, especially in the reprocessing of nuclear fuels and in the storage of spent nuclear fuel. The production of tritium is not a goal, but rather a side effect. It is discharged to the atmosphere in small quantities by some nuclear power plants.

Tritium is the only *radioactive isotope* of hydrogen and is commonly represented by the chemical symbol T or H_3. While the most common form of the hydrogen atom has a nucleus consisting of a single proton, the nucleus of tritium is comprised of three particles: two neutrons and a single proton (Fig. 3.2).

In fact, tritium atoms have a tendency to replace one of the stable hydrogen atoms in water, H_2O, thus becoming part of the water molecule. The resulting compound is known as tritiated water, with the chemical formula HTO or T_2O. Tritiated water is colorless and odorless, just like regular water, and can exist alongside regular water

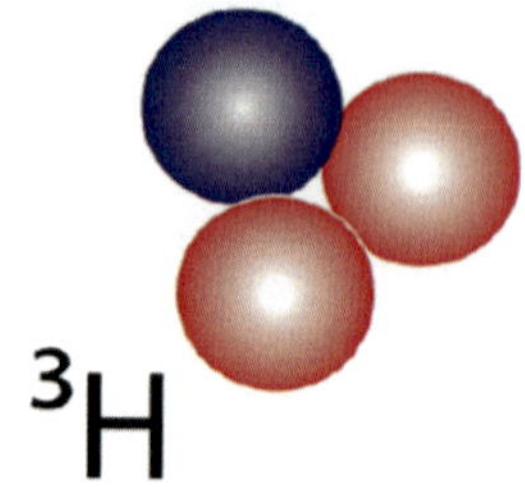

Fig. 3.2 Tritium isotope atom

molecules. Given the chemical properties of tritium and the fact that roughly two-thirds of human body mass is composed of water, it is very common for tritium to exist within the human body.

Energy released from thermonuclear fusion reactions between the two isotopes of hydrogen are obtained from two very different experimental innovative approaches. First, the fusion reaction takes place when the deuterium nuclei ($_1D^2$) alone is involved. Second, and seemingly very common these days, both from an inertial confinement fusion (ICF) [2] and magnetic confinement fusion (MCF) [3] point of view, is the fusion reaction between the deuterium nuclei ($_1D^2$) and tritium ($_1T^3$). In the past decade these have studied among scientists in respect to the lateral fusion reaction and it was determined that the reaction between these two isotopes of hydrogen are very promising at a very high temperature reaction and their cross-sections were determined by laboratory measurements (depicted in Fig. 3.3).

By means of the quantum mechanics theory of Coulomb barriers penetration, it is much more convenient to make use of cross-sections obtained experimentally, as plotted in Fig. 3.9, for reactions such as D–D, D–T, and D–He^3 by bombarding targets containing deuterium, tritium, or helium-3 with deuterons of known energies. Technically, for the purpose of marginal safety measurements of the cross-section, order-of-magnitude estimation is usually performed, at least of the rates or cross-section of thermonuclear reactions obtained experimentally.

It will be observed that the D–T curve demonstrates a maximum at energy of 110 keV, which is an example of the resonance phenomenon, which often occurs in nuclear reactions [1].

However, the appreciable cross-sections for energies well below the top of the Coulomb barriers for each of the reactions studies provides an experimental illustration of the reality of the barrier penetration effect [1].

The other way in which the production of nuclear fusion energy has been achieved on Earth is the Manhattan Project towards the end of World War II (WWII), when the dawn of hydrogen bombs started. Devices known as hydrogen bombs are thermonuclear weapons in which a "fission reaction" is produced at high temperatures and then the nuclei of isotopes of hydrogen undergo a "fusion reaction," liberating uncontrollable and tremendous amounts of energy. Under these circumstances, nuclear reactions are propagated so rapidly that the energy is released in this uncontrolled fashion, just as one can observe in a typical TNT explosive material, yet with much more pressure and the release of energy.

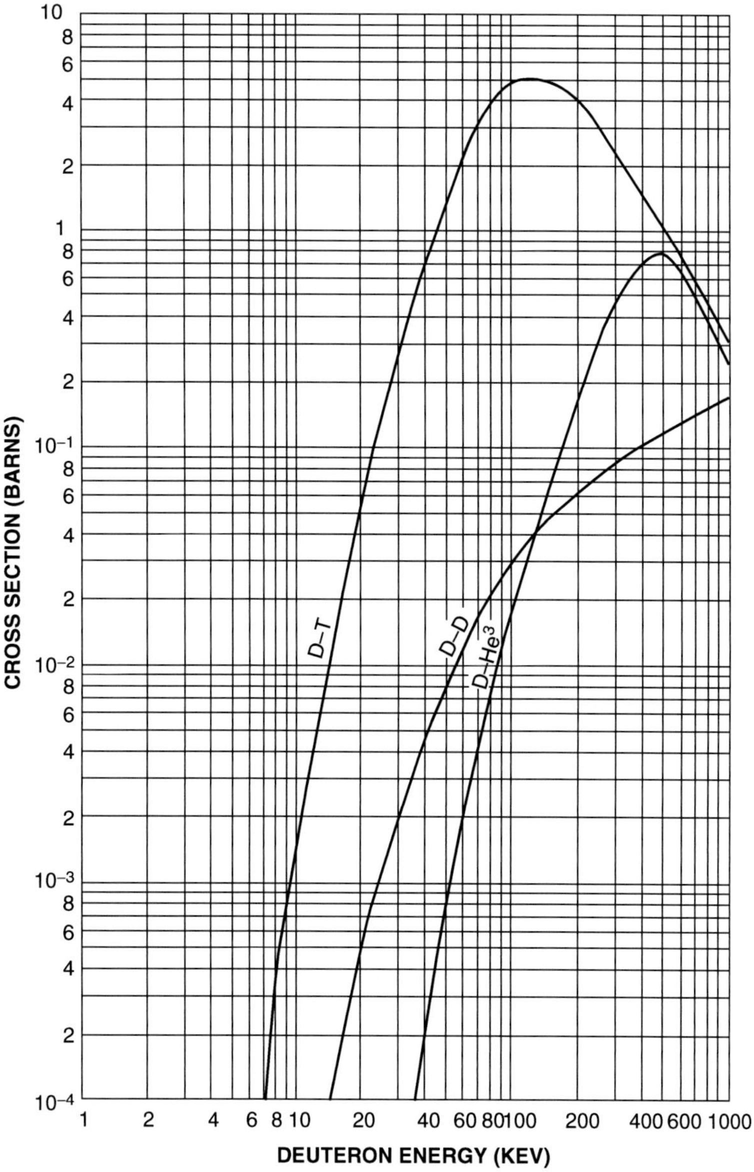

Fig. 3.3 Cross-sections for D–T, D–D total, and D–He^3 reactions. D deuterium, He^3 helium-3, T tritium

3.3 Hydrogen Isotopes

The common hydrogen isotope—protium—consists of only one proton and one electron, while deuterium has one proton and one neutron in its nucleus, and tritium has one proton and two neutrons in its nucleus (Fig. 3.4).

Both deuterium and tritium have identical chemical properties to protium. Protium and deuterium are not radioactive, but tritium is radioactive due to the instability of the larger nuclei. When the nucleus breaks apart, energy is released and radiates out from the atom.

3.4 Thermonuclear Fusion Reactions

As noted earlier, in order to bring about fusion reactions, it is necessary that the interacting nuclei collide with sufficient energy to overcome the forces of electrostatic repulsion, which tend to keep them apart (i.e., Coulombs' barrier) [1].

In a thermonuclear fusion reaction, two light nuclei are forced together and then fuse with an energy yield, as depicted in Fig. 3.5. The reason behind the energy yield

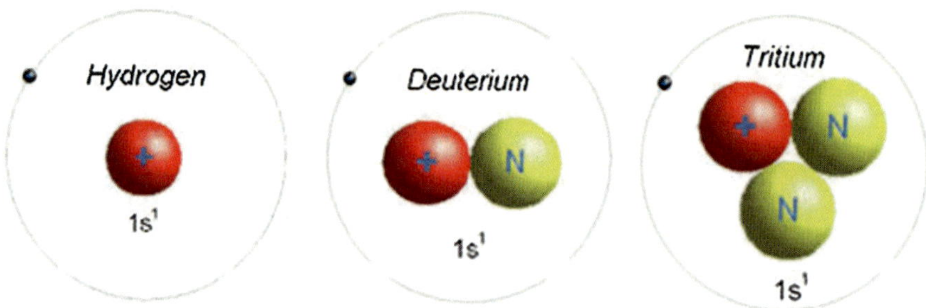

Fig. 3.4 Hydrogen isotopes

Fig. 3.5 Thermonuclear fusion reaction

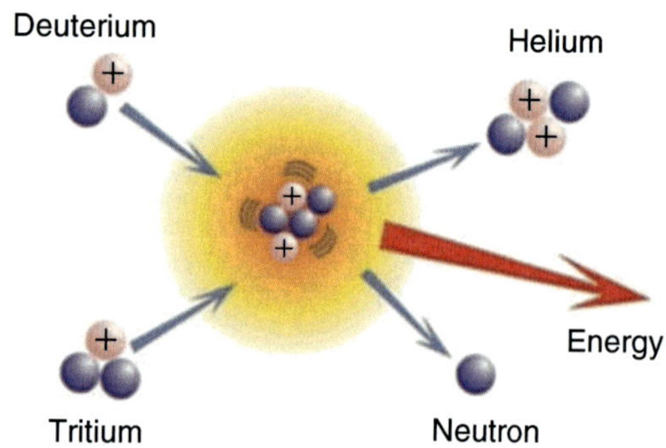

is the fact that the mass of the combined fusion reaction is less than the sum of the masses of the individual nuclei.

If the combined nuclear mass is less than that of iron at the peak of the binding energy curve, then the nuclear particles will be more tightly bound than they were in the lighter nuclei, and that decrease in mass comes off in the form of energy according to the Einstein relationship, known as the theory of relativity. However, for elements heavier than iron, the fission reaction will yield energy.

The Einstein relationship indicates that relativistic energy is presented as follows:

$$E = mc^2 \tag{3.1}$$

where m is the effective relativistic mass of a particle traveling at the speed of light (c). Kinetic energy (KE) can be calculated using and total energy (mc^2) and rest mass energy (m_0) for a particle:

$$\mathrm{KE} = mc^2 - m_0c^2 \tag{3.2}$$

Further analysis of Einstein's relativity theory allows us to blend Eq. 3.2 into the relativistic momentum (p) expression as follows:

$$p = \frac{m_0 \mathrm{v}}{\sqrt{1 - \frac{\mathrm{v}^2}{c^2}}} \tag{3.3}$$

The combination of relativistic momentum (p) and particle speed (c) shows up often in relativistic quantum mechanics and relativistic mechanics as the multiplication pc and can be manipulated as follows (and illustrated in Fig. 3.6):

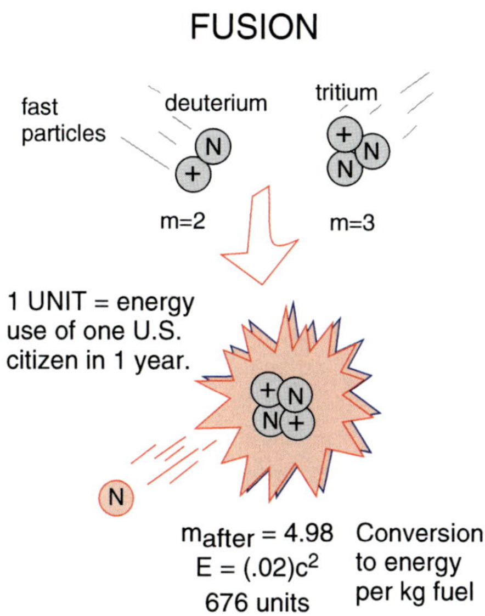

Fig. 3.6 Deuterium–tritium fusion reaction

$$p^2c^2 = \frac{m_0^2 v^2 c^2}{1-\frac{v^2}{c^2}} = \frac{m_0^2 \frac{v^2}{c^2} c^4}{1-\frac{v^2}{c^2}} \tag{3.4a}$$

By adding and subtracting a term, it can be put in the following form:

$$p^2c^2 = \frac{m_0^2 c^4 \left[\frac{v^2}{c^2}-1\right]}{1-\frac{v^2}{c^2}} + \frac{m_0^2 c^4}{1-\frac{v^2}{c^2}} = -m_0^2 c^4 + \left(mc^2\right)^2 \tag{3.4b}$$

which may be rearranged to give the following expression for energy:

$$E = \sqrt{p^2c^2 + (m_0c^2)^2} \tag{3.4c}$$

Note that again, m_0 is the rest mass and m is the effective relativistic mass of the particle of interest at very high speed c.

As per Eq. 3.4c, the relativistic energy of a particle can also be expressed in terms of its momentum in the expression:

$$E = mc^2 = \sqrt{p^2c^2 + m_0^2c^4} \tag{3.5}$$

The relativistic energy expression is the tool used to calculate binding energies of nuclei and energy yields of both nuclear fission and thermonuclear fusion reactions.

Bear in mind that the nuclear binding energy rises from the fact that nuclei are made up of protons and neutrons, but the mass of a nucleus is always less than the sum of the individual masses of the protons and neutrons which constitute it. The difference is a measure of the nuclear binding energy, which holds the nucleus together. This binding energy can be calculated from the Einstein relationship:

$$\text{Nuclear Binding Energy} = \Delta mc^2 \tag{3.6}$$

Now that we have a better understanding of the physics of thermonuclear fusion reactions and have explained what the Coulomb barriers and energy is all about, our attention is turned to the thermonuclear fusion reaction of hydrogen, which is a fundamental chemical element of generating energy-driven controlled fusion.

According to Glasstone and Lovberg [4], "because of the increased height of the Coulomb energy barrier with increasing atomic number, it is generally true that, at a given temperature, reactions involving the nuclei of hydrogen isotopes take place more readily than do analogous reactions with heavier nuclei. In view of the greater abundance of the lightest isotope of the hydrogen, with mass number 1, it is natural to see if thermonuclear fusion reactions involving this isotope could be used for the release of energy" [1].

Unfortunately, the three possible reactions between hydrogen (H) nuclei alone or with deuterium (D) or tritium (T) nuclei are known to have cross-sections that are too small to permit a net gain of energy at temperatures which may be regarded as attainable [1]. The three reactions are as follows:

$$_{1}H^{1} + _{1}H^{1} \rightarrow _{1}D^{2} + _{1}e^{0}$$
$$_{1}H^{1} + _{1}D^{2} \rightarrow _{1}D^{2} + \gamma$$
$$_{1}H^{1} + _{1}T^{3} \rightarrow _{2}He^{4} + \gamma$$

Consequently, recourse must be had to the next most abundant isotope—deuterium—and two reactions that occur at approximately the same rate over a considerable range of energies are of interest. These are the D–D reactions:

$$_{1}D^{2} + _{1}D^{2} \rightarrow _{2}He^{3} + _{1}n^{0} + 3.27\ \mathrm{MeV}$$

and

$$_{1}D^{2} + _{1}D^{2} \rightarrow _{2}T^{3} + _{1}H^{0} + 4.03\ \mathrm{MeV}$$

These are called the "neutron branch" and the "proton branch," respectively. The tritium produced in the proton branch or obtained in another way, as explained below, can then react, at a considerably faster rate, with deuterium nuclei in the D–T reaction, as follows:

$$_{1}D^{2} + _{1}T^{3} \rightarrow _{2}He^{4} + _{1}n^{0} + 17.60\ \mathrm{MeV}$$

The He^{3} formed in the first D–D reaction can also react with deuterium. Thus:

$$_{1}D^{2} + _{1}He^{3} \rightarrow _{2}He^{4} + _{1}H^{0} + 18.30\ \mathrm{MeV}$$

This reaction is of interest because, as in the D–T reaction, there is a large energy release; the D–He^{3} reaction is, however, slower than the others at low thermonuclear temperatures, but its rate approached that of the D–D reactions at 100 KeV, as demonstrated in Fig. 2.9.

In the methods currently being considered for production of thermonuclear power, the fast neutrons produced in neutron branch of the D–D reactions and in the D–T reactions would most probably escape from the immediate reactions environment. Thus, a suitable moderator to slow down these neutrons would be water, lithium, or beryllium, and the heat liberated as a result of their kinetic energy can be utilized.

$$_{3}Li^{6} + _{0}n^{1} \rightarrow _{2}He^{4} + _{1}T^{3} + 4.6\ \mathrm{MeV}$$

The slow neutrons can then be captured in lithium-6, which constitutes 7.5 atomic percent of natural lithium, using the reaction above, leading to the production of tritium. The energy released can be used as heat, and the tritium can, in principle, be transferred to the thermonuclear system to react with deuterium.

If we produce a sufficient initial ignition temperature for the four thermonuclear reactions above, all four fusion processes will take place, and the two neutrons produced would subsequently be captured by lithium-6, as can be seen in the chemical reaction between $_{3}Li^{6}$ and the neutron.

The analytical expression for σ and $\overline{\sigma v}$ (Fig. 3.8) for the D–D and D–T fusion reactions can be obtained using Eq. 3.7, as given here in a somewhat modified form [1]:

$$R(E) \approx \frac{\mathrm{C}}{E^{3/2}T^{3/2}} \exp\left[-\frac{2^{3/2}\pi^2 M^{1/2} Z_1 Z_2 e^2}{hE^{1/2}} - \frac{E}{kT}\right] \tag{3.7}$$

The relative kinetic energy E of the nuclei is given as follows:

$$E = \frac{1}{2}Mv^2 \tag{3.8}$$

where v is the relative velocity, and the deuteron energy (E_D), in terms of which cross-section is expressed, is $m_\mathrm{D}v^2/2$, where m_D is the mass of the deuteron. Hence, $(M/E)^{1/2}$ in Eq. 3.7 may be replaced by $(m_\mathrm{D}/E_\mathrm{D})^{1/2}$ since Z_1 and Z_2 are both unity. The result then is:

$$\begin{aligned}\sigma(E_\mathrm{D}) &= \frac{\mathrm{C}}{E_\mathrm{D}} \exp\left[-\frac{2^{3/2}\pi^2 {m_\mathrm{D}}^{1/2} e^2}{h{E_\mathrm{D}}^{1/2}}\right] \\ &= \frac{\mathrm{C}}{E_\mathrm{D}} \exp\left[-\frac{44.24}{E_\mathrm{D}^{1/2}}\right]\end{aligned} \tag{3.9}$$

where E_D is expressed in kilo-electron volts. Note that the potential factor is the same for both the D–D and D–T thermonuclear fusion reactions, with the deuteron as the projectile particle. The factor preceding the exponential will, however, be different in the two cases (Fig. 3.7) [1].

Now if we are interested in mean free path reaction λ, in a system containing n nuclei/cm^3 of a particular reacting species, then λ is the average distance traveled by a nucleus before it undergoes a reaction, and is equal to $1/n\sigma$, where σ is the cross-section for the given reaction [1].

We replace σ with $\overline{\sigma}$, if we consider a Maxwellian distribution, and in this case the averaged cross-section $\overline{\sigma}$ is taken over all energies from zero to infinity, at a given kinetic temperature.

Figure 3.8 is an illustration of the mean free path values for a deuteron in centimeters as a function of the deuteron particle density n, in nuclei/cm^3, for the D–D and D–T reactions at two kinetic temperatures, 10 and 100 KeV, in each case. Temperatures of this order of magnitude would be required in a controlled thermonuclear fusion reactor.

The particle of interest for the possible fusion reaction for the controlled thermonuclear process has a possible density of about 10^{15} deuterons/cm^3, and the mean free path at 100 KeV for the D–D reactions, according to Fig. 3.8, is about 2×10^{16} cm. This statement translates to the fact that, at the specified temperature and particle density, a deuteron would travel, on average, a distance of 120,000 miles before reacting. For the D–T reaction, the mean free paths are shorter, because of the large

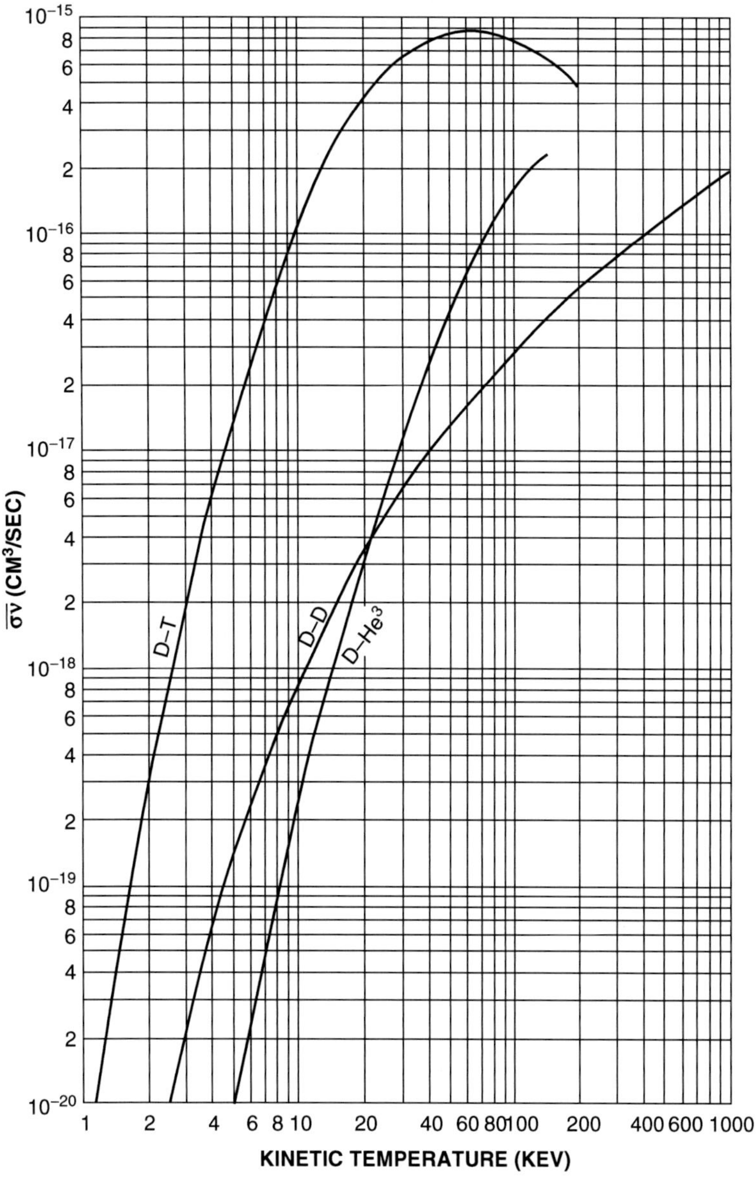

Fig. 3.7 Values of $\overline{\sigma v}$ based on Maxwellian distribution for D–T, D–D (total), and D–He^3 reactions. D deuterium, He^3 helium-3, T tritium

cross-sections for deuterons of given energies, but they are still large in comparison with the dimensions of normal equipment. All these results play a part in the problem of confinement of the particles in a thermonuclear fusion reacting system such as a Tokamak machine or any other means.

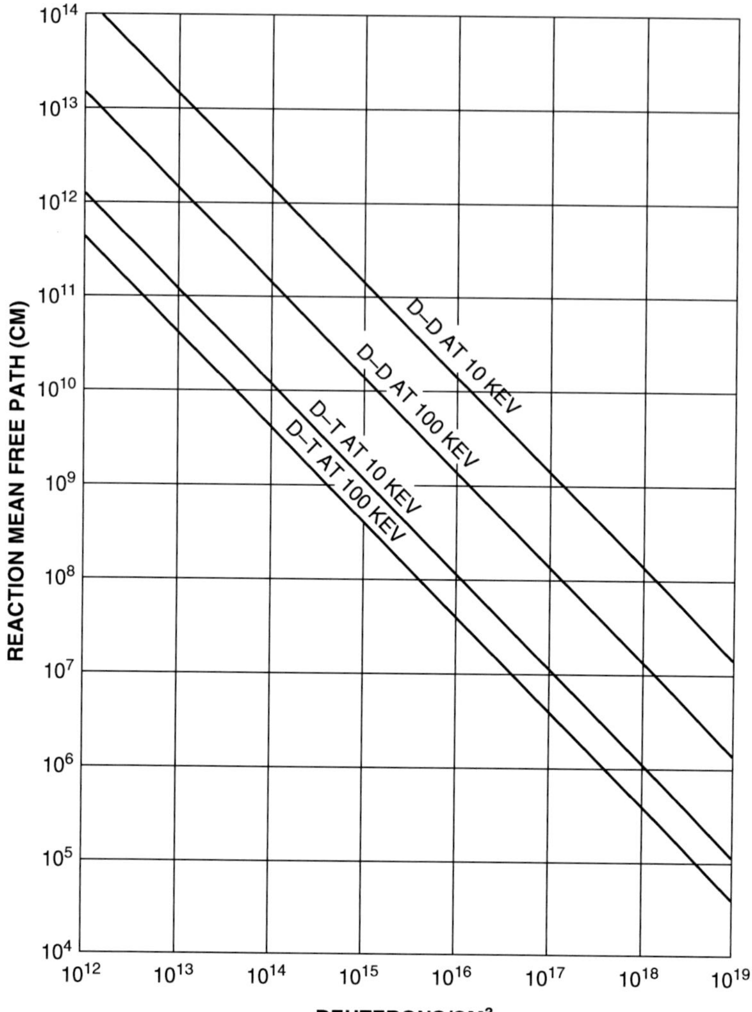

Fig. 3.8 Mean free path for D–T and D–D (total) thermonuclear reactions. D deuterium, T tritium

For purpose of obtaining a power density P_{DD} for a thermonuclear fusion reaction, such as D–D, we use either Eqs. 3.19 or 3.21 to calculate the rate of thermonuclear energy production. If we assume an amount of average energy Q in erg is produced per nuclear interaction, then using Eq. 3.21 it follows that:

$$\text{Rate of energy release} = \frac{1}{2} n_D^2 \overline{\sigma v} Q \ \text{ergs}/\left(\text{cm}^3\right)(\text{s}) \tag{3.10}$$

If the dimension of power density P_{DD} is given in Watts/cm^3, which is equal to 10^7 ergs/(cm^3) (s), then we can write:

$$P_{\mathrm{DD}} = \frac{1}{2} n_{\mathrm{D}}^2 \overline{\sigma v} Q \times 10^{-7} \tag{3.11}$$

with n_{D} in deuterons/cm^3, $\overline{\sigma v}$ in cm^3/s, and average energy Q in erg.

For every two D–D interactions, an average of 8.3 MeV of energy is deposited within the reacting system. The energy Q per interaction is thus (1/2) $\times$ 8.3 $\times$ 1.60 $\times 10^{-6} = 6.6 \times 10^{-6}$ erg, and upon substitution into Eq. 3.11, it yields that:

$$P_{\mathrm{DD}} = 3.3 \times 10^{-13} n_{\mathrm{D}}^2 \overline{\sigma v} \quad \mathrm{Watts/cm^3} \tag{3.12}$$

Utilizing Eq. 3.11, the example of a D–D reaction at 10 KeV can be given. From Fig. 3.7 or, for a given kinetic temperature, we see that $\overline{\sigma v}$ is equal to 8.6×10^{-19} cm^3 /sec, and therefore the power density is:

$$P_{\mathrm{DD}}(10\ \mathrm{KeV}) = 2.8 \times 10^{-31} n_{\mathrm{D}}^2\, \mathrm{Watts/cm^3} \tag{3.13}$$

and at 100 KeV, when $\overline{\sigma v}$ is equal to 3.0×10^{-17} cm^3/s, the power density will be as follows:

$$P_{\mathrm{DD}}(100\ \mathrm{KeV}) = 10^{-29} n_{\mathrm{D}}^2\, \mathrm{Watts/cm^3} \tag{3.14}$$

Similar analysis can be performed for the thermonuclear fusion reaction of D–T, knowing that if the energy remaining in the system per interaction is 3.5 MeV, that is, 3.5 $\times$ 1.6 $\times 10^{-6}$ erg, then the reaction rate is given by Eq. 3.20 in Sect. 3.5, and, therefore, the thermonuclear reactor density power is as follows:

$$P_{\mathrm{DT}} = \frac{1}{2} n_{\mathrm{D}} n_{\mathrm{T}} \overline{\sigma v} Q \times 10^{-7} \tag{3.15}$$

where, in this case, the average energy Q is 5.6×10^{-6} erg; hence:

$$P_{\mathrm{DT}}(10\ \mathrm{KeV}) = 6.2 \times 10^{-13} n_{\mathrm{D}} n_{\mathrm{T}}\, \mathrm{Watts/cm^3} \tag{3.16}$$

and

$$P_{\mathrm{DT}}(100\ \mathrm{KeV}) = 4.5 \times 10^{-28} n_{\mathrm{D}} n_{\mathrm{T}}\, \mathrm{Watts/cm^3} \tag{3.17}$$

There is no exact parallel correlation between the conditions of heat transfer and operating pressures, which limit the power density of a fission reactor, and those that might apply to a thermonuclear fusion reactor. Nevertheless, similar limitations must apply to power transfer in a continuously operating thermonuclear reactor as in other electrical power systems.

A large steam-powered electricity-generating plant has about 500 megawatts of power, that is, 5×10^8 watts. Figure 3.9 illustrates that 100 KeV in a D–D reactor with a power of 5×10^8 watts would provide a reacting volume of only 0.03 cm^3, with deuteron particle densities equivalent to those at a standard temperature.

Meanwhile, the gas kinetic pressure exerted by the thermonuclear fuel would be about 10^7 atm or 1.5×10^8 psi. Since the mean reaction lifetime is only a few

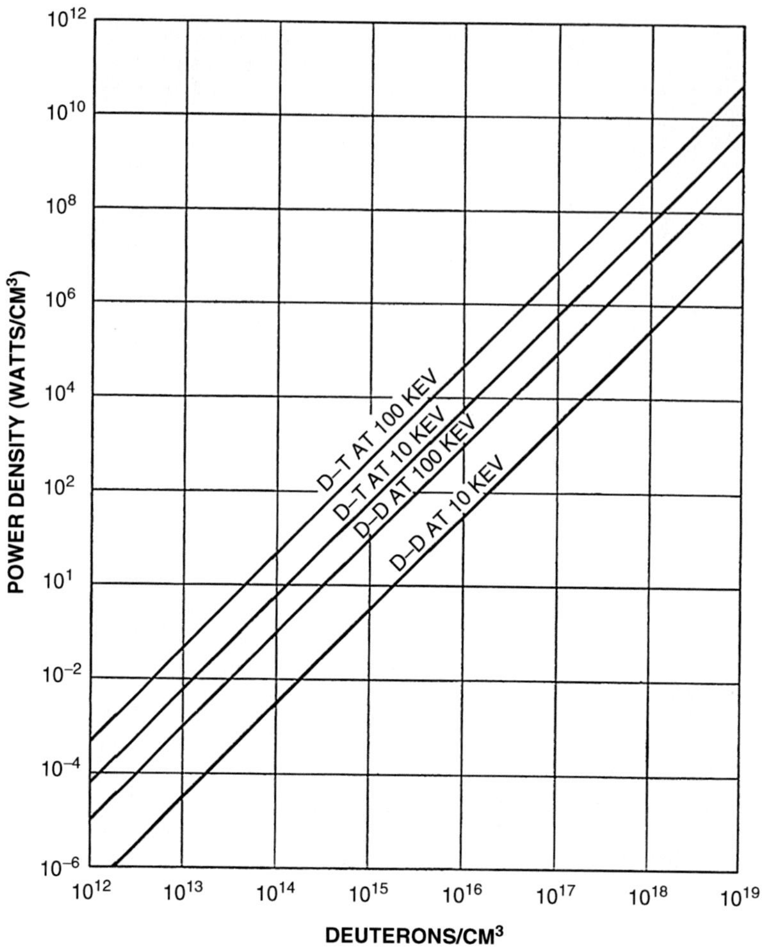

Fig. 3.9 Power densities for D–T and D–D (total) thermonuclear reactions. D deuterium, T tritium

milliseconds under the conditions specified, it is obvious that the situation would be completely impractical [1].

From what have discussed so far, it seems that the particle density in a practical thermonuclear reactor must be near 10^{15} nuclei/cm^3. Other problems associated with the controlled thermonuclear fusion reaction for plasma confinement and this is, why the density cannot be much larger, and it can be explained via stability requirements that is frequently restricted by dimensionless ratio, β. This ratio is defined as part of the plasma confinement driven by the magnetic field, which is equal to the kinetic pressure of the particles in plasma in terms of its ratio to the external magnetic pressure or energy density β, which is defined by Eq. 3.15.

Details of this dimensionless parameter are also defined toward the end of this chapter (Sect. 3.5).

3.5 Rates of Thermonuclear Reactions

Consider a binary reaction in a system containing n_1 nuclei/cm^3 of one reacting element and n_2 of the other. To determine the rate at which the two nuclear elements interact, it may be supposed that the nuclei of the first element form a stationary lattice within the nuclei of the second element and move at random with a constant velocity υ cm/s, equal to the relative velocity of the nuclei. The total cross-section for all the stationary nuclei in 1 cm^3 is then $n_1\sigma$ nuclei/cm. This gives the number of nuclei of the first element with which each nucleus of the second element will react while traveling a distance of 1 cm. The total distance traversed in 1 s by all the nuclei of the latter type present in 1 cm^3 is equal to $n_2\upsilon$ nuclei/(cm^2) (s). Hence, the nuclear reaction rate R_{12} is equal to the product of $n_1\sigma$ and $n_2\upsilon$. Thus:

$$R_{12} = n_1 n_2 \sigma\upsilon \text{ Interaction}/\left(\text{cm}^2\right)(\text{s}) \tag{3.18}$$

If the reaction occurs between two nuclei of the same type, for example two deuterons, so that n_1 and n_2 are equal, the expression for the nuclear reaction rate, represented by R_{11}, becomes:

$$R_{11} = \frac{1}{2} n^2 \sigma\upsilon \text{ Interaction}/\left(\text{cm}^2\right)(\text{s}) \tag{3.19}$$

where n is the number of reactant nuclei/cm^3 (see Fig. 3.6).

So that each interaction between identical nuclei is not counted twice, the factor of 1/2 is introduced into Eq. 3.19.

Going forward, Eqs. 3.18 and 3.19 are applicable when the relative velocity of the interacting nuclei is constant, as is true, approximately at least, for high-energy particles from an accelerator. However, for a thermonuclear reaction, there would also be a distribution of velocities and energies over a wide range.

As depicted in Fig. 3.10 (right side), the cross-section of the reaction is dependent on the energy or velocity and, generally speaking, it follows that the product $\sigma\upsilon$ in

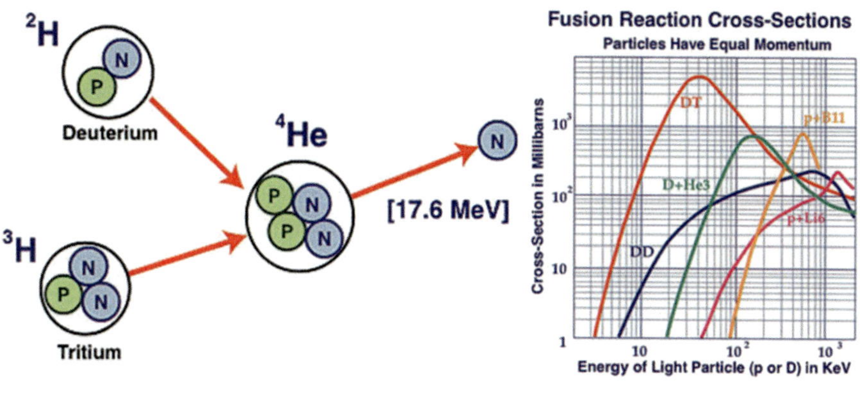

Fig. 3.10 Depiction of all hydrogen isotopes' thermonuclear reactions

Eqs. 3.18 and 3.19 needs to be replaced with a value such as $\overline{\sigma v}$, which is averaged over the whole range of relative velocities. Thus, Eq. 2.27 is written as follows:

$$R_{12} = n_1 n_2 \overline{\sigma v} \text{ Interaction}/\left(\text{cm}^2\right)(\text{s}) \tag{3.20}$$

Accordingly, Eq. 3.19 becomes:

$$R_{11} = \frac{1}{2} n^2 \overline{\sigma v} \tag{3.21}$$

I have demonstrated [1] that the Maxwellian distribution, in terms of mass M and kinetic energy $E = Mv^2/2$, is given by:

$$dn = n \left(\frac{M}{2\pi kT}\right)^{3/2} \exp\left(-\frac{Mv^2}{2kT}\right) v^2 dv \tag{3.22}$$

where k is the Boltzmann constant, T is temperature, and dn is the number of particles whose velocities relative to that of a given particle lie in the range from v to $v + dv$. Hence, it follows that:

$$\overline{\sigma v} = \frac{\int_0^\infty \sigma v dn}{\int_0^\infty dn} = \frac{\int_0^\infty \sigma v \left[\exp\left(-\frac{Mv^2}{2kT}\right) v^2 dv\right]}{\int_0^\infty \exp\left(-\frac{Mv^2}{2kT}\right) v^2 dv} \tag{3.23}$$

The integral in the denominator of Eq. 3.23 is equal to $[(2kT/M)^{3/2}](\pi^{1/2}/4)$, and so this equation becomes:

$$\overline{\sigma v} = \frac{4}{\pi^{1/2}} \left(\frac{Mv^2}{2kT}\right) \int_0^\infty \sigma \exp\left(-\frac{Mv^2}{2kT}\right) v^2 dv \tag{3.24}$$

The integral term in Eq. 3.24 can be evaluated by changing the variable. Since nuclear cross-sections are always determined and expressed as a function of the energy of the bombarding particle, the bombarded particle being essentially at rest in the target, the actual velocity of the bombarding nucleus is also its relative velocity. Hence, if E is the actual energy, in the laboratory system, of the bombarding nucleus of mass m, then E is written as follows:

$$E = \frac{1}{2} mv^2 \tag{3.25a}$$

So that:

$$v = \left(\frac{2E}{m}\right)^{1/2} \tag{3.25b}$$

And, differentiating both sides of Eq. 3.25b, we get:

$$\upsilon^2 d\upsilon = \frac{2E}{m^2} dE \tag{3.25c}$$

Substitution of Eq. 3.25c into Eq. 3.24 yields:

$$\overline{\sigma\upsilon} = \frac{4}{\pi^{1/2}} \left(\frac{M}{2kT}\right)^{3/2} \frac{1}{m^2} \int_0^\infty \sigma \exp\left(-\frac{ME}{mkT}\right) E dE \tag{3.26}$$

where σ in the integrand is the cross-section for a bombarding nucleus of mass m and energy E.

If temperature T in Eq. 3.26 is expressed in kilo-electron volts, and the values of E are in the same units, it is convenient to rewrite Eq. 3.26 in a new form as follows:

$$\overline{\sigma\upsilon} = \left(\frac{8}{\pi^{1/2}}\right)^{1/2} \frac{M^{3/2}}{m^2} \int_0^\infty \sigma \exp\left(-\frac{ME}{mT}\right) \frac{E}{T} dE \tag{3.27}$$

where the quantity E/T is dimensionless. If σ, determined experimentally, can be expressed as a relatively simple function of E, as is sometimes the case, the integration in Eq. 3.27 may be performed analytically. Alternatively, numerical methods, for example, Simpson's rule, may be employed.

In any event, the values of $\overline{\sigma\upsilon}$ for various kinetic temperatures can be derived from Eq. 3.27, based on a Maxwellian distribution of energies or velocities, and the results can be inserted in Eqs. 3.20 or 3.21 to give the rate of a thermonuclear reaction at a specified temperature.

3.6 Critical Ignition Temperature for the Fusion Reaction

The fusion temperature obtained by setting the average thermal energy equal to the Coulomb barrier gives too high a temperature because fusion can be initiated by those particles which are out on the high-energy tail of the Maxwellian distribution of particle energies. The critical ignition temperature is lowered further by the fact that some particles that have energies below the Coulomb barrier can tunnel through the barrier.

The presumed height of the Coulomb barrier is based on the distance at which the nuclear strong force could overcome the Coulomb repulsion. The required temperature may be overestimated if the classical radii of the nuclei are used for this distance, since the range of the strong interaction is significantly greater than a classical proton radius. With all these considerations, the critical temperatures for the two most important cases are approximately:

$$\text{Deuterium} - \text{Deuterium Fusion} : 40 \times 10^7 \text{K}$$

Fig. 3.11 Physical shape of the Tokamak fusion test reactor (TFTR) at the Princeton Plasma Physics Laboratory

$$\text{Deuterium} - \text{Tritium Fusion}: 4.5 \times 10^7 \text{K}$$

The Tokamak fusion test reactor (TFTR), for example, reached a temperature of 5.1×10^8 K, well above the critical ignition temperature for D–T fusion. The TFTR was the world's first magnetic fusion device to perform extensive scientific experiments with plasmas composed of 50/50 deuterium/tritium (D–T), the fuel mix required for practical fusion power production, and also the first to produce more than 10 million watts of fusion power.

The TFTR (see Fig. 3.11) was an experimental Tokamak built at the Princeton Plasma Physics Laboratory (Princeton, New Jersey, USA) circa 1980. Following on from the Poloidal Diverter Experiment (PDX) and Princeton Large Torus (PLT) devices, it was hoped that the TFTR would finally achieve fusion energy break-even; unfortunately, the TFTR never achieved this goal. However, it did produce major advances in confinement time and energy density, which ultimately contributed to the knowledge base necessary to build the International Thermonuclear Experimental Reactor (ITER). The TFTR operated from 1982 to 1997.

The ITER is an international nuclear fusion research and engineering megaproject, which will be the world's largest magnetic confinement plasma experiment. It is an experimental Tokamak nuclear fusion reactor, which is being built next to the Cadarache facility in Saint-Paul-lès-Durance in the south of France. Figure 3.12 depicts a sectional view of the ITER, providing a scale comparison to the man standing to the lower right of the picture.

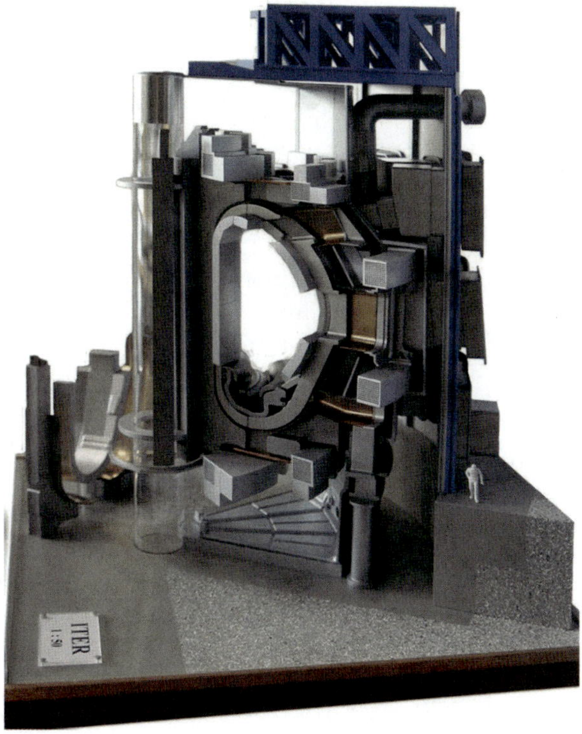

Fig. 3.12 Sectional view of the International Thermonuclear Experimental Reactor's (ITER) Tokamak

Therefore, in summary, the temperature required for fusion to occur to overcome the Coulomb barrier is so high as to require extraordinary means for its achievement.

In the Sun, the proton–proton cycle of fusion is presumed to proceed at a much lower temperature because of the extremely high density and high population of particles.

$$\text{Interior of the sun, proton cycle}: 1.5 \times 10^7\text{K}$$

3.7 Ideal Ignition Temperature for Controlled Thermonuclear Fusion

The minimum operating temperature for a self-sustaining thermonuclear fusion reactor of the MCF type is that at which the energy deposited by nuclear fusion within the reacting system just exceeds that lost from the system as a result of Bremsstrahlung emission, which is thoroughly described in this section.

To determine this value, the rates of thermonuclear energy production at a number of temperatures needs to be calculated, using Eqs. 3.12 and 3.15 together with Fig. 3.7, for *charged-particle products only*, and the results need to be compared with the rates of energy loss as Bremsstrahlung emissions derived using the following equations (Eqs. 3.28 and 3.29):

$$P_{\mathrm{DD(br)}} = 5.5 \times 10^{-31} n_{\mathrm{D}}^2 T_e^{1/2} \quad \mathrm{Watts/cm^3} \tag{3.28}$$

and

$$P_{\mathrm{DT(br)}} = 2.14 \times 10^{-30} n_{\mathrm{D}} n_{\mathrm{T}} T_e^{1/2} \quad \mathrm{Watts/cm^3} \tag{3.29}$$

Note that these two equations are established with the assumption that, for a plasma consisting only of hydrogen isotopes, $Z = 1$ and n_i and n_e are equal, so that the factor $n_e \sum (n_i Z^2)$ may be replaced by n^2 where n is the particle density of either electrons or nuclei.

Note that the factor $n_e \sum (n_i Z^2)$ is sometimes written in the form $n_e^2 \left(\sum n_e Z^2 / \sum n_i Z \right)$, since n_e is equal to $\sum n_i Z$.

The assumption that we have made here and utilizing both Eqs. 3.28 and 3.29 arises from the fact that the kinetic ion (nuclear) temperature and the electron temperature are the same in the plasma.

To illustrate the ideal ignition temperature schematically, we take n_{D} to be as 10^{15} nuclei/cm^3 for the D–D reactions, whereas n_{D} and n_{T} are both 0.5×10^{15} nuclei/cm^3 for the D–T reaction. This makes Bremsstrahlung losses the same for the two cases. The results of the calculations are shown in Fig. 3.13.

The energy rates are expressed in terms of the respective power densities, that is, the energy produced or lost per unit of time per unit of volume of reaching system. It seems that the curve for the rate of Bremsstrahlung energy loss intersects the D–T and D–D energy production curves at the temperatures of 4 and 36 KeV, that is, 4.6×10^7 and 4.1×10^8 K, respectively. These are sometime called the *ideal ignition temperatures*.

If we assume a Maxwellian distribution of electron velocities, then for the rate of Bremsstrahlung energy emission per unit volume, it provides an accurate equation for the total power radiation (P_{br}):

$$P_{\mathrm{br}} = g \frac{32\pi}{3\sqrt{3}} \cdot \frac{(2\pi kT)^{1/2} e^6}{m_e^{3/2} c^3 h} n_e \sum n_i Z^2 \tag{3.30}$$

This equation is explained in more detail in Case I. The ideal ignition temperature values defined earlier are the lowest possible operating temperatures for a self-sustaining thermonuclear fusion reactor. For temperatures lower than the ideal ignition values, the Bremsstrahlung loss would exceed the rate of thermonuclear energy deposition by charged particles in the reacting system [1].

Two other factors also exist which require the actual plasma kinetic temperature to exceed the ideal ignition temperature vales given earlier. These are in addition to the various losses beside the Bremsstrahlung radiation losses that possibly can be minimized, but not completely eliminated, in a thermonuclear fusion power plant reactor [1].

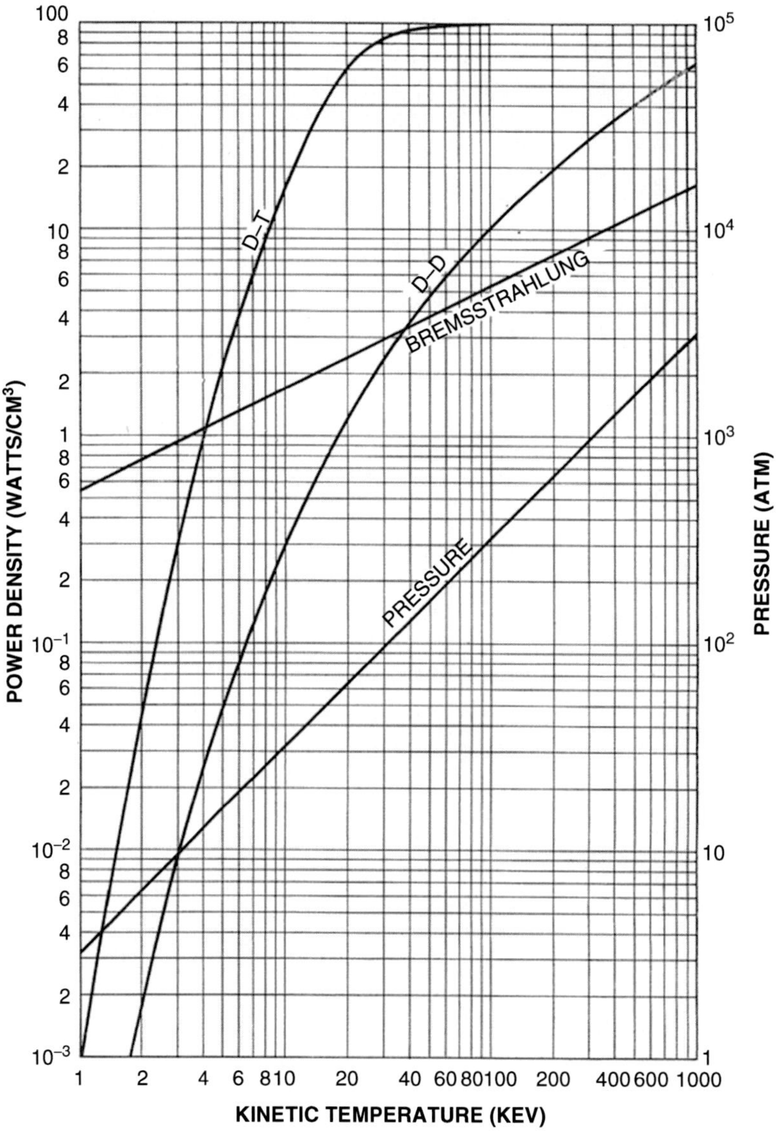

Fig. 3.13 Characteristics of thermonuclear fusion reactions and the ideal ignition temperature [1, 4]

1. We have not yet considered the Bremsstrahlung emission as described in Sect. 2.11 of Zohuri [1] arising from Coulomb interaction of electrons with the helium nuclei produced in the thermonuclear fusion reactions (as shown in Figure 2.20 of Zohuri [1]). Since they carry two-unit charges, the loss of energy will be greater than for the same concentration of hydrogen isotope ions.

2. At the high temperatures present in a thermonuclear fusion reaction, the production of Bremsstrahlung emissions due to electron–electron interactions are very distinct from those resulting from the electron–ion interactions considered above. This is a concern, as the relativistic effects should not occur and there should not be any electron–electron Bremsstrahlung emissions, but at high electron velocities this is not the case and appreciable losses can occur from this form of radiation.

In addition to power densities, Fig. 3.13 reveals the pressures at the various temperature stages, based on the ideal gas equation $p = (n_i + n_e)kT$, where $(n_i + n_e)$ is the total number of particles of nuclei and electrons, respectively, per cm^3 and T is the presentation of kinetic temperature in Kelvin. Under the present condition, $n_i = n_e = 10^{15}$ particles/cm^3, so that $(n_i + n_e) = 2 \times 10^{15}$.

With k having dimension of erg/^{0}K, the values are found in the dimension of dynes/cm^2, and the results have been converted to atmospheres assuming 1 atm $= 1.01 \times 10^6$ dynes/cm^2 and then plotted in Fig. 2.15. This figure also shows that the thermonuclear power densities near the ideal ignition temperatures are in the range of 100–1000 watts/cm^3, which would be reasonable for continuous reactor operation of a thermonuclear fusion reaction, and that is the reason behind choosing the density values of 10^{15} nuclei/cm^3 for the purpose of illustrating reacting particles [1].

It should be noted that although the energy emitted as Bremsstrahlung emissions may be lost as far as maintaining the temperature of the thermonuclear reacting system is concerned, it would not be a complete loss in the operating fusion reactor. Zohuri [1] and Glasstone and Lovberg [4] demonstrate that the energy distribution of the electron velocities is Maxwellian or approximately so and dependence of the Bremsstrahlung energy emission on the wavelength or photon energy and related equation can be derived as well.

3.8 Confinement Systems for Controlled Thermonuclear Fusion

Fusion power is the generation of energy by nuclear fusion. Fusion reactions are high-energy reactions in which two lighter atomic nuclei fuse to form a heavier nucleus. This major area of plasma physics research is concerned with harnessing this reaction as a source of large-scale sustainable energy. There is no question of fusion's scientific feasibility, since stellar nucleosynthesis is the process in which stars transmute matter into energy emitted as radiation. Conversion of mass of matter to energy is very well-understood, and demonstrated by Einstein's theory of relativity and his famous formula (see Eq. 3.1), in which E is the kinetic energy produced by M, which is the reduced mass of two individual particles interacting with each other and is multiplied by C, which is the speed of light. Figure 3.14 shows such to energy production which takes place at the surface of the Sun in our solar system, which is a natural fusion reactor.

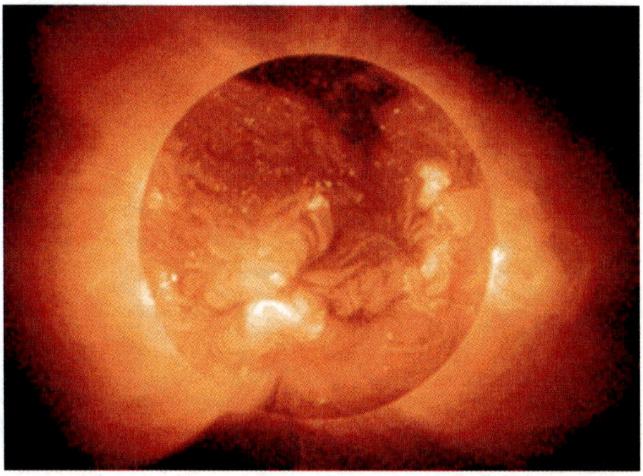

Fig. 3.14 Sun is a natural fusion reactor

For reduced mass M to exist and the relationship in Eq. 3.1 to take place, the particles must come within range of the nuclear forces and surpass the Coulomb barrier via driven kinetic energy available in the center of the mass system of the colliding particles. As we observed in Chap. 2, it was realized that bombardment of light element targets with high-energy particle beams could not produce enough power, unless the energy imparted to outer shell electrons in the collision process was utilized.

What the preceding text implies is that the reacting particles must be confined at high density for a time sufficiently long for energy transfer to the nuclei to take place—what is called the "break-even" condition, also known as the Lawson criterion [5].

The Lawson criterion is an important general measure of a system that defines the conditions needed for a fusion reactor to reach what is known as *ignition temperature*, which is the heating of plasma by the products of the fusion reactions to be sufficient to maintain the temperature of the plasma against all losses with external power input. As formulated Chapter 2 in Zohuri [1], the Lawson criterion gives a minimum required value for the product of the particle plasma density such as electron n_e and the *energy confinement time* τ_E.

Figure 3.15 shows a typical Lawson criterion, or minimum value of electron density multiplied by energy confinement time required for self-heating, for fusion reactions. For the D–T reaction, $n_e\tau_E$ minimizes near the temperature 25 KeV, or roughly 300 million Kelvin, as can be seen in the figure.

Note that although the discussion in this section is based on the MCF approach, similar reasoning would apply to ICF by multiplying the density of plasma particles by the radius of the pellet containing the D–T for the fusion reaction, which is shown later in this chapter.

To summarize, the temperature at which the reaction rate takes place is proportional to the square of the density; the time during which confinement can be secured turns out to be limited to a small fraction of a second and, therefore, the density needed in order to achieve a useful power output is very high (see Sect. 2.5 of Chap. 2 in Zohuri [1]).

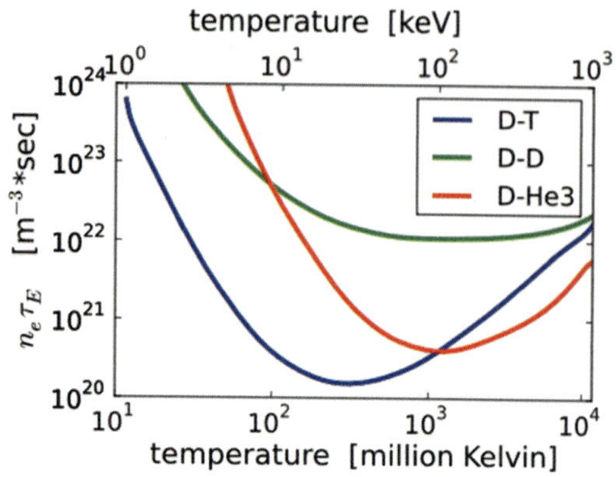

Fig. 3.15 Depiction of Lawson criterion for three fusion reactions

In addition, the temperature required for barrier penetration and the density required (see Sect. 2.2 of Chap. 2 in Zohuri [1]) for a practical device will be determined from data concerning reaction cross-sections, and they represent conditions of matter known to exist in the terrestrial galaxy surrounding us. The concept behind such phenomena on Earth was first produced in thermonuclear weapons technology, and similar conditions were used to trigger the most devastating weapons known to human beings.

Although the first release of manmade thermonuclear energy via a hydrogen bomb ("H-bomb") took place in 1952, the problem of how to control this sudden release of energy for the purpose of generating electric power is still unresolved.

3.9 Magnetic Confinement Fusion

The major magnetic fusion concepts that are under consideration in the quest to confine plasma for magnetic fusion concepts are the:

1. Tokamak
2. Stellarator
3. Reversed Field Pinch
4. Spheromak
5. Field Reversed Configuration
6. Levitated Dipole.

All these magnetic fusion concepts, except the Stellarator, are two-dimensional (2D) axisymmetric toroidal configurations; however, the Stellarator is an inherently three-dimensional (3D) configuration. In this section we discuss only the Tokamak and Stellarator, and their configurations are discussed in Sect. 3.10. These two concepts are primarily discussed from the point of view of macroscopic magneto hydrodynamics (MHD) equilibrium and stability.

Magnetic confinement of plasma is an attempt to prevent particles of moderate density of around 10^{14}–10^{15} cm^{-3} in plasma from escaping the reaction volume by thermal velocity for long periods (i.e., $\tau \geq 1$ s). The concept is based on the foundation that a charged particle path generally forms a spiral along magnetic field lines, which is created by the Lorentz force acting on plasma particles with a charge q, and moving with velocity of $\vec{v}$ in a magnetic field with induction of $\vec{B}$, as explained in Chap. 2 in Zohuri [1].

The above approach is based on a single particle and its motion, depending on the density of charged particles of plasma and their behavior; they present a fluid, either with collective effects being dominant or as collective individual particles. In dense plasmas, the electrical forces between particles couple them to each other and to the electromagnetic fields, which affects their motion.

To gain a better concept, for the single particle approach, and understand what it means, we look at the rarefied plasmas. Under these circumstances, the charged particles do not interact with one another and their motion does not govern a large enough current to significantly affect the electromagnetic fields. Therefore, under these conditions, the motion of each particle, classically, can be treated independent of any other, by solving the Lorentz force equation for prescribed electric and magnetic field. This procedure is known as a single-particle approach and is valid for investigating high-energy particles in the Earth's radiation belts (i.e., Van Allen radiation belt), the solar corona, and in practical devices such as cathode ray tubes or traveling-wave amplifiers, as a few examples. Figure 3.16 is a conceptual image of the cross-section of the Van Allen belt, which is an imaginary belt created by the radiation layer of energetic charged particles that is held in place around a magnetized planet, such as Earth, by its magnetic field.

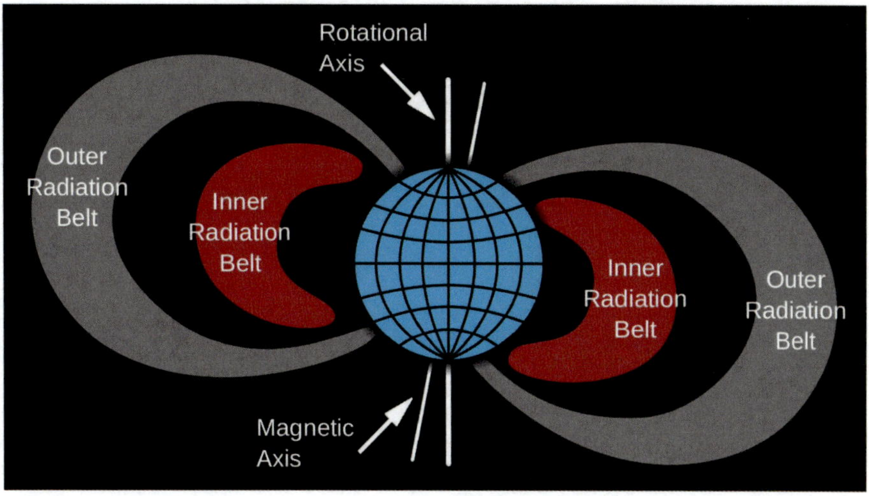

Fig. 3.16 Conceptual cross-section of the Van Allen belt around the earth

In magnetized plasmas under the influence of an external static force or a slowly varying magnetic field produced by the electric field, the single-particle approach is the only applicable classical solution for studying charged particle motion utilizing the Lorentz force equation, which in general defined as:

$$\vec{F} = m\,\vec{a} = m \cdot \frac{d\,\vec{v}}{dt} = q\left(\vec{E} + \vec{v} \times \vec{B}\right) \tag{3.31}$$

The equation given (Eq. 3.31) for the motion of charged particles in magnetized plasmas is relevant if the external magnetic field is quite strong, as opposed to when the magnetic field produced by the electric current arises from the motion of charged particles, an event that is explained by physics or the theory of electromagnetism.

Note that here we are only concerned with non-relativistic motion of charged particles obeying Newtonian classical mechanics rules and the second law of motions. Equation 3.2 is valid for the relativistic case if we simply replace particle mass m with Einstein's formula of relativity in terms of $m = m_0(1 - v^2/C^2)^{-1/2}$, where m_0 is the resting mass of the particle. More commonly, the relativistic form of Eq. 3.2 is written in terms of particle momentum $\vec{P} = m\,\vec{v}$, rather than velocity $\vec{v}$.

Case I: Uniform $\vec{E}$ and $\vec{B}$ Fields, and $\vec{E} = 0$

As stated earlier for the simplest cases of motion in uniform fields, when a particle is under the domination of a static electric field, which is uniform in space, the Lorentz force $\vec{F}$ is expressed in the following form, with only a static and uniform magnetic field present [1]:

$$\vec{F} = q\,\vec{v} \times \vec{B} \tag{3.32}$$

In this case, the particle moves with a constant acceleration along the direction of the field, and the case does not warrant further study.

From a classical mechanics point of view, Lorentz force is also equal to the mass of the particle of interest multiplied by its own mass, so:

$$\vec{F} = m\,\vec{a} = m \cdot \frac{d\,\vec{v}}{dt} \tag{3.33}$$

Combining Eqs. 3.32 and 3.33, the momentum balance equation for this type of particle is as follows:

$$m \cdot \frac{d\,\vec{v}}{dt} = q\,\vec{v} \times \vec{B} \tag{3.34}$$

For further analysis, we can decompose the particle velocity vector $\vec{v}$ into its two components, namely parallel $\vec{v}_{\parallel}$ and perpendicular $\vec{v}_{\perp}$, respective to the magnetic field:

$$\vec{v} = \vec{v}_{\parallel} + \vec{v}_{\perp} \tag{3.35}$$

Lorentz force $\vec{F}$ is proportional to the vector product $\vec{v} \times \vec{B}$, and is vertical to the plane of vector velocity $\vec{v}$ and magnetic $\vec{B}$ and because $\vec{\upsilon} \times \vec{B} = \vec{\upsilon}_{\perp} \times \vec{B}$ it is a function only of the velocity component $\vec{v}_{\perp}$ which is vertical to $\vec{B}$. Note that $\vec{v}_{\perp}$ is the vertical component of vector velocity $\vec{v}$. As far as the parallel component of velocity is concerned, it has no effect because $\vec{v}_{\parallel} \times \vec{B} = 0$, the component $\vec{v}_{\parallel}$ of the particle velocity being parallel to $\vec{B}$, and does not lead to any force influencing the particle.

Using our knowledge of vector analyses and taking the dot-product of Eq. 3.34 with vector $\vec{v}$, we have the following:

$$\begin{aligned}
&\vec{v} \cdot m\frac{d\vec{v}}{dt} = \vec{v} \cdot q\left(\vec{v} \times \vec{B}\right) \\
&m\frac{1}{2}\left\{\frac{d\left(\vec{v} \cdot \vec{v}\right)}{dt}\right\} = q\left[\vec{v} \cdot \left(\vec{v} \times \vec{B}\right)\right] \\
&\frac{d}{dt}\left(\frac{m\upsilon^2}{2}\right) = 0
\end{aligned} \tag{3.36}$$

where $\upsilon = \left|\vec{v}\right|$ is the speed of particle and, as noted before, $\left(\vec{v} \times \vec{B}\right)$ is perpendicular to $\vec{v}$ so the right-hand side is zero.

Obviously, from this we can see that the static magnetic field cannot change the kinetic energy of the particle, since the force is always perpendicular to the direction of motion and this is true even for a spatially non-uniform field. This is because the deviation given here did not use the fact that the field is uniform in space.

Using Eq. 3.35 and rewriting Eq. 3.34, we have:

$$\frac{d\vec{v}_{\parallel}}{dt} + \frac{d\vec{v}_{\perp}}{dt} = \frac{q}{m}\left(\vec{v}_{\perp} \times \vec{B}\right) \tag{3.37}$$

However, as stated above, $\vec{v}_{\parallel} \times \vec{B} = 0$. Equation 3.8 can be split into two equations in terms of $\vec{v}_{\parallel}$ and, $\vec{v}_{\perp}$, and thus we have:

$$\begin{aligned}
&\frac{d\vec{v}_{\parallel}}{dt} = 0 \rightarrow \vec{v}_{\parallel} = \text{constant} \\
&\frac{d\vec{v}_{\perp}}{dt} = \frac{q}{m}\left(\vec{v}_{\perp} \times \vec{B}\right)
\end{aligned} \tag{3.38}$$

Further investigation of Eq. 3.38 reveals that the magnetic field $\vec{B}$ has no effect on the motion of the particle in the direction along it, and that it only affects the particle velocity in the direction perpendicular to it.

We now consider a static magnetic field oriented along the z-axis in vector form as $\vec{B} = \hat{z}B$ in order to be able to examine the characteristic of the perpendicular further on. Equation 3.34 can then be written in component form as follows:

$$m\frac{dv_x}{dt} = qBv_y \tag{3.39a}$$

$$m\frac{dv_y}{dt} = -qBv_x \tag{3.39b}$$

$$m\frac{dv_z}{dt} = 0 \tag{3.39c}$$

The parallel component of particle velocity $\vec{v}_{\parallel}$ to the magnetic field is usually denoted as v_z and is constant, since the Lorentz force $q(\vec{v} \times \vec{B})$ is perpendicular to $\hat{z}$. To determine the time variations of v_x and v_y, we refer to Eq. 3.39a and 3.39b by taking the second derivative of these equations in respect to time t to obtain the following sets of equations:

$$\frac{d^2v_x}{dt^2} + \omega_c^2 v_x = 0 \tag{3.40a}$$

$$\frac{d^2v_y}{dt^2} + \omega_c^2 v_y = 0 \tag{3.40b}$$

where $\omega_c = -qB/m$ is the *gyrofrequency* or cyclotron frequency, shown as the following equation:

$$\text{Cyclotron Frequency} \quad \boxed{\omega_c \equiv \frac{qB}{m}} \tag{3.41}$$

The dimension of ω_c as an angular frequency is rad/m and can be a positive or negative value which is drive by the sign of charge q.

Figure 3.17 presents a cylindrical coordinate with an azimuthal angle of ϕ with a right-hand sense of rotation in a positive direction from the x-axis and also shows the motion of particles, where the z-axis is indicated with the symbol $\otimes$.

The solution to the linear differential sets in Eq. 3.40a and 3.40b in the form of harmonic motion is provided as follows, assuming that $\vec{v}_{\perp} = v_{\perp}$ and $\vec{v}_{\parallel} = v_z = v_{\parallel}$:

$$v_x = v_{\perp}\cos(\omega_c + \psi) = v_{\perp}e^{i\omega_c t} = \frac{dx}{dt} = \dot{x} \tag{3.42a}$$

$$v_y = v_{\perp}\cos(\omega_c + \psi) = \frac{m}{qB}\dot{v}_x = \pm\frac{1}{\omega_c}\dot{v}_x = \pm iv_{\perp}e^{iw_c t} = \dot{y} \tag{3.42b}$$

$$v_z = v_{\parallel} \tag{3.42c}$$

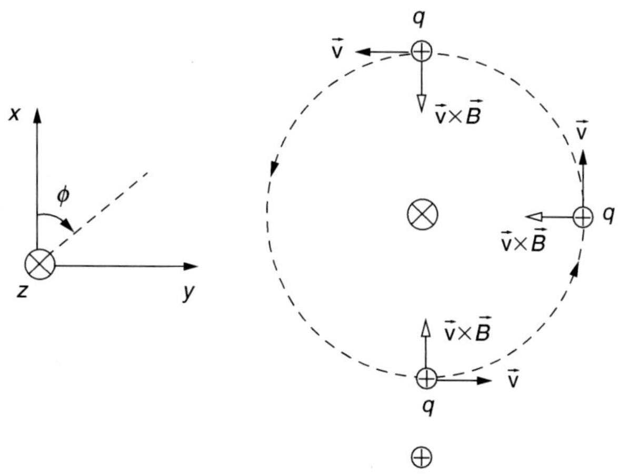

Fig. 3.17 Motion of particles in a magnetic field

where ψ is an arbitrary phase angle, which defines the orientation of the particle velocity at$t = 0$, and $v_\perp = \sqrt{v_x^2 + v_y^2}$ is the constant speed in the plane perpendicular to magnetic field $\vec{B}$.

Considering Fig. 3.4 and assumption of a positive charge q in motion, at different points along its orbit we can clearly see that the particles experience a force of $\vec{F} \propto \vec{v} \times \vec{B}$ directed inward at all times at any given point, which balances the centrifugal force, driven by circular motion of the particle. For a magnetic field in z-direction, in case of electrons, the particle rotation follows the right-hand thumb rule in electromagnetism.

The right-hand rule (RHR) for magnetic force describes how the interactions between the current, flow of electrons, and magnets can be used to do useful work, such as power motors, and will continue to be important in the future because they can be used for things such as wireless energy transfer. This simple demonstration shows how strongly and quickly they interact with each other (see Fig. 3.18).

A more complicated RHR is Fleming's RHR, which describes the motion or force in which something moves. It is useful for understanding the direction of various players in electromagnetism since they interact at right angles. The direction of the thumb is the direction of the force, the direction of the index finger indicates the direction of the magnetic field, and the direction of the middle finger is the direction of the electric current.

From what we have so far, we can easily find the radius of the circular trajectory, which can be found by considering the fact that the $\vec{v} \times \vec{B}$ force is balanced by the centripetal force; therefore, we have:

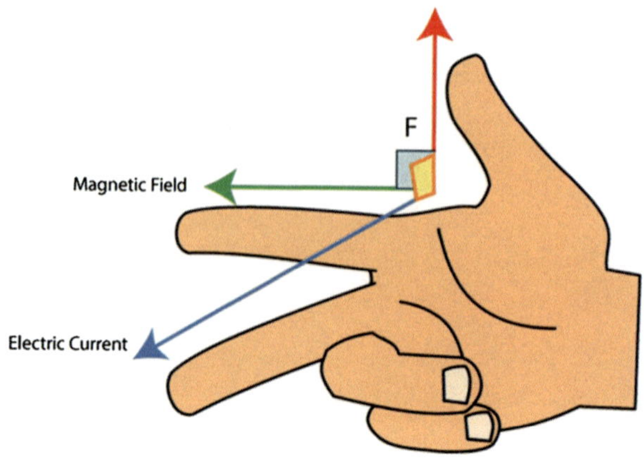

Fig. 3.18 The right-hand rule for magnetic force

$$-\frac{m\mathrm{v}_{\perp}^{2}}{r} = q\,\vec{\mathrm{v}} \times \vec{B} = qv_{\perp}B \tag{3.43}$$

Using Eq. 3.41 and substituting it into Eq. 3.43, we get the result for the final form of trajectory radius of gyroradius, which also known as the *Larmor radius* and is written as:

$$r_c = \frac{-mv_{\perp}}{qB} = \frac{v_{\perp}}{\omega_c} \tag{3.44}$$

Note that the magnitude of the particle velocity remains constant, since the magnetic field force is at all times perpendicular to the motion, as can be seen in Fig. 3.4. Additionally, by the convention, the gyroradius is written as r_c and can take a negative value. This is a mathematical formulation that allows the expression for the particle trajectory to be written for either positive or negative charges in compact form. The gyroradius should always be interpreted as a real physical distance [2].

Note that the magnetic field has no influence over changing the kinetic energy of the particle; however, it does change the direction of its momentum. It is important to note that the gyrofrequency ω_c of the charged particle does not depend on its velocity or kinetic energy and is only a function of the intensity of the magnetic field.

Further analyses can be performed to show the particle position as a function of time by integrating Eqs. 3.42a, 3.42b, and 3.42c sets to find the following information:

$$x = r_c \sin(\omega_c t + \psi) + (x_0 - r_c \sin\psi) \tag{3.45a}$$

$$y = -r_c \cos(\omega_c t + \psi) + (y_0 - r_c \cos\psi) \tag{3.45b}$$

$$z = z_0 + v_{\parallel} t \tag{3.45c}$$

where x_0, y_0, and z_0 are the coordinates of the location of the particle at $t = 0$, and ψ is simply the phase with respect to a particular time of origin.

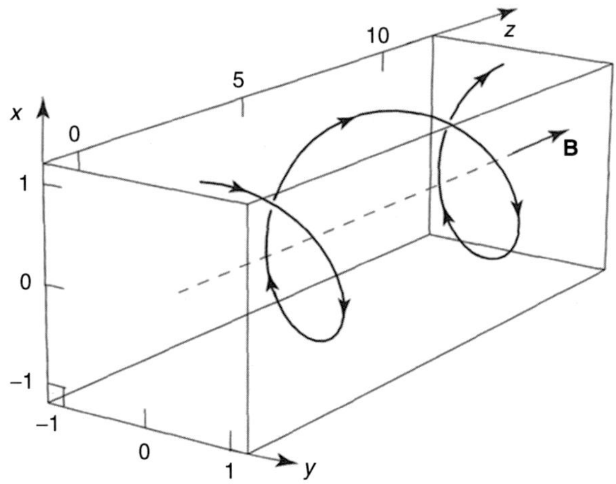

Fig. 3.19 Electron guiding center motion in a magnetic field $\vec{B} = B\hat{z}$. (Courtesy of Inan and Golkowski [6])

Plotting the trajectory function of sets of Eq. 3.45 shows that the particle moves in a circular orbit perpendicular to magnetic field $\vec{B}$ with an angular frequency ω_c and radius r_c about a *guiding center* $\vec{r}_g = x_0\hat{x} + y_0\hat{y} + (z_0 + v_{\|}t)\hat{z}$.

If we are considering particle motion (i.e., electron) in an inhomogeneous field, then the concept of a guiding center is very useful, since the gyration is often much more rapid than the motion of the guiding center. Considering the sets of Eqs. 3.42a, 3.42b, and 3.42c, in their present form, influences the guiding center to simply move linearly along z-axis at a uniform speed $v_{\|}$, as depicted in Fig. 3.6, although the particle motion itself is helical.

From Fig. 3.19, the *pitch angle* of the helix is defined as:

$$\alpha = \tan^{-1}\left(\frac{v_{\perp}}{v_{\|}}\right) \tag{3.46}$$

Noticeably, for both positive and negative charges such as protons or electrons, respectively, the particle gyration constitutes an electric current in the $-\phi$ direction (i.e., opposite to the direction of the fingers of the right-hand, when the thumb points in the direction of the +z-axis). The conceptual direction using the RHR is depicted in Fig. 3.20, and in that case, magnetic moment μ associated with such a current loop is given by the current multiplied by the area, or mathematically presented as follows:

$$\mu = \underbrace{\left(\left|\frac{q\omega_c}{2\pi}\right|\right)}_{\text{current}} \underbrace{\left(\pi r_c^2\right)}_{\text{area}} = \frac{mv_{\perp}^2}{2B} \tag{3.47}$$

Similarly, if we are interested in the torque $\vec{\tau}$ at this stage, it is defined by first expressing the rate of change of angular momentum $\vec{L}$, which is as follows:

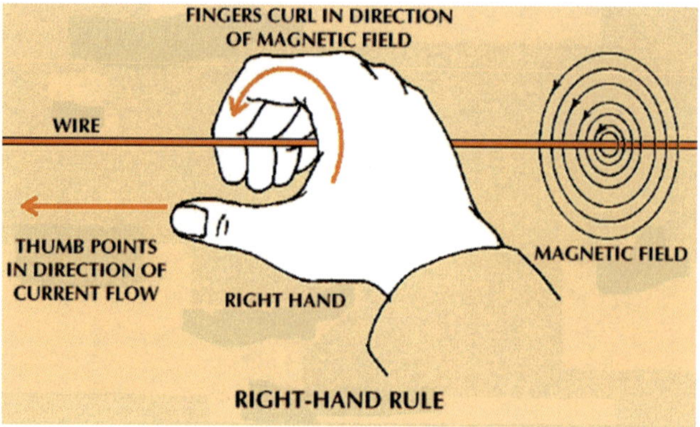

Fig. 3.20 Right-hand rule direction

$$\vec{\tau} = \left(\frac{d}{dt}\right)\vec{L} = \vec{r_c} \times \vec{F} \tag{3.48}$$

The angular momentum in term of linear momentum $\vec{p}$ of the particle in motion is expressed as:

$$\vec{L} = \vec{r_c} \times \vec{p} \tag{3.49}$$

For more details of derivation, refer to Chap. 1 of this book under the same subject.

Note as well that the direction of the magnetic field generated by the gyration is opposite to that of the external field. Thus, the plasma particles that are freely mobile will respond to an external magnetic field with some tendency to reduce the total magnetic field. In other words, plasma is a *diamagnetic* medium and has a tendency to exclude magnetic fields.

As a summary of single particle motion, this has been covered so far by applying the general form of the Lorenz force given in Eq. 3.31 in uniform electric field $\vec{E}$ and magnetic field $\vec{B}$ and reducing it to the form of Eq. 3.32, and used it to find the result for a simple harmonic oscillator and, consequently, the cyclotron frequency as well. In addition, Eq. 3.44 has been used to determine the Larmor radius, and, finally, the trajectory of particle function as sets of Eq. 3.45 have been found and showed the concept of the guiding center.

I now turn to a quest for all possible forms of the general Lorenz force function that will reduce to different categories based on conditions of electric and magnetic fields as combined elements the Lorenz formula.

The sets of Eqs. 3.45a, 3.45b, and 3.45c can also be written in the following format as a complete set:

$$m\dot{\upsilon}_x = qB\upsilon_y \quad m\dot{\upsilon}_y = -qB\upsilon_x \quad m\dot{\upsilon}_Z = 0$$

$$\ddot{\upsilon}_x = \frac{qB}{m}\dot{\upsilon}_y = -\left(\frac{qB}{m}\right)^2 \upsilon_x$$
$$\ddot{\upsilon}_y = \frac{qB}{m}\dot{\upsilon}_x = -\left(\frac{qB}{m}\right)^2 \upsilon_y \tag{3.50a}$$

The circular orbit around a guiding center (x_0, y_0), which is a fixed point, can be written as follows [3]:

$$x - x_0 = r_L \sin \omega_c t$$
$$y - y_0 = \pm r_L \cos \omega_c t \tag{3.50b}$$

Case II: Finite $\vec{E}$

In this case, we allow an electric field to be present and the motion will be found as a summation of the two motions and the usual circular Larmor gyration plus a drift of the guiding center is taking place. In this scenario, electric field $\vec{E}$ is taken to be lying in the $x - z$ plane, thus $E_y = 0$. However, the z component of velocity is unrelated to the transverse components, as in Case I, and can be treated separately. Then, the general Lorenz force equation function of motion applies:

$$\vec{F} = q\left(\vec{E} + \vec{v} \times \vec{B}\right) \tag{3.51a}$$

and

$$m\frac{d\vec{v}}{dt} = q\left(\vec{E} + \vec{v} \times \vec{B}\right) \tag{3.51b}$$

which has the z component velocity as:

$$\frac{d\upsilon_Z}{dt} = \frac{q}{m}E_Z \tag{3.52a}$$

Or, integration of Eq. 3.52a in respect to time t provides the following:

$$\upsilon_z = \frac{qE_z}{m}t + \upsilon_0 \tag{3.53b}$$

These relationships reveal straightforward acceleration along magnetic field $\vec{B}$ and the transverse components of Eqs. 3.51a and 3.51b will be as follows:

$$\begin{aligned}\frac{dv_x}{dt} &= \frac{q}{m}E_x \pm \omega_c v_y \\ \frac{dv_y}{dt} &= 0 \mp \omega_c v_x\end{aligned} \tag{3.53}$$

Differentiating, we have for constant$\vec{E}$:

$$\begin{aligned}\ddot{v}_x &= -\omega_c^2 v_x \\ \ddot{v}_y &= \mp\omega_c\frac{q}{m}\left(E_x \pm \omega_c v_y\right) = -\omega_c^2\left(\frac{E_x}{B} + v_y\right)\end{aligned} \tag{3.54}$$

We can then write the following for this case:

$$\frac{d^2}{dt^2}\left(v_y + \frac{E_x}{B}\right) = -\omega_c^2\left(v_y + \frac{E_x}{B}\right) \tag{3.55}$$

Comparing this equation with Eqs. 3.50a and 3.50b, we can easily see that Eq. 3.55 is a reduced version of Eqs. 3.50a and 3.50b as in Case I if we replace v_y by v_y + (E_x/B). However, Eqs. 3.42a and 3.42b can therefore be replaced by the following:

$$\begin{aligned}v_x &= v_\perp e^{i\omega_c t} \\ v_y &= \pm v_\perp e^{i\omega_c t} - \frac{E_x}{B}\end{aligned} \tag{3.56}$$

We can find the general form of Larmor motion as before with the help of superimposition of a drift of the guiding center velocity $\vec{\text{v}}_{gc}$ in the $-y$ direction for $E_x > 0$, which is illustrated in Fig. 3.21.

Thus, by eliminating the term $md\,\vec{\text{v}}\,/dt$ in Eq. 3.51a and doing our algebraic homework by taking the vector cross product with the magnetic field, we get the general formula:

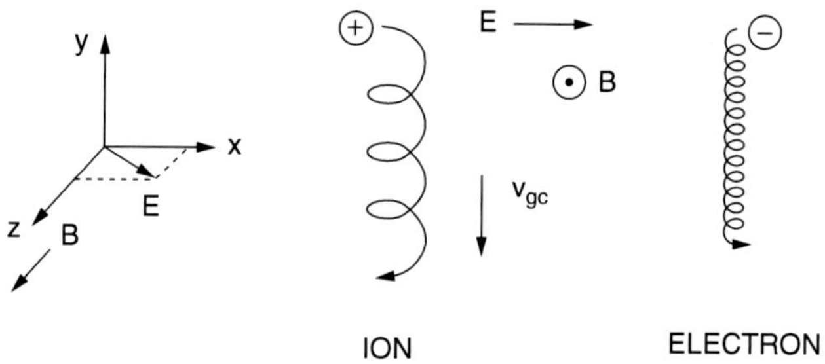

Fig. 3.21 Particle drifts in crossed electric and magnetic fields. (Courtesy of Springer Publishing Company [3])

$$\vec{E} \times \vec{B} = \vec{B} \times (\vec{v} \times \vec{B}) = vB^2 - B(\vec{v} \cdot \vec{B}) \tag{3.57}$$

The transverse components of this equation (i.e., Eq. 3.51a) are as follows:

$$\underset{\perp gc}{\vec{v}} = \frac{\vec{E} \times \vec{B}}{B^2} \equiv \underset{E}{\vec{v}} \tag{3.58}$$

The magnitude of electric field drift v_E of the guiding center is then given by the following equation:

$$v_E = \frac{E(V/m)}{B(\text{tesla})} \frac{m}{s} \tag{3.59}$$

More detailed information and discussion can be found in the book by Chen [7].

Case III: Non-Uniform $\vec{B}$ Field

Cases I and II firmly established the concept of the guiding center, and now we need to have a concept and understanding of particle motion in inhomogeneous electric $\vec{E}$ and magnetic $\vec{B}$ fields where they vary in space or time. Expression of the guiding center for uniform fields has been able to be established; however, the problem of the guiding center becomes too complicated to deal with and be able to find an exact solution to the problem as soon as we introduce an inhomogeneity condition to it.

An approximate answer to the question in Case III can be found if a customary approach to expanding the small fraction r_L/L, for orbit radius of r_L, where L is the scale length of inhomogeneity. Seeking a solution using this type of theory is called "*orbital theory*," and is extremely complex and involved, but, for the sake of argument, we study only the simplest cases here, where only one inhomogeneity for either the electric or magnetic field takes place at a time.

Case III-1: $\nabla \vec{B} \perp \vec{B}$, Gradient $\vec{B}$ Drift

In this case, the magnetic field lines are often called "lines of force," although they are not lines of force but are straight lines, and their density increases as an example in y-direction, as illustrated in Fig. 3.22.

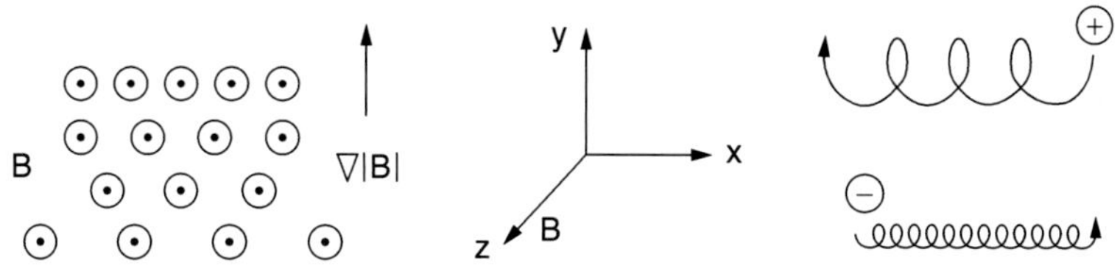

Fig. 3.22 Drift of a gyrating particle in a non-uniform magnetic field. (Courtesy of Springer Publishing Company [3])

The solution to this simple case (Case III-1) is expressed by Chen [7] and readers should refer to this book; however, for sake of this discussion, the relate equations are summarized here, considering Fig. 3.22. The gradient in $\left|\vec{B}\right|$ causes the Larmor radius to be larger at the bottom of the orbit than at the top, and this leads to a drift, in opposite directions for ion and electron particles, perpendicular to both $\vec{B}$ and $\nabla\vec{B}$. Under this situation, the drift velocity is proportional to r_L/L and to $v_\perp$.

For the purpose of this analysis, the Lorentz force $\vec{F}=q\vec{\mathrm{v}}\times\vec{B}$ is averaged over a gyration and, clearly, since the particle spends more time moving up and down, $F_x=0$, as shown in Fig. 3.22. The Lorentz force in y-direction namely v, can be calculated in an approximation method utilizing the *undisturbed orbit* of the particle using Eqs. 3.42a, 3.42b, and 3.42c to find the average for a uniform magnetic field $\vec{B}$.

Real part of complex form of Eqs. 3.42a and 3.42b is given as follows:

$$F_y=-qv_xB_z(y)=-qv_\perp(\cos\omega_c t)\left[B_0\pm r_L(\cos\omega_c t)\frac{\partial B}{\partial y}\right] \tag{3.60}$$

Equation 3.60, along with utilization of Taylor series approximation of $\vec{B}$ field about the point $x_0=0$ and $y_0=0$, gives:

$$\begin{aligned}\vec{B}&=B_0+\left(\vec{r}\cdot\nabla\right)\vec{B}+\cdots\\ B_z&=B_0+y(\partial B_z/\partial y)+\cdots\end{aligned} \tag{3.61}$$

For this expansion the required condition $(r_L/L)\ll 1$ needs to hold, where L is the scale length of $\partial B_z/\partial y$. The first term of Eq. 3.60 averages zero in a gyration and the average of $\cos\omega_c t=1/2$, giving:

$$F_y=\mp qv_\perp r_L(\partial B_z/\partial y)/2 \tag{3.62}$$

The guiding center drift velocity is then:

$$\vec{\mathrm{v}}_{gc}=\frac{1}{q}\frac{\vec{F}\times\vec{B}}{B^2}=\frac{1}{q}\frac{F_y}{\left|\vec{B}\right|}\hat{x}=\mp\left(\frac{v_\perp r_L}{\vec{B}}\right)\left(\frac{1}{2}\frac{\partial B}{\partial y}\hat{x}\right) \tag{3.63}$$

where the following equation is used in the presence of gravitational force by replacing $q\vec{E}$ in the motion of Eq. 3.51a by the forgoing result that can be applied to the other forces:

$$\vec{\mathrm{v}}_f=\frac{1}{q}\frac{\vec{F}\times\vec{B}}{B^2} \tag{3.64}$$

Therefore, we can write the general form as follows:

$$\vec{\mathrm{v}}_{\nabla\vec{B}}=\pm\frac{1}{2}v_\perp r_L\frac{\vec{B}\times\nabla\vec{B}}{B^2} \tag{3.65}$$

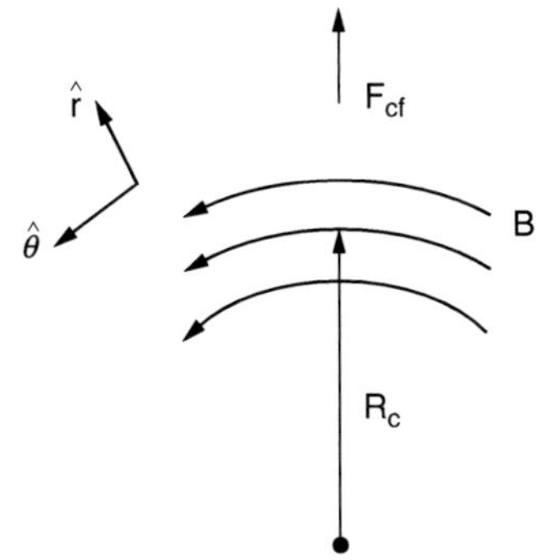

Fig. 3.23 Curved magnetic field. (Courtesy of Springer Publishing Company [3])

This equation has all the dependences expected from the physical picture minus the factor 1/2, which arises from the averaging.

Case III-2: Curved $\vec{B}$, Curvature Drift

In this case, the lines of force are assumed to be curved with a constant radius of curvature R_c and $\left|\vec{B}\right|$ is constant (see Fig. 3.23), and the average square of the component of random velocity $v_{\parallel}^2$ along with centrifugal force F_{cf} is given as follows:

$$F_{\mathrm{cf}} = \frac{mv_{\parallel}}{R_c} = mv_{\parallel}^2 \frac{R_c}{R_c^2} \tag{3.66}$$

According to Eq. 3.64, this gives rise to a drift:

$$\vec{\mathrm{v}}_R = \frac{1}{q}\frac{F_{\mathrm{cf}} \times \vec{B}}{B^2} = \frac{mv_{\parallel}^2}{qB^2}\frac{\vec{R}_c \times \vec{B}}{\frac{\vec{R}_c}{R_c^2}} \tag{3.67}$$

The general form of total drift in a curved vacuum field is as follows:

$$\vec{\mathrm{v}}_R + \vec{\mathrm{v}}_{\nabla\vec{B}} = \frac{m}{q}\frac{\vec{R}_c \times \vec{B}}{R_c^2 B^2}\left(v_{\parallel}^2 + \frac{1}{2}v_{\perp}^2\right) \tag{3.68}$$

Adding these drifts means that if one bends a magnetic field into a torus for the purpose of confining a thermonuclear plasma, the particles will drift out of the torus no matter how one juggles the temperatures and magnetic fields.

For more details and further analysis, readers should refer to Chen's textbook [7].

Case III-3: $\nabla \vec{B} \| \vec{B}$, Magnetic Mirrors

Now consider magnetic field $\vec{B}$ primarily lies in z-direction, varies in that direction, and is axisymmetric with $B_\theta = 0$ and $\partial/\partial\theta = 0$. Figure 3.24 shows the drift of a

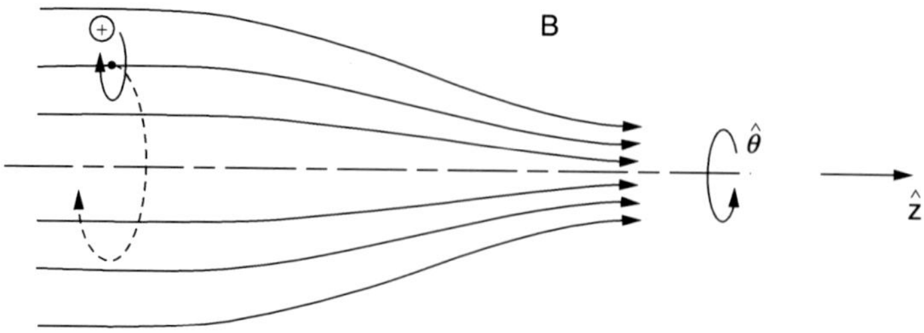

Fig. 3.24 Drift of a particle in a magnetic mirror field. (Courtesy of Springer Publishing Company [3])

particle in the magnetic mirror field, where the lines of force coverage and diverge with a component of magnetic field B_r in direction r of the cylindrical coordinate. This scenario will give rise to a force that traps a particle in the magnetic field.

We are able to obtain the B_r and $\nabla \cdot \vec{B} = 0$ using the following calculations, by means of the cylindrical coordinate systems with the assumption of axisymmetric around angle θ:

$$\frac{1}{r}\frac{\partial}{\partial r}(rB_r) + \frac{\partial B_z}{\partial z} = 0 \tag{3.69}$$

If $\partial B_z/\partial z$ is given at $r = 0$ and does not vary much with r, we have approximately the following:

$$\begin{aligned} rB_r &= -\int_0^r r\frac{\partial B_z}{\partial z}dr \simeq -\frac{1}{2}r^2\left[\frac{\partial B_z}{\partial z}\right]_{r=0} \\ B_r &= -\frac{1}{2}r\left[\frac{\partial B_z}{\partial z}\right]_{r=0} \end{aligned} \tag{3.70}$$

The variation of $\left|\vec{B}\right|$ with r causes a gradient $\vec{B}$ drift of guiding centers about the axis of symmetry with no radial gradient of magnetic field $\vec{B}$ drift due to $\partial B_\theta/\partial\theta = 0$. Therefore, the components of the Lorentz force are as follows:

$$\begin{aligned} F_r &= q(v_\theta B_z - v_z B_\theta) \\ F_\theta &= q(-v_r B_z + v_z B_r) \\ F_z &= q(v_r B_\theta - v_\theta B_r) \end{aligned} \tag{3.71}$$

Moreover, we are interested in the following terms of Eq. 3.71:

$$F_z = \frac{1}{2}qv_\theta r\left(\frac{\partial B_z}{\partial z}\right) \tag{3.72}$$

Averaging out this equation over one gyration by considering a particle whose guiding center lies on the axis, then v_θ is a constant during a gyration; depending on the sign of particle charge q, v_θ is $\mp v_\perp$. Since $r = r_L$, the average force is then:

$$F_z = \mp\frac{1}{2}qv_\perp r_L\frac{\partial B_z}{\partial z} = \mp\frac{1}{2}q\frac{v_\perp^2}{\omega_c}\frac{\partial B_z}{\partial z} = -\frac{1}{2}\frac{mv_\perp^2}{B}\frac{\partial B_z}{\partial z} \tag{3.73}$$

Defining the "*magnetic moment*" of the gyrating particle, which is the same as the definition for the magnetic moment of a current loop with area A and current I, showing it as $\mu = IA$, we have:

$$\mu \equiv \frac{1}{2}\frac{mv_\perp^2}{B} \tag{3.74}$$

So that:

$$F_z = -\mu\left(\frac{\partial B_z}{\partial z}\right) \tag{3.75}$$

Then, the general form of force on a diamagnetic particle is as follows:

$$F_\parallel = -\mu\left(\frac{\partial B}{\partial s}\right) = -\mu\nabla_\parallel B \tag{3.76}$$

where ds is a line element along magnetic field $\vec{B}$.

In any case, from the definition of Eq. 3.74 and from a single particle charge such as that from an ion, I is generated by a charge e coming around $\omega_c/2\pi$ times a second as $I = e\omega_c/2\pi$ and the area A is calculated based on $\pi r_L^2 = \pi v_\perp^2/\omega_c^2$; thus, we can write;

$$\mu = \frac{\pi v_\perp^2}{\omega_c^2}\frac{e\omega_c}{2\pi} = \frac{1}{2}\frac{ev_\perp^2}{\omega_c} = \frac{1}{2}\frac{mv_\perp^2}{B} \tag{3.77}$$

The Larmor radius varies as the particle goes through regions of stronger or weaker magnetic field $\vec{B}$; however, the magnetic moment μ does remain *invariant* and the proof can be seen in Chen's textbook [7].

The invariance of magnetic moment μ is the foundation for one of initial schemes for an plasma confinement approach using a magnetic device called the *magnetic mirror*.

Figure 3.25 shows a simplistic illustration of such a device, where the non-uniform field of a simple pair of coils forms two magnetic mirrors between which the plasma can be trapped and consequently confined. This effect works on both ions and electrons, holding either positive or negative charges, respectively.

Conservation of energy requires that:

$$\frac{1}{2}\frac{mv_{\perp 0}^2}{B_0} = \frac{1}{2}\frac{mv_\perp'^2}{B'} \tag{3.78}$$

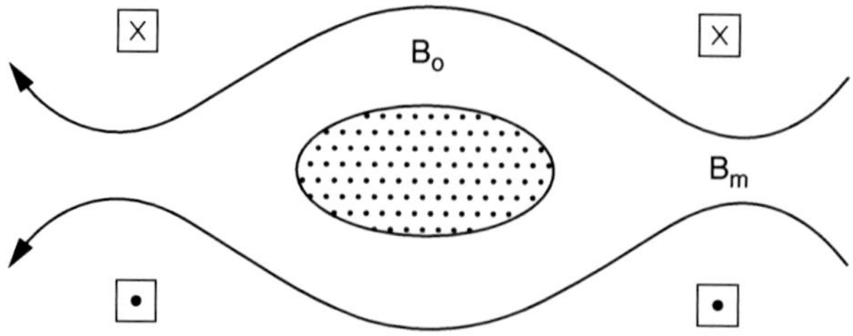

Fig. 3.25 Plasma trapped between magnetic mirrors. (Courtesy of Springer Publishing Company [3])

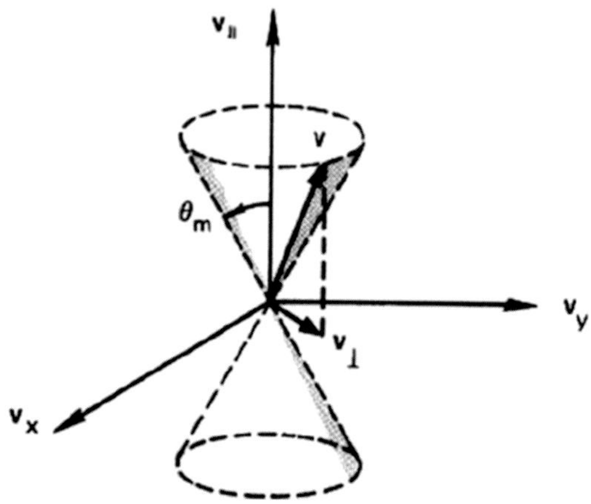

Fig. 3.26 The loss cone. (Courtesy of Springer Publishing Company [3])

where:

$$v_{\perp}^{\prime 2} = v_{\perp 0}^{2} + v_{\| 0}^{2} \tag{3.79}$$

Combining Eq. 3.78 with Eq. 3.79, we can write:

$$\frac{B_0}{B'} = \frac{v_{\perp 0}^2}{v_{\perp}^{\prime 2}} = \frac{v_{\perp 0}^2}{v_0^2} \equiv \sin^2\theta \tag{3.80}$$

where θ is the pitch angle of the orbit in the weak-field region and with a smaller value of this angle, the particle will mirror regions of higher magnetic field B; however, if the this angle is too small, B' exceeds B_m and the particle does not mirror at all. If we replace B' with B_m in Eqs. 3.50a and 3.50b, we observe that the smallest pitch angle θ of a confined particle is provided by:

$$\sin^2\theta_m = \frac{B_0}{B_m} \equiv \frac{1}{R_m} \tag{3.81}$$

where R_m is the *mirror ratio.*

Figure 3.26 illustrates the *loss cone*, where Eq. 3.81 defines the boundary of a region in velocity space in the shape of a cone.

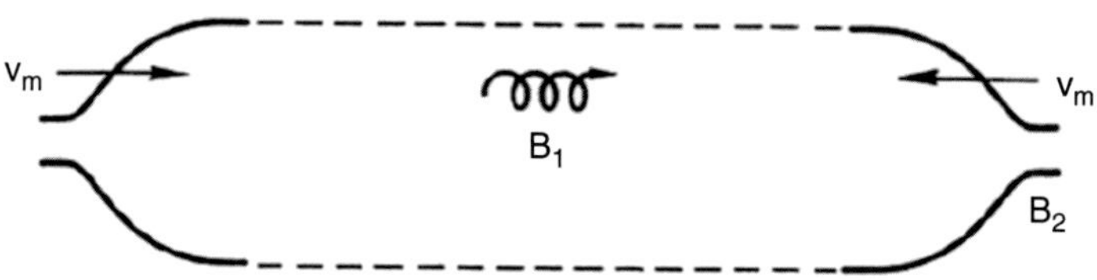

Fig. 3.27 Cosmic ray proton trap device. (Courtesy of Springer Publishing Company [3])

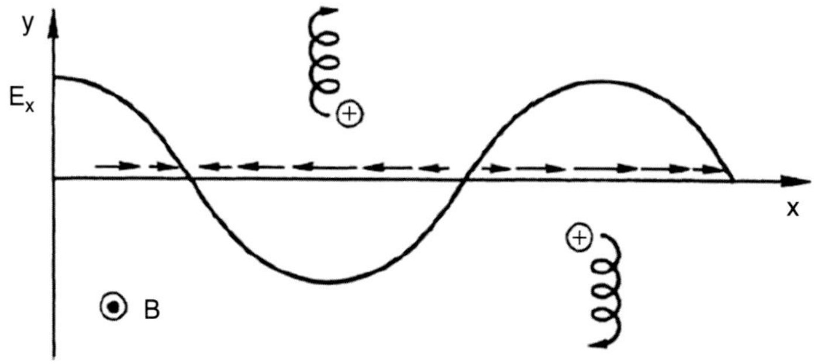

Fig. 3.28 Drift of a gyrating particle in a non-uniform electric field. (Courtesy of Springer Publishing Company [3])

The magnetic mirror was first configured and proposed by Enrico Fermi as an instrument/machine for the acceleration of cosmic rays. His configuration is depicted in Fig. 3.27, where protons are bouncing between magnetic fields.

As stated previously, a further example of the mirror effect confinement of particles can be observed in Van Allen belts, as shown in Fig. 3.3.

Case IV: Non-Uniform $\vec{E}$ Field

In this case, it is assumed that the magnetic field is uniform and the electric field is in a non-uniform condition, and for simplicity of the problem in hand, we assume electric field $\vec{E}$ is in x-direction and varies in that direction sinusoidally, as shown in Fig. 3.28 and presented with the following equation:

$$\vec{E} = E_0(\cos kx)\hat{x} \tag{3.82}$$

The associated field distribution has a wavelength $\lambda = 2\pi/k$ and is the result of a sinusoidal distribution of charges, which we do not specify. Practically, such distribution can arise in plasma during a wave motion. Therefore, the equation of motion is as follows:

$$m\left(\frac{d\vec{\text{v}}}{dt}\right) = q\left[\vec{E}(x) + \vec{\text{v}} \times \vec{B}\right] \tag{3.83}$$

The transverse components of this are:

$$\begin{aligned}\dot{v}_x &= \frac{qB}{m}v_y + \frac{q}{m}E_x(x) \\ \dot{v}_y &= -\frac{qB}{m}v_x\end{aligned} \tag{3.84}$$

and

$$\ddot{v}_x = -\omega_c^2 v_x \pm \omega_c \frac{\dot{E}_x(x)}{B} \tag{3.85}$$

$$\ddot{v}_y = -\omega_c^2 v_y = \omega_c^2 \frac{\dot{E}_x(x)}{B} \tag{3.86}$$

The component of electric field $E_x(x)$ in x-direction in these equations is presented in the field at the position of particle and can be evaluated if we know the particle's orbit, which we need to solve in the first place. However, for a weak electric field, we use an approximation of *undisturbed orbit* to assess $E_x(x)$. The orbit in the absence of the electric field given by Eq. 3.50b is written as follows:

$$x = x_0 + r_L \sin \omega_c^t \tag{3.87}$$

From Eqs. 3.86 and 3.82, we obtain:

$$\ddot{v}_y = -\omega_c^2 v_y - \omega_c^2 \frac{\dot{E}_x(x)}{B} \cos k(x_0 + r_L \sin \omega_c t) \tag{3.88}$$

The solution to Eq. 3.89 can be found as follows [3]:

$$\overline{\ddot{v}}_y = 0 = -\omega_c^2 \overline{v}_y - \omega_c^2 \frac{E_0}{B} \overline{\cos k(x_0 + r_L \sin \omega_c t)} \tag{3.89}$$

Expanding the cosine, we have:

$$\begin{aligned}\cos k(x_0 + r_L \sin \omega_c t) = {} & \cos(kx_0)\cos(kr_L \sin \omega_c t) \\ & - \sin(kx_0)\sin(kr_L \sin \omega_c t)\end{aligned} \tag{3.90}$$

It will suffice to treat the small Larmor radius case, $kr_L \ll 1$. The Taylor expansions are:

$$\begin{aligned}\cos \varepsilon &= 1 - \frac{1}{2}\varepsilon^2 + \cdots \\ \sin \varepsilon &= \varepsilon + \cdots\end{aligned} \tag{3.91}$$

This allows us to write:

$$\begin{aligned}\cos k(x_0 + r_L \sin \omega_c t) \approx {} & (\cos kx_0)\left(1 - \frac{1}{2}k^2 r_L^2 \sin^2 \omega_c t\right) \\ & -(\sin kx_0) k r_L \sin \omega_c t\end{aligned} \tag{3.92}$$

The last term of Eq. 3.92 vanishes upon averaging over time, and then Eq. 3.59 reduces to the following form:

$$\overline{\upsilon}_y = -\frac{E_0}{B}(\cos kx_0)\left(1 - \frac{1}{4}k^2 r_L^2\right) = -\frac{E_x(x_0)}{B}\left(1 - \frac{1}{4}k^2 r_L^2\right) \tag{3.93}$$

Thus, the usual $\vec{E} \times \vec{B}$ drift is modified by the inhomogeneity to read:

$$\vec{\mathrm{v}}_E = \frac{\vec{E} \times \vec{B}}{B^2}\left(1 - \frac{1}{4}k^2 r_L^2\right) \tag{3.94}$$

Chen [7] argues that the *finite-Larmor-radius effect* should be found using the expansion of Eq. 3.94 in the following form, and readers should refer to Chen's book:

$$\vec{\mathrm{v}}_E = \left(1 - \frac{1}{4}r_L^2 \nabla^2\right)\frac{\vec{E} \times \vec{B}}{B^2} \tag{3.95}$$

Case V: Time-Varying $\vec{E}$ Field

In this case, we provide the equation related to the case and leave the reader to see the proof of the details in Chen [7] and other plasma-related books. The condition that we consider in this case calls for both electric and magnetic fields to be uniform in space but varying in time.

$$\vec{E} = E_0 e^{i\omega t}\widehat{x} \tag{3.96}$$

Since $\dot{E}_x = i\omega E_x$, we can write Eq. 3.55 as follows:

$$\theta \tag{3.97}$$

The rest of the equations related to this case are provided here just as they are, without detailed explanation:

$$\tan\left(\frac{\theta}{2}\right) = \frac{Ze^2}{m_e \upsilon^2 b} \tag{3.98}$$

$$\begin{aligned} \ddot{\upsilon}_x &= -\omega_c^2\left(\upsilon_x - \widetilde{\upsilon}_p\right) \\ \ddot{\upsilon}_y &= -\omega_c^2\left(\upsilon_y - \widetilde{\upsilon}_E\right) \end{aligned} \tag{3.99}$$

The solution of Eq. 3.99 is as follows:

$$\begin{aligned} \upsilon_x &= \upsilon_\perp e^{i\omega_c t} + \widetilde{\upsilon}_p \\ \upsilon_y &= \pm i\upsilon_\perp e^{i\omega_c t} + \widetilde{\upsilon}_E \end{aligned} \tag{3.100}$$

Twice differentiation of Eq. 3.101 in respect to time results in:

$$\begin{aligned}\ddot{\upsilon}_x &= -\omega_c^2\upsilon_x + \left(\omega_c^2 - \omega^2\right)\widetilde{\upsilon}_p \\ \widetilde{\upsilon}_y &= -\omega_c^2\upsilon_y + \left(\omega_c^2 - \omega^2\right)\widetilde{\upsilon}_E\end{aligned} \tag{3.101}$$

The polarization drift for the x-component along the direction of $\vec{E}$ field is given as follows:

$$\vec{\mathrm{v}}_p = \pm\frac{1}{\omega_c \vec{B}}\frac{d\vec{E}}{dt} \tag{3.102}$$

In addition, the polarization current is:

$$\vec{j}_p = ne\left(\upsilon_{ip} - \upsilon_p\right) = \frac{ne}{eB^2}(M+m)\frac{d\vec{E}}{dt} = \frac{\rho}{B^2}\frac{d\vec{E}}{dt} \tag{3.103}$$

where ρ is the mass density while M and m are the particle masses involved and are defined as before.

If a field $\vec{E}$ is suddenly applied, the first thing the ion does is to move in the direction of $\vec{E}$. Only after picking up a velocity $\vec{\mathrm{v}}$ does the ion feel a Lorentz force $e\vec{\mathrm{v}} \times \vec{B}$ and begin to move downward, as illustrated in Fig. 3.29.

Case VI: Time-Varying $\vec{B}$ Field

For this case we let the magnetic field vary in time, and due to the fact that the Lorentz force is perpendicular to $\vec{\mathrm{v}}$, a magnetic field by itself does not have any impact on the energy of a charged particle. However, an electric field $\vec{E}$ exists that is associated with magnetic field $\vec{B}$, that can accelerate the particle, as given here:

$$\frac{\nabla\times\vec{E} = -\vec{B}}{B^2} \tag{3.104}$$

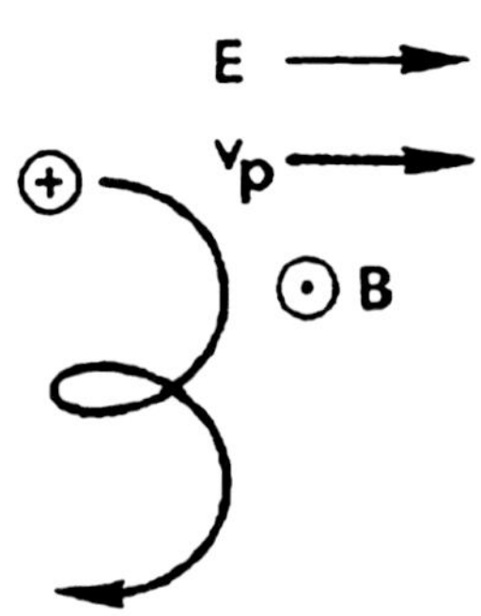

Fig. 3.29 The polarization drift. (Courtesy of Springer Publishing Company [3])

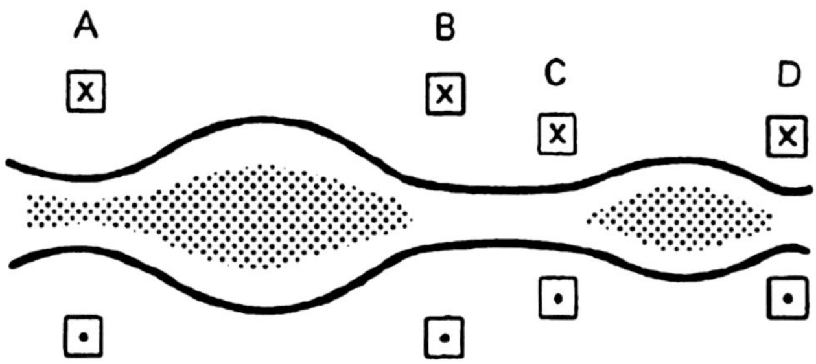

Fig. 3.30 Two-stage adiabatic compression of plasma. (Courtesy of Springer Publishing Company [3])

Details of this analysis are as per Zohuri [3] and all related equations are briefly shown here, including the magnetic moment μ that is invariant and slowly varying in magnetic fields and magnetic flux Φ through a Larmor orbit that is constant as follows:

$$\begin{aligned} &\delta\left(\frac{1}{2}mv_{\perp}^2\right) = \mu\delta B \\ &\delta\mu = 0 \\ &\Phi = B\pi\frac{v_{\perp}^2}{\omega_c^2} = B\pi\frac{v_{\perp}^2}{q^2B^2} = \frac{2\pi m}{q^2}\frac{\frac{1}{2}mv_{\perp}^2}{B} = \frac{2\pi m}{q^2}\mu \end{aligned} \tag{3.105}$$

This property is used in a method of plasma heating known as adiabatic compression. Figure 3.30 shows a schematic of how this is done. Plasma is injected into the region between mirrors A and B. Coils A and B are then pulsed to increase B and hence $v_{\perp}^2$. The heated plasma can then be transferred to the region C–D by a further pulse in A, increasing the mirror ratio there. Coils C and D are then pulsed to further compress and heat the plasma. Early magnetic mirror fusion devices employed this type of heating [7].

3.9.1 Summary of Guiding Center Drift

$$\text{General force } \vec{F}: \quad \vec{v}_f = \frac{1}{q}\frac{\vec{F}\times\vec{B}}{B^2} \tag{3.106}$$

$$\text{Electric field } \vec{E}: \quad \vec{v}_E = \frac{\vec{E}\times\vec{B}}{B^2} \tag{3.107}$$

$$\text{Gravitational field } \vec{v}_g: \quad \vec{v}_g = \frac{m}{q}\frac{\vec{g}\times\vec{B}}{B^2} \tag{3.108}$$

$$\text{Non-uniform } \vec{E}: \quad \vec{\mathrm{v}}_E = \left(1 + \frac{1}{4} r_L^2 \nabla^2\right) \frac{\vec{E} \times \vec{B}}{B^2} \tag{3.109}$$

Non-uniform Magnetic Field $\vec{B}$

$$\text{Grad-}\vec{B}\text{ drift}: \quad \vec{\mathrm{v}}_{\nabla \vec{B}} = \pm \frac{1}{2} v_\perp r_L \frac{\vec{B} \times \nabla \vec{B}}{B^2} \tag{3.110}$$

$$\text{Curvature drift}: \quad \vec{\mathrm{v}}_R = \frac{m v_\parallel^2}{q} \frac{\vec{R}_c \times B}{R_c^2 B^2} \tag{3.111}$$

$$\text{Curved vacuum field}: \quad \vec{\mathrm{v}}_R + \vec{\mathrm{v}}_{\nabla \vec{B}} = \frac{m}{q} \left(v_\parallel^2 + \frac{1}{2} v_\perp^2\right) \frac{\vec{R}_c \times B}{R_c^2 B^2} \tag{3.112}$$

$$\text{Polarization drift}: \quad \vec{\mathrm{v}}_p = \pm \frac{1}{\omega_c \vec{B}} \frac{d\vec{E}}{dt} \tag{3.113}$$

More details can be found in many standard plasma textbooks.

3.10 How the Tokamak Reactors Works

The Tokamak was invented in the old Soviet Union by Andrei Sakharov and Igor Tamm. A conceptual sketch of its configuration is shown in Fig. 3.31.

As of 2008, the United States Department of Energy (DOE) and other federal agencies have spent approximately US$18 billion on energy devices using the fusion reaction between deuterium and tritium (D–T fusion; bottom left of Fig. 3.19). In this reaction the hydrogen isotope deuterium (with one "extra" neutron) collides with the

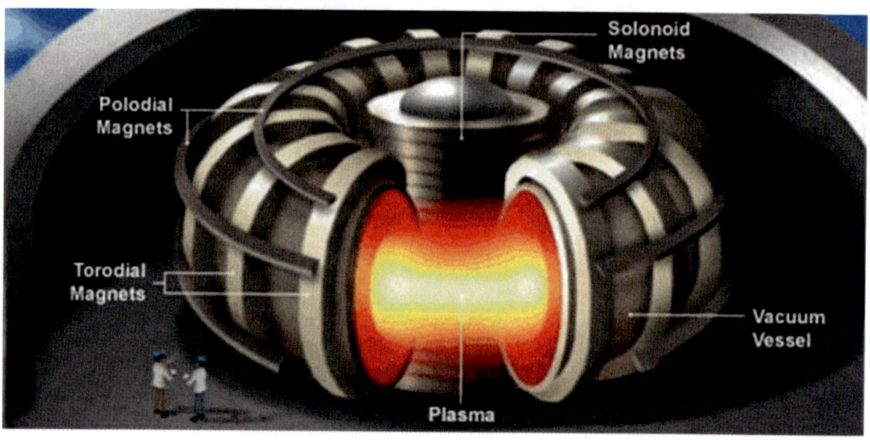

Fig. 3.31 Conceptual sketch of the Tokamak

hydrogen isotope tritium (with two "extra" neutrons) to form an alpha particle (a helium nucleus) and a neutron. This is a nuclear reaction: between them, the new alpha and the neutron possess 17.6 MeV (million electron volts) of energy.

In Fig. 3.10 (top right), the red D–T (deuterium–tritium) curve peaks at about 40 KeV (40,000 electron volts). This means that the optimum activation energy required for the D–T fusion reaction is only about 40 KeV. The curves for the other reactions peak at much higher energies. The energy required to make the D–T reaction happen is lower (in KeV) than the energy required for any other nuclear fusion reaction. In addition, the height of the D–T curve (cross-section in millibarns) indicates that the deuterium and tritium isotopes "see" each other as being relatively large, compared with the isotopes in the other reactions shown. Thus, at the proper activation energy, this reaction is much more likely to happen than any other fusion reaction. The DOE and many other entities pursue the D–T reaction because it requires less energy to initiate, and because it is more probable.

Unfortunately, there are several serious disadvantages to this reaction:

1. Tritium is both radioactive and expensive.
2. The neutrons released can harm living things and damage any other materials surrounding them.
3. The neutrons can make some materials radioactive.

At this time, the device preferred for making this reaction happen is the Tokamak. The DOE, the European Union, Japan, Russia, China, and India are all part of the ITER program, which is working on it. Their dream is that the Tokamak will heat a plasma-containing tritium and deuterium nuclei. The hotter these nuclei get, the faster they will move. When the plasma is hot enough, some of the nuclei will be moving fast enough to react when they collide. The energy of the newly produced, highly energetic helium nuclei (alphas) will be used to keep the plasma hot; and the energy of the new neutrons will be released to a lithium metal blanket that lines the Tokamak. Water lines will run through the lithium. The hot lithium will heat the water to steam, and the steam will be used to spin turbines, which will spin generators to make electricity.

However, there is a substantial gap between this dream and its fulfillment. For at least 50 years, the practical use of Tokamaks and other D–T devices to make electricity has been forecast to be about thirty years in the future. To be commercially useful, a controlled fusion reaction must produce more energy than the energy required to cause the reaction in the first place (i.e., the 40 KeV activation energy as mentioned earlier). The point at which the energy produced exceeds the energy required is called "net power" or "break-even." Various organizations in different parts of the world have been working to produce "net power" nuclear fusion for about 50 years. Many billions of rubles, dollars, yen, and euros have been spent on this endeavor, but no one has been successful yet.

Many of the efforts have involved the idea of heating plasma of deuterium (D) and tritium (T) gases until the nuclei fuse. When the heat of the plasma increases, the average energy (speed) of the particles increases; but there is an enormous variation in the energies of the individual particles within the plasma. This set of

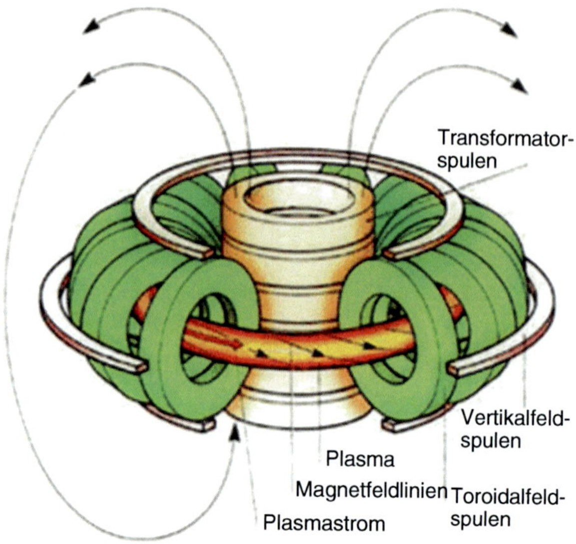

Fig. 3.32 Tokamak donut hole shape

all the different energies of the particles in plasma or a gas is called a Maxwellian distribution. Unfortunately, in the typical Maxwellian distribution, only a few of the nuclei have the 40 KeV of energy required to react—all the other particles are just along for the ride. If the temperature is increased to the point where an adequate number of nuclei have enough energy, then other problems develop which can compromise the integrity of the containment.

Both the Tokamak and the Stellarator use magnetic fields to manipulate the D–T plasma. However, the distinguishing feature of the Tokamak is its "step-down" transformer. The transformer's primary is the stack of beige coils in the center of the Tokamak's torus (in the donut's hole below; Fig. 3.32). The transformer's secondary is the ring of plasma—the orange skinny donut. An increasing current in the many-coiled primary induces a much larger current in the single-coiled plasma "donut" secondary.

Two magnetic fields combine to produce the resultant magnetic field (labeled left in Fig. 3.32) that spirals helically around the Tokamak's torus (orange skinny donut). This *resultant* field contains and controls the plasma. The two magnetic fields that combine vectorially to make the resultant field are (1) the *toroidal* field, generated by the green *toroidal* coils; and (2) the *poloidal* field generated by the orange plasma current in the torus. The vertical coils (the large rings around the outside of the Tokamak, and above and below it) can create a vertical magnetic field for controlling the position of the plasma inside the torus.

The transformer coils also cause "ohmic" (RI [2]) heating in the plasma, which contributes to raising its temperature. However, since the electrical resistance of plasma decreases as its temperature increases, the upper limit on the "ohmic" heating turns out to be about 20–30 million degrees Celsius, which is not high enough for fusion.

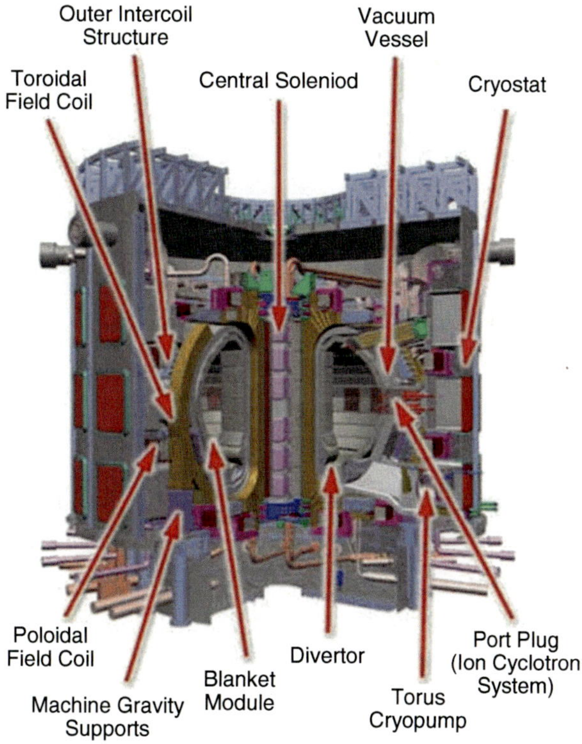

Fig. 3.33 France's International Thermonuclear Experimental Reactor (ITER) Tokamak machine

Thus, it is necessary to further increase the temperature by three additional strategies: radio frequency heating, magnetic compression, and neutral beam injection.

The proposed ITER Tokamak, to be built in France, is pictured in Fig. 3.33. To get an idea of the scale that is involved, notice the tiny little laboratory technician in the blue coat standing on the floor, near the machine.

A somewhat similar fusion effort is the Stellarator, also known as the Wendelstein 7-X, in Germany (see Fig. 3.34).

Both the Stellarator and the Tokamak use magnetic containment to control the fuel. A distinguishing feature of the Stellarator is the use of odd-shaped coils to manipulate the shape of the plasma donut within the coils. To have a better concept of how the Stellarator works, we introduce the plasma beta, β, which is the ratio of plasma pressure to magnetic pressure and is defined as follows:

$$\beta = \frac{p_{\text{Plasma}}}{p_{\text{Magnetic}}} = \frac{nk_B T}{B^2/(2\mu_0)} \tag{3.114}$$

where $n$is the plasma density, k_B is the Boltzmann constant, T is the plasma temperature, B is the magnetic field, and μ_0 is the magnetic moment.

Fig. 3.34 Stellarator, also known as the Wendelstein 7-X, under construction in Germany

Given that the magnets are a dominant factor in MCF reactor design, and that density and temperature combine to produce pressure, the ratio of the pressure of the plasma to the magnetic energy density naturally becomes a useful figure of merit when comparing MCF designs. In effect, the ratio illustrates how effectively a design confines its plasma.

β is normally measured in terms of the total magnetic field and the term is commonly used in studies of the Sun and Earth's magnetic field, and in the field of magnetic fusion power designs. However, in any real-world design, the strength of the field varies over the volume of the plasma, so, to be specific, the average beta is sometimes, referred to as the "beta toroidal". In the Tokamak design, the total field is a combination of the external toroidal field and the current-induced poloidal one, so the "beta poloidal" is sometimes used to compare the relative strengths of these fields. In addition, as the external magnetic field is the driver of reactor cost, "beta external"is used to consider just this contribution.

In the magnetic fusion power field, plasma is often confined using large superconducting magnets that are very expensive. Since the temperature of the fuel scales with pressure, reactors attempt to reach the highest pressures possible. The costs of large magnets roughly scale as ß½. Therefore, beta can be thought of as a ratio of money out to money in for a reactor, and beta can be thought of (very approximately) as an economic indicator of reactor efficiency. To make an economically useful reactor, betas better than 5% are needed.

The same term is also used when discussing the interactions of solar wind with various magnetic fields. For example, beta in the corona of the Sun is about 0.01.

Tokamaks have been studied the most and have achieved the best overall performance for MCF purposes, followed by the Stellarator and then the spherical Tokamak (see Fig. 3.35), which is actually a very tight aspect ratio Tokamak.

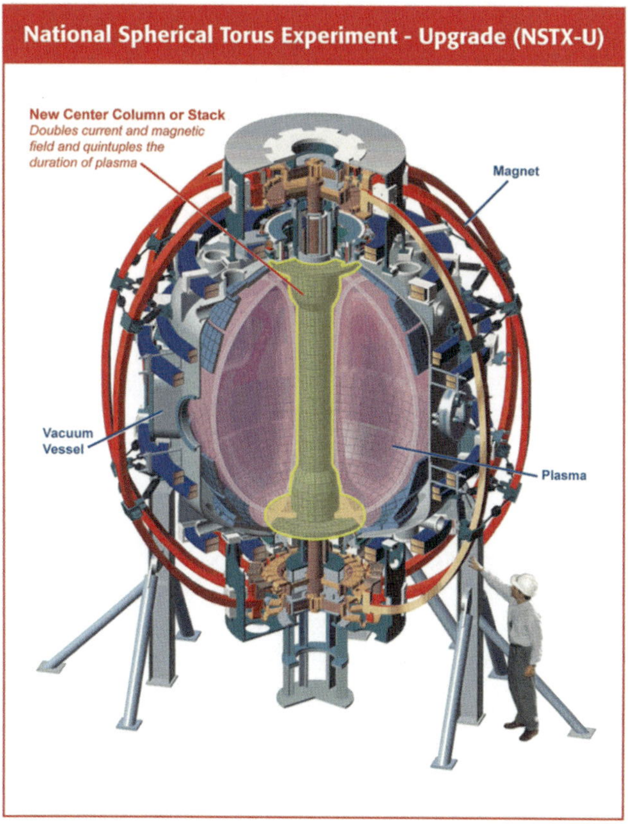

Fig. 3.35 Spherical tokamak

These configurations (i.e., Tokamak and Stellarator) all have relatively strong toroidal magnetic fields and reasonable transport losses. Each is capable of stable MHD operation at acceptable values of β, without the need for a conducting wall close to the plasma.

The advantage of the Stellarator is that it is the only concept that does not require a toroidal current device in a magnetic plasma fusion reactor, but it has a noticeably more complicated magnetic configuration which increases the complexity and cost.

3.11 Inertial Confinement Fusion (ICF)

In recent years, ICF has raised a lot of interest beyond just the national laboratories in the United States and abroad. ICF is aimed towards producing clean energy, using high-energy laser beams or for that matter a particle beam (i.e., the particle beam may consist of heavy or light ion beam) to drive a pellet of two isotopes of hydrogen

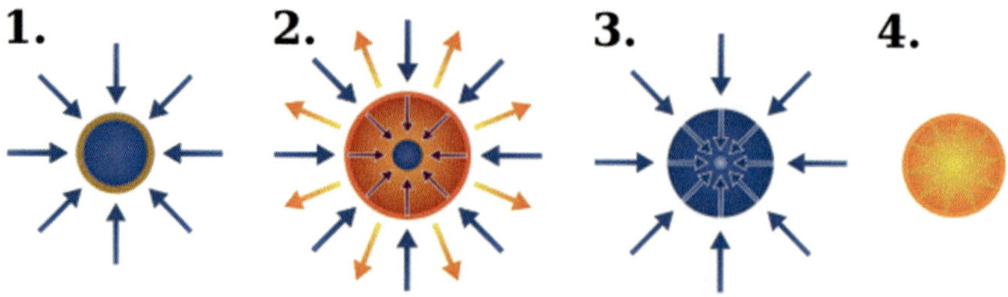

Fig. 3.36 Direct laser-driven compression of a fusion pellet

to fuse and release energy. See the D–T fusion process in Eq. 3.115, where n is the neutron and α a particle such as helium $\left({}_2^4\mathrm{H}^2\mathrm{e}\right)$.

$$D + T \rightarrow n(14.06\,\mathrm{MeV}) + \alpha(3.5\mathrm{MeV}) \tag{3.115}$$

These two isotopes of hydrogen are known as deuterium ($\mathrm{D} = {}^2\mathrm{H}$) and tritium ($\mathrm{T} = {}^3\mathrm{H}$) as part of fuel, to ignition temperature in order to satisfy the confinement criterion of $\rho r \geq 1$ gram/cm^2, where ρ and r are the compressed fuel density and radius pellet, respectively. In order for the confinement criteria, also known as the Lawson criterion, to be satisfied, it needs to take place before occurrence of Rayleigh-Taylor hydrodynamics instability would happen for uniform illumination of the target's surface, namely pellets of deuterium and tritium.

In a direct laser-driven pellet approach in order to overcome Raleigh-Taylor instability, we require a large number of laser beams (see Fig. 3.36).

Figure 3.36 provides a schematic of the stages of ICF using lasers. The blue arrows represent radiation; orange is blow-off; purple is inwardly transported thermal energy.

1. Laser beams or laser-produced x-rays rapidly heat the surface of the fusion target, forming a surrounding plasma envelope.
2. Fuel is compressed by the rocket-like blow-off of the hot surface of the material.
3. During the final part of the capsule implosion, the fuel core reaches 20 times the density of lead and ignites at 100,000,000 °C.
4. Thermonuclear burn spreads rapidly through the compressed fuel, yielding many times the input energy.

In the indirect illuminating target approach, the laser light is converted into soft x-rays, which are trapped inside a hohlraum chamber, surrounding the fusion fuel and irradiating it uniformly. In this approach, in order to archive fusion inertial confinement, the energy source that drives the ablation and compression, is soft x-rays. This is produced by the conversion of a non-thermal, directed energy source, such as lasers or ion beams, into thermal radiation inside a high-opacity enclosure that is referred to as a hohlraum (see Fig. 3.37).

In Fig. 3.38, a schematic of the stages of ICF using lasers driving the pellets, the compression proceeds along several steps from left to right, as follows:

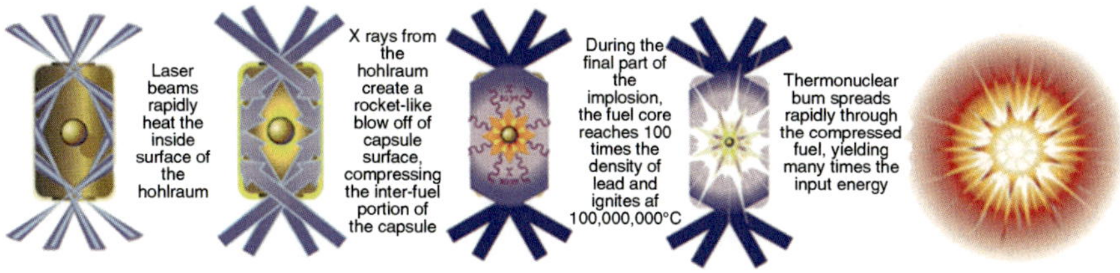

Fig. 3.37 Indirect soft x-ray hohlraum drive compression of a fusion pellet

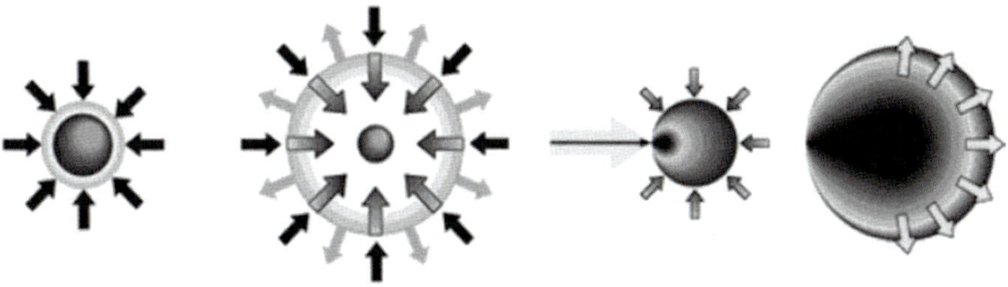

Fig. 3.38 Single-beam igniter concept for fusion pellets

1. *Laser illumination*: Laser beams rapidly heat the inside surface of the hohlraum.
2. *Indirect drive illumination*: The walls of the hohlraum create an inverse rocket effect from the blow-off of the fusion pellet surface, compressing the inner fuel portion of the pellet.
3. *Fuel pellet compression*: During the final part of the implosion process the fuel core reaches a high density and temperature.
4. *Fuel ignition and burn*: The thermonuclear burn propagates through the compressed fusion fuel, amplifying the input energy in fusion fuel burn.

In addition to these approaches, there is third approach, as depicted in Fig. 3.26; this is a single-beam direct approach, where a single beam is used for the compression using the following steps:

1. *Atmospheric formation*: A laser or a particle beam rapidly heats up the surface of the fusion pellet, surrounding it with a plasma envelope.
2. *Compression*: The fuel is compressed by the inverse rocket blow-off of the pellet surface, imploding it inwards.
3. *Beam fuel ignition*: At the instant of maximum compression, a short, high-intensity pulse ignites the compressed core. An intensity of 10^{19} [Watts/cm^2] is contemplated with a pulse duration of 1–10 microseconds.
4. *Burn phase*: The thermonuclear burn propagates through the compressed fusion fuel, yielding several times the driver input energy.

In either of these approaches, Lawson criterion for simple cases of physics of ICF can easily be calculated as shown below.

The Lawson criterion applies to ICF as well as to MCF, but is more usefully expressed in a different form. A good approximation for the inertial confinement time τ_E is the time that it takes an ion to travel over a distance r at its thermal speed $\mathrm{v}_{\text{Thermal}}$.

$$\mathrm{v}_{\text{Thermal}} = \sqrt{\frac{k_B T}{m_i}} \tag{3.116}$$

where k_B is the Boltzmann constant, m_i is the mean ionic mass, and T is the temperature.

Equation 3.116 is derived from kinetic energy theory and the gas pressure relationship.

The inertial confinement time τ_E can thus be approximated as:

$$\tau_E \approx \frac{r}{\mathrm{v}_{\text{Thermal}}} \tag{3.117}$$

Substituting Eq. 3.116 into Eq. 3.117 results in:

$$\begin{aligned}\tau_E &\approx \frac{r}{\mathrm{v}_{\text{Thermal}}}\\ &= \frac{r}{\sqrt{\frac{k_B T}{m_i}}}\\ &= r \cdot \sqrt{\frac{m_i}{k_B T}}\end{aligned} \tag{3.118}$$

However, the Lawson criterion requires that fusion heating fE_{ch} exceeds the power losses P_{loss} as written here:

$$fE_{\text{ch}} \geq P_{\text{loss}} \tag{3.119}$$

In this equation the *volume rate* f, which is the reactions per volume time of fusion reaction, is written as follows:

$$f = n_{\text{Deuterium}} n_{\text{Tritium}} < \sigma \mathrm{v} > = \frac{1}{4} n^2 < \sigma \mathrm{v} > \tag{3.120}$$

Moreover, E_{ch} is the energy of the charged fusion products, and in case of deuterium–tritium reaction is equal to 3.5 MeV.

In addition, power loss density P_{loss} is the rate of emery loss per unit volume and is written as follows:

$$P_{\text{loss}} = \frac{W}{\tau_E} \tag{3.121}$$

where W is the energy density or energy per unit volume and is given by:

$$W = 3nk_BT \tag{3.122}$$

In these equations, the variables that are used are define as follows:

k_B = Boltzmann constant
n = particle density
$n_{\text{Deuterium}}$ = deuterium particle density
n_{Tritium} = tritium particle density
τ_E = confinement time, which measures the rate at which a system loses energy to its surrounding environment
σ = fusion cross-section
v = relative velocity
$<\sigma\text{v}>$ = average over the Maxwellian velocity distribution at temperature T
T = temperature

Substituting for all the quantities in Eq. 3.89, the result is written as follows:

$$n\tau_E \geq \frac{12}{E_{\text{ch}}} \frac{k_BT}{<\sigma\text{v}>} \equiv L \tag{3.123}$$

Equation 3.123 is known as Lawson criterion and for the deuterium and tritium reaction is at least $n\tau_E \geq 1.5 \times 10^{20}$ s/m^3, where the minimum of the product occurs near $T = 25$ keV. The quantity $T/<\sigma\text{v}>$ is a function of temperature with an absolute minimum. Replacing the function with its minimum value provides an absolute lower limit for the product $n\tau_E$.

Substituting Eq. 3.123 into Eq. 3.118, we obtain:

$$n\tau_E \approx n \cdot r \cdot \sqrt{\frac{m_i}{k_BT}} \geq \frac{12}{E_{\text{ch}}} \frac{k_BT}{<\sigma\text{v}>} \tag{3.124}$$

Or:

$$n \cdot r \geq \frac{12}{E_{ch}} \cdot \frac{(k_BT)^{3/2}}{<\sigma\text{v}> \cdot m_i^{1/2}} \tag{3.125}$$

Equation 3.125 could be approximated to the following form:

$$n \cdot r \geq \frac{(k_BT)^{3/2}}{<\sigma\text{v}>} \tag{3.126}$$

This product must be greater than a minimum value of $T^{3/2}/<\sigma\text{v}>$. The same requirement is traditionally expressed in terms of mass density $\rho = <nm_i>$, as follows:

$$\rho r \geq 1\,\text{g/cm}^2 \tag{3.127}$$

Satisfaction of this criterion at the density of solid deuterium–tritium (0.2 g/cm^3) would require a laser pulse of implausibly large energy. Assuming the energy required scales with the mass of the fusion plasma ($E_{\text{Laser}} \approx \rho r^3 \approx \rho^{-2}$), compressing the fuel to 10^3 or 10^4 times solid density would reduce the energy required by a factor of 106 or 108, bringing it into a realistic range. With a compression by 10^3, the compressed density will be 200 g/cm^3, and the compressed radius can be as small as 0.05 mm. The radius of the fuel before compression would be 0.5 mm. The initial pellet will be perhaps twice as large, since most of the mass will be ablated during the compression.

The fusion power density is a figure of good merit to determine the optimum temperature for magnetic confinement, but for inertial confinement the fractional burn-up of the fuel is probably more useful. The burn-up should be proportional to the specific reaction rate ($n^2 < \sigma v>$) times the confinement time (which scales as $T^{-1/2}$) divided by the particle density n:

$$\text{burn-up fraction} \Rightarrow \begin{cases} \propto & n^2 < \sigma v > T^{-1/2}/n \\ \propto & (nT) < \sigma v > /T^{3/2} \end{cases} \tag{3.128}$$

Thus, the optimum temperature for ICF maximizes $<\sigma v > /T^{3/2}$, which is slightly higher than the optimum temperature for magnetic confinement.

Note that one of the key issues for a laser to drive pellets of micro-balloon-containing deuterium and tritium to achieve fusion is symmetrical homogenous compression, which means aiming for perfectly spherical implosions and explosions. However, in reality, this ideal situation will never take place perfectly and as result there are a number of physics problems and consequences, as follows:

- Instabilities and mixing
  - Rayleigh–Taylor unstable compression [1]
  - Break in symmetry destroys confinement
- How to improve energy coupling into the target (i.e., pellet of D–T), which requires the conversion of kinetic energy from the implosion into internal energy of the fuel, which is not perfect. Additionally, we need to prevent reduction of the maximum compression.
- Severe perturbing of a spherical homogeneous and symmetric implosion can result in small-scale turbulences and even in break-up of the target shell.
- The hot spot area at the ablation surface is increased or has a large surface due to the perturbed structure, which leads to reduction of the ignition temperature to achieve fusion reactions in the corona of the pellet, and it causes the α-particle created in Eq. 3.115 to escape the hot spot area. This also lowers the self-heating (see Figs. 3.39 and 3.40).
- Finally, what is the best material for the first wall of the pellet of D–T as the target?

In summary, the Rayleigh–Taylor instabilities occur when a lower-density fluid such as oil underlies a higher-density fluid such as water. In inertial confinement where the implosion and explosion process takes place in a sequence, the higher-density fluid is the pellet surface and the lower-density fluid is the plasma

Fig. 3.39 Reduced self-heating of the hot-spot area from prematurely escaping α-particles

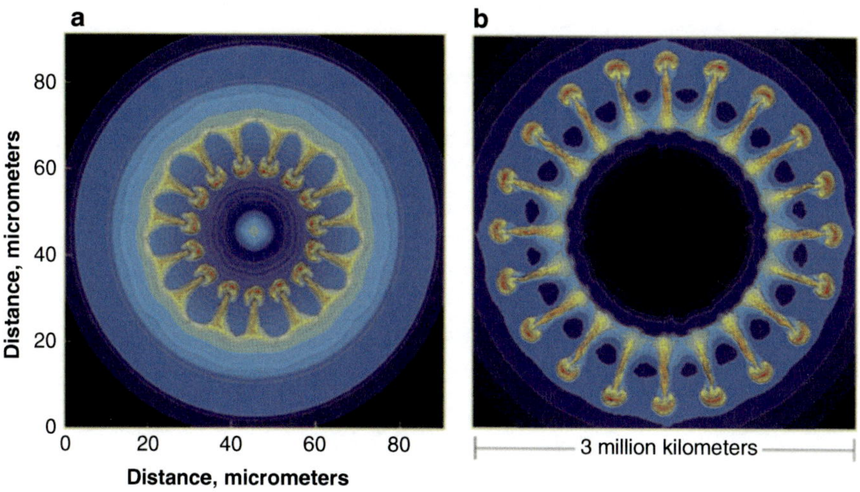

Fig. 3.40 Striking similarities exist between hydrodynamic instabilities in (**a**) inertial confinement fusion capsule implosions; and (**b**) core-collapse supernova explosions

surrounding it, and it compressing the pellet through the inverse rocket action (i.e., inertial) of the implosion process.

All approaches stated here for the inertial confinement via laser or particle beams imploding and exploding target pellets in a symmetrical and homogeneous mode are mainly influenced by Rayleigh–Taylor instabilities at the ablation surface.

As stated, the impact and effect of the Rayleigh–Taylor instabilities is because they initially grow exponentially, so that even very small and insignificant disturbances can grow to a size that has an adverse effect on the entire compression in the homogeneous and symmetrical mode, as observed in Fig. 3.41. In this figure, the major instability is again because of a heavy material pushed on a low-density one.

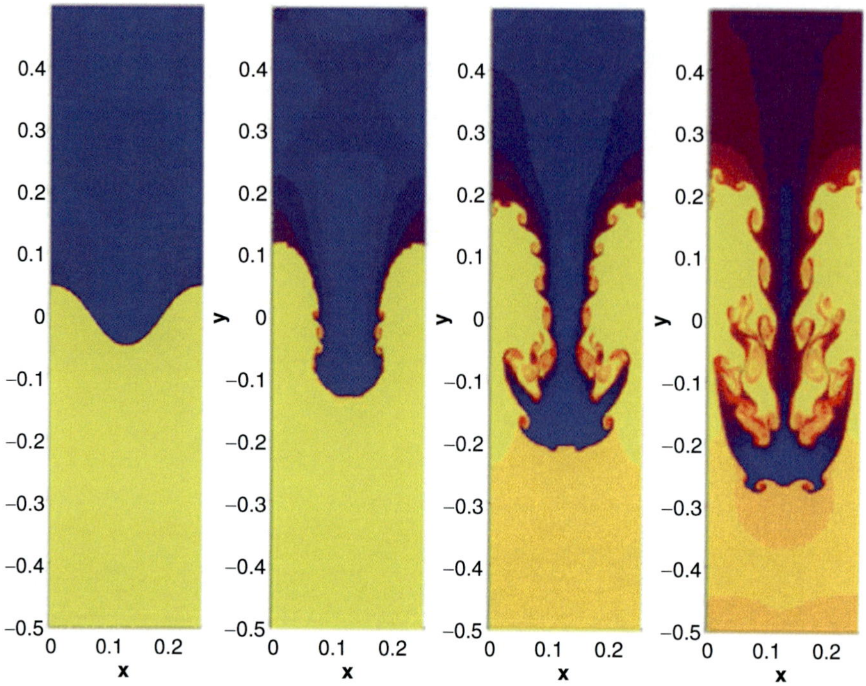

Fig. 3.41 Growth of Rayleigh–Taylor instability

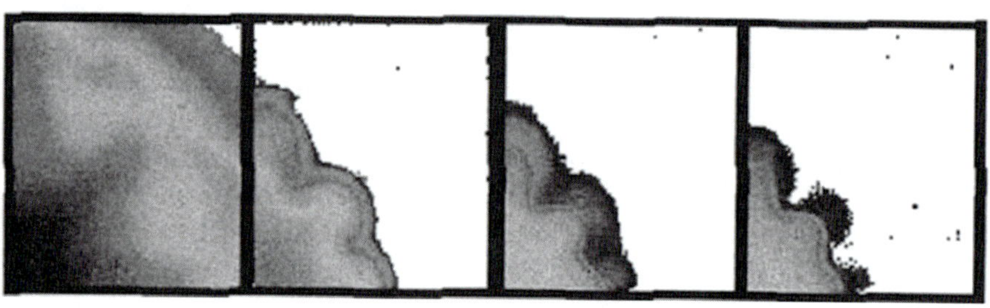

Fig. 3.42 Growth of Rayleigh–Taylor instabilities during pellet implosion

This instability always occurs, since the laser or particle beam as a driver of the deuterium–tritium pellet is never 100% homogeneous and symmetric, and, consequently, the Rayleigh–Taylor instability is always growing.

The growth rate of the Rayleigh–Taylor instability can be measured in a wavelength range not previously accessible and is a very important factor that needs to be paid attention to during the implosion and explosion of the pellet. Moreover, it is important for the purpose of delivering energy to the corona of the pellet as symmetrically and homogenously as possible before the plasma frequency generated at the ablation surface in order to reach the same beam wavelength frequency as the driver (see Fig. 3.42).

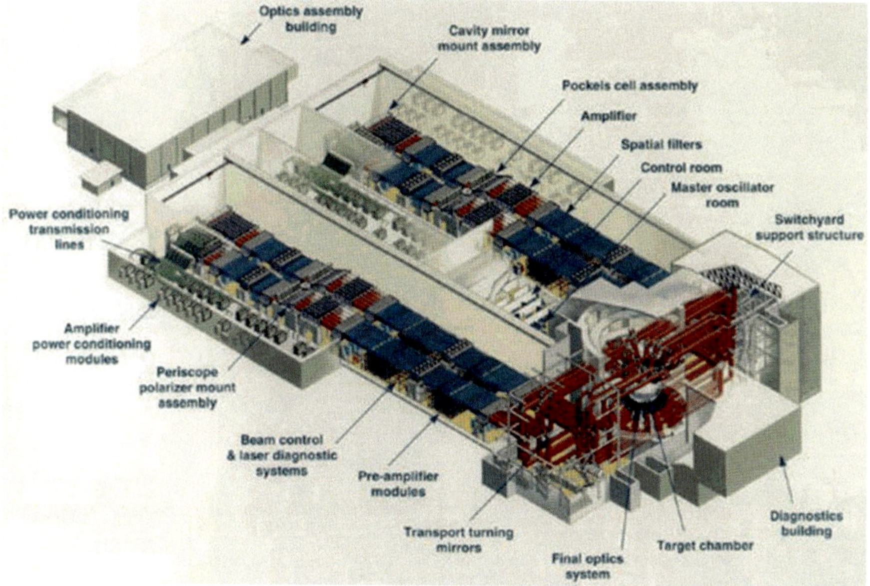

Fig. 3.43 Top view of national ignition facility (NIF) at Lawrence Livermore Laboratory, Livermore, California, USA

Thus, in conclusion, the fusion targets can be illuminated with the energy of different drivers. The primary efforts in inertial confinement exist in the United States, France, and Japan.

The National Ignition Facility (NIF) is the world's largest and most energetic laser ever built. The NIF is also the most precise and reproducible laser as well as the largest optical instrument. The giant laser has nearly 40,000 optics, which precisely guide, reflect, amplify, and focus 192 laser beams onto a fusion target about the size of a pencil eraser. The NIF became operational in March 2009. The NIF is the size of a sports stadium—three football fields could fit inside. Figure 3.43 provides an image of the top view of this facility at the Lawrence Livermore National Laboratory (LLNL) in Livermore, California, USA.

The NIF is making important advances toward achieving fusion ignition in the laboratory for the first time. Its goal is to focus the intense energy of 192 giant laser beams on a BB-sized target (see Fig. 3.34) filled with hydrogen fuel, fusing the nuclei of the hydrogen atoms and releasing many times more energy than it took to initiate the fusion reaction. The BB-sized target is a hohlraum cylinder, which contains the NIF fusion fuel capsule, and is just a few millimeters wide, about the size of a pencil eraser, with beam entrance holes at either end. The fuel capsule is the size of a small pea.

In addition to energy for the future, the NIF's primary missions include national security and understanding the universe. Moses noted that there are many benefits to being able to create conditions to study fusion reactions in lieu of weapons testing. The NIF target chamber will also allow study of the cosmos.

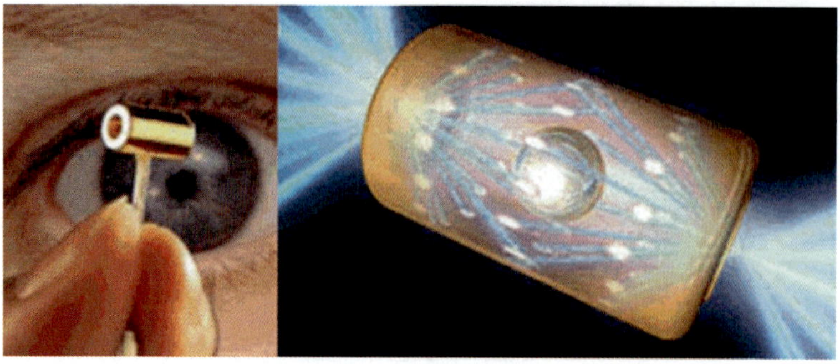

Fig. 3.44 Illustration of BB-sized target

References

1. B. Zohuri, *Plasma Physics and Controlled Thermonuclear Reactions Drive Fusion Energy*, 1st edn. (Springer Publishing Company, Cham, 2016)
2. B. Zohuri, *Inertial Confinement Fusion Driven Thermonuclear Energy*, 1stn edn. (Springer Publishing Company, Cham, 2016)
3. B. Zohuri, *Magnetic Confinement Fusion Driven Thermonuclear Energy*, 1st edn. (Springer Publishing Company, Cham, 2016)
4. S. Glasstone, R.H. Lovberg, *Controlled Thermonuclear Reactions* (D. Van Nostrand Company, Inc, Huntington, 1960)
5. J.D. Lawson, Some criteria for a power producing thermonuclear reactor. Proc Phys Soc **70**, 6 (1957)
6. U.S. Inan, M. Golkowski, *Principles of Plasma Physics for Engineers and Scientists* (Cambridge University Press, Cambridge, 2011)
7. F.F. Chen, *Introduction to Plasma Physics and Controlled Fusion*, 3rd edn. (Springer Publishing Company, Cham, 2015)

Chapter 4
Cryogenics and Liquid Hydrogen Storage

Cryogenics is the science that addresses the production and effects of very low temperatures. The word originates from the Greek words *kryos* meaning "frost" and *genic* meaning "to produce." Using this definition, the term could be used to include all temperatures below the freezing point of water (0 °C). However, Professor Kamerlingh Onnes of the University of Leiden in the Netherlands first used the word in 1894 to describe the art and science of producing much lower temperatures. He used the word in reference to the liquefaction of permanent gases such as oxygen, nitrogen, hydrogen, and helium. Oxygen had been liquefied at −183 °C a few years earlier in 1887 and a race was in progress to liquefy the remaining permanent gases at even lower temperatures. The techniques employed in producing such low temperatures were quite different from those used somewhat earlier in the production of artificial ice. In particular, efficient heat exchangers are required to reach very low temperatures. Over the years the term cryogenics has generally been used to refer to temperatures below approximately −150 °C (123.15 K, −238.00 °F). Cryogenic applications extends beyond its present day-to-day usage, and one important aspect of it is storage of high-density liquid hydrogen. To liquefy hydrogen, it must be cooled to cryogenic temperatures through a liquefaction process. Hydrogen is most commonly transported and delivered as a liquid when high-volume transport is needed in the absence of pipelines. Trucks transporting liquid hydrogen are referred to as liquid tankers [1].

4.1 Introduction

A workshop on "Advanced Composite Materials for Cold and Cryogenic Hydrogen Storage Applications in Fuel Cell Electric Vehicles" was hosted by the United States Department of Energy (DOE) Office of Energy Efficiency and Renewable Energy's Fuel Cell Technologies Office and Pacific Northwest National Laboratory in Dallas, Texas, on October 29, 2015 [10]. The objectives of the workshop were to (1) gather

© Springer International Publishing AG, part of Springer Nature 2019

B. Zohuri, *Hydrogen Energy*, https://doi.org/10.1007/978-3-319-93461-7_4

input and discuss the state of knowledge on composite materials and processing for use at sub-ambient temperatures; and (2) identify research needs and recommended development pathways for use of composite materials at sub-ambient temperature high-pressure applications. This input will be used to help guide future activities for the DOE hydrogen storage program (see Chap. 8).

A key technology requirement for the widespread commercialization of fuel cell electric vehicles (FCEVs) and other hydrogen fuel cell applications is compact, reliable, safe, and cost-effective storage of hydrogen. However, while some light-duty FCEVs with a driving range of about 300 miles are emerging in limited markets, achieving sufficient affordable onboard storage that does not sacrifice passenger and cargo space remains a barrier to commercialization beyond limited vehicle platforms and niche markets. While the energy per mass of hydrogen is much greater than that of most other fuels, its energy by volume is significantly less than that of liquid fuels such as gasoline. The current state of the art is to store hydrogen in composite overwrapped pressure vessels (COPVs) with polymer liners at 700 bar pressures.

To make the systems more compact, longer-term research is focusing on developing advanced hydrogen storage technologies that can provide greater energy density than 700 bar compressed hydrogen, at a competitive cost. This is where we need liquefaction of gaseous hydrogen by cooling it below −253 °C (−423 °F) with help from our knowledge of the physics of cryogenics as an ultra-low temperature phenomenon [1].

Research is now being performed for high-pressure hydrogen storage at cold (e.g., ~200 K) and cryogenic (e.g., ≤ 200 K) temperatures. Cold and cryogenic-compressed hydrogen storage systems allow the same quantity of hydrogen to be stored—either in smaller volumes at similar pressures or in similar volumes at lower pressures—thus providing higher hydrogen densities for the widespread commercialization of FCEVs. The workshop aimed to help identify the implications of using composite materials in these low-temperature, high-pressure, long-cycle-life applications and any knowledge gaps that need to be addressed.

Figure 4.1 illustrates liquid hydrogen storage tanks, where liquefaction using the cryogenics process has stored the hydrogen for future consumption in liquid form.

As already stated, in the cryogenic process, gaseous hydrogen is liquefied by cooling it to below −253 °C (−423 °F). Once hydrogen is liquefied it can be stored at the liquefaction plant in large insulated tanks. It takes energy to liquefy hydrogen—using today's technology, liquefaction consumes more than 30% of the energy content of the hydrogen and is expensive. In addition, some amount of stored hydrogen will be lost through evaporation, or "boil off," of liquefied hydrogen, especially when using small tanks with large surface:volume ratios. Research to improve liquefaction technology, as well as improved economies of scale, could help lower the energy required and the cost.

Currently, for longer distances, hydrogen is transported as a liquid in super-insulated, cryogenic tanker trucks, as shown in Fig. 4.2. After liquefaction, the liquid hydrogen is dispensed to delivery trucks and transported to distribution sites where it is vaporized to a high-pressure gaseous product for dispensing.

Fig. 4.1 Hydrogen liquefier. (Courtesy of Linde Kryotechnik AG, Switzerland)

Fig. 4.2 Liquid hydrogen tanker. (Courtesy of Air Products)

Over long distances, trucking liquid hydrogen (LH_2) is more economical than trucking gaseous hydrogen because a liquid tanker truck can hold a much larger mass of hydrogen than a gaseous tube trailer can. Challenges with liquid transportation include the potential for boil-off during delivery.

Figure 4.2 shows a liquid tanker installed on the back of transport truck for consumption by a space shuttle, which is on its launch pad, getting prepared for takeoff. The cryogenics process keeps the hydrogen in liquid form by cooling the hydrogen gas to an ultra-low temperature below −253 °C (−423 °F).

However, one main issue is the type of materials required, which need to be advanced enough to withstand cold hydrogen liquid at cryogenic temperature stage and yet not to be fragile based on their crystal from solid-state point of view, and have the physical infrastructure to stand such low temperatures.

It should be noted that tank manufacturers tend to do their own modeling for their own specific tank designs. This implies that there is a lack of agreement on design principles for composites in cryogenic temperature environments. Thus, we need to identify the main groups of materials that can be used for extreme temperature modeling of composite materials, either at low or high temperatures. It also needs to be discussed how aerospace original equipment manufacturers (OEMs) typically perform their extreme temperature modeling of composite materials, relying strongly on empirical relationships to correlate modeling assumptions with experimental observations.

A discussion on material choice and processing at a low cost and high production volume in the automotive industry, where a compact shape is needed for these storage tanks in vehicles, should take place among all researchers in this field, since it is very difficult to obtain material properties at cryogenic temperatures. Thus, the question that needs to be answered is whether one can adapt room temperature modeling tools to the cryogenic regime? Would room temperature assumptions still be valid for low temperature modeling/how far off would they be? Can one trust ambient temperature relationships at cryogenic temperatures?

4.2 Physics of Cryogenics

This section provides a high-level discussion of the physics of cryogenics. Readers are encouraged to refer to the book by Zohuri [1] for more detail on this subject.

According to the laws of thermodynamics, a limit to the lowest temperature that can be achieved exists, which is known as absolute zero. Molecules are in their lowest, but finite, energy state at absolute zero. Such a temperature is impossible to reach because the input power required approaches infinity. However, temperatures within a few billionths of a degree above absolute zero have been achieved. Absolute zero is the zero of the absolute or thermodynamic temperature scale and is equal to −273.15 °C or − 459.67 °F. The metric or SI (International System of Units) absolute scale is known as the Kelvin scale, the unit of which is the kelvin (not Kelvin), which has the same magnitude as degrees Celsius. The symbol for the

Table 4.1 Normal boiling, triple, and critical points

Cryogen	K	°C	°R	°F	Triple point	Critical point
Methane	111.7	−161.5	201.1	−258.6	90.7	190.5
Oxygen	90.2	−183.0	162.4	−297.3	54.4	154.6
Nitrogen	77.4	−195.8	139.3	−320.4	63.1	126.2
Hydrogen	20.3	−252.9	36.5	−423.2	13.8	33.2
Helium	4.2	−269.0	7.6	−452.1	2.2	5.2
Absolute zero	0.0	−273.15	0.0	−459.67	–	–

Kelvin scale is K, as adopted by the 13th General Council on Weights and Measures (CGPM) in 1968. Thus, 0 °C equals 273.15 K. The English absolute scale, known as the Rankine scale, uses the symbol R and has an increment the same as that of the Fahrenheit scale. In terms of the Kelvin scale, the cryogenic region is often considered to be that below approximately 120 K (−153 °C). The common permanent gases referred to earlier change from gas to liquid at atmospheric pressure at the temperatures shown in Table 4.1, called the normal boiling point (NBP). Table 4.1 also includes the triple point and critical point. Such liquids are known as cryogenic liquids or cryogens. When liquid helium is cooled further to 2.17 K or below, it becomes a super-fluid with very unusual properties that are associated with being in the quantum mechanical ground state. For example, it has zero viscosity and produces a film that can creep up and over the walls of an open container, such as a beaker, and drip off the bottom as long as the temperature of the container remains below 2.17 K.

The measurement of cryogenic temperatures requires methods that may not be familiar to the general public. Normal mercury or alcohol thermometers freeze at such low temperatures and become useless. A platinum resistance thermometer has a well-defined behavior of electrical resistance versus temperature. It is commonly used to measure cryogenic temperatures down to about 20 K accurately. Certain semiconducting materials, such as doped germanium, are also useful as electrical resistance thermometers for temperatures down to 1 K and below, as long as they are calibrated over the range they are to be used in. Such secondary thermometers are calibrated against primary thermometers that utilize fundamental laws of physics in which a physical variable changes in a well-known theoretical way with temperature.

The production of cryogenic temperatures usually utilizes the compression and expansion of gases. In a typical air liquefaction process the air is compressed, causing it to heat; it is allowed to cool back to room temperature while still pressurized. The compressed air is further cooled in a heat exchanger before it is allowed to expand back to atmospheric pressure. The expansion causes the air to cool and a portion of it to liquefy. The remaining cooled gaseous portion is returned through the other side of the heat exchanger where it pre-cools the incoming high-pressure air before returning to the compressor. The liquid portion is usually distilled to produce liquid oxygen, liquid nitrogen, and liquid argon. Other gases, such as helium, are used in a similar process to produce even lower temperatures, but several stages of expansion are necessary.

Cryogenics has many applications. Cryogenic liquids, such as oxygen, nitrogen, and argon, are often used in industrial and medical applications. The electrical resistance of most metals decreases as temperature decreases, and certain metals lose all electrical resistance below some transition temperatures and become superconductors. An electromagnet wound with a wire made from such a metal can produce extremely high magnetic fields with no generation of heat and no consumption of electric power once the field is established and the metal remains cold. These metals, typically niobium alloys cooled to 4.2 K, are used for the magnets of magnetic resonance imaging (MRI) systems in most hospitals. Superconductivity in some metals was first discovered in 1911 by Onnes, but since 1986 another class of materials, known as high-temperature superconductors, has been found to be superconducting at much higher temperatures, currently up to about 145 K. They are a type of ceramic, and because of their brittle nature are more difficult to fabricate into wires for magnets.

Other applications of cryogenics include fast freezing of some foods and the preservation of some biological materials such as livestock semen as well as human blood, tissue, and embryos. The practice of freezing an entire human body after death in the hope of later restoring life is known as cryonics, but it is not an accepted scientific application of cryogenics. The freezing of portions of the body to destroy unwanted or malfunctioning tissue is known as cryosurgery. It is used to treat cancers and abnormalities of the skin, cervix, uterus, prostate gland, and liver.

4.2.1 Low Temperatures in Science and Technology

Cryogenics, as described in the previous section, is defined as "that branch of physics, which deals with the production of very low temperatures and their effect on matter" [2], a definition which addresses both aspects of attaining low temperatures that do not naturally occur on Earth, and of using them for the study of nature or human industry. In a more operational way [3], it is also defined as *the science and technology of temperatures below 120 K*. The reason for this latter definition can be understood by examining characteristic temperatures of cryogenic fluids as shown in Table 4.1.

The temperature limit of 120 K comprehensively includes the NBPs of the main atmospheric gases, as well as of methane, which constitutes the principal component of natural gas. Today, liquid natural gas (LNG) represents one of the largest—and fastest-growing—industrial domains of the application of cryogenics (see Fig. 4.3), together with the liquefaction and separation of air gases (see Fig. 4.4). Densification by condensation and separation via distillation of gases was historically—and remains today—the main driving force for the cryogenic industry. This is exemplified not only by liquid oxygen, and by nitrogen used in chemical as well as metallurgical processes, but also by the cryogenic liquid propellants of rocket engines (see Fig. 4.5) where the proposed use of hydrogen is as a "clean" energy vector in transportation (see Fig. 4.6) [1].

Fig. 4.3 130,000 m^3 liquid natural gas (LNG) carrier with integrated Invar tank

Fig. 4.4 Cryogenic air separation plant with heat exchanger and distillation column towers

As already stated, cryogenic technology has the need for smaller cryocoolers because of the advances in the miniaturization of electrical and optical devices and the need for cooling and conducting efficiency (see Fig. 4.7). Cryogenic technology deals with materials at low temperatures and the physics of their behavior at these

Fig. 4.5 Rockets using cryogenic liquid propellants: (**a**) Ariane 5 (25 t liquid hydrogen, 130 t liquid oxygen); and (**b**) space shuttle (100 t liquid hydrogen, 600 t liquid oxygen)

Fig. 4.6 Automotive liquid hydrogen fuel tank

temperatures. New applications are being discovered for cryocooled electrical and optical sensors and devices, with particular emphasis on high-end commercial applications in medical and scientific fields as well as in the aerospace and military industries.

Refrigerators, cryocoolers, and micro-coolers are needed by various commercial, industrial, space, and military systems. Cryogenic cooling plays an important role in

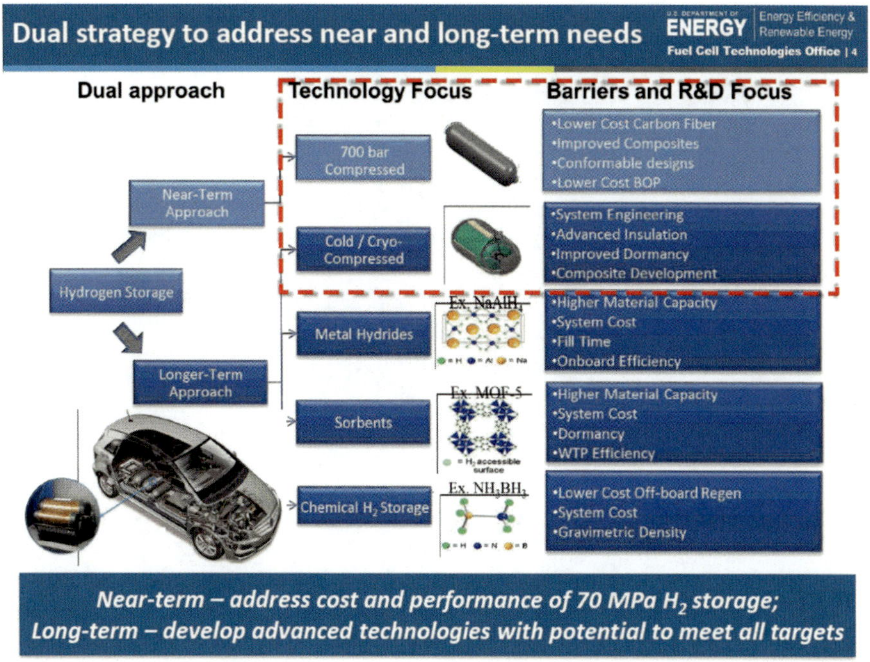

Fig. 4.7 Strategy for near- and long-term storage of liquid hydrogen at 70 Mpa. (Courtesy of the US Department of Energy)

unmanned aerial vehicle systems, infrared search and track sensors, missile warning receivers, satellite tracking systems, and a host of other commercial and military systems [1].

Now with new-generation nuclear power plants that are known as GEN-IV, a lot of attention is focused toward making them more efficient and cost effective [3] as well as using cryogenics techniques to implement energy storage in nuclear power plants [4]. Energy storage in nuclear power plants relies on a novel method of integration of nuclear power generation with cryogenic energy storage (CES) to achieve an effective time shift of the electrical power output. CES stores excess electricity in the form of cryogen (liquid air/nitrogen) through an air liquefaction process at off-peak hours and recovers the stored power by expanding the cryogen during peak hours [5].

The quest for low temperatures, however, finds its origin in early thermodynamics, with Amontons's gas pressure thermometer (1703) opening the way for the concept of absolute zero that was inferred a century later by Charles and Gay-Lussac, and eventually formulated by Kelvin. It was, however, with the advent of Boltzmann's statistical thermodynamics in the late nineteenth century that temperature—a phenomenological quantity—could be explained in terms of microscopic structure and dynamics. Consider a thermodynamic system in a macrostate, which

Fig. 4.8 Ludwig Boltzmann's grave in the Zentralfriedhof, Vienna, Austria, bearing the entropy formula

can be obtained by a multiplicity (W) of microstates. The entropy S of the system was postulated by Boltzmann as follows [1]:

$$S = k_B \log W \tag{4.1}$$

with $k_B \simeq 1.38 \times 10^{-23}$ J/K. This formula, which founded statistical thermodynamics, is displayed on Boltzmann's grave in Vienna (see Fig. 4.8).

Adding reversibly heat dQ to the system produces a change of its entropy dS, with a proportionality factor T that is precisely temperature:

$$T = \frac{dQ}{dS} \tag{4.2}$$

Thus, a low-temperature system can be defined as one to which a minute addition of heat produces a large change in entropy, that is, a large change in its range of possible microscopic configurations. Boltzmann also found that the average thermal energy of a particle in a system in equilibrium at temperature T is as follows:

$$E \sim k_B T \tag{4.3}$$

Consequently, a temperature of 1 K is equivalent to a thermal energy of 10^{-4} eV or 10^{-23} J per particle.

Table 4.2 Characteristic temperature of low-energy phenomena

Phenomenon	Temperature (K)
Debye temperature of metals	Few
High-temperature superconductors	~100
Low temperature superconductors	~10
Intrinsic transport properties of metals	<10
Cryopumping	Few
Cosmic microwave background	2.7
Super fluid helium 4	2.2
Bolometers for cosmic radiation	<1
Low-density atomic Bose–Einstein condensates	$\sim 10^{-6}$

A temperature is therefore low for a given physical process when k_BT is small compared to the characteristic energy of the process that is considered.

Cryogenic temperatures thus reveal phenomena with low characteristic energy (Table 4.2) and enable their application when significantly lower than the characteristic energy of the phenomenon of interest. From Tables 4.1 and 4.2, it is clear that "low temperature" superconductivity requires helium cryogenics: several examples of helium-cooled superconducting devices are shown in Fig. 4.9. Considering vapor pressures of gases at low temperature (see Fig. 4.10), it is also clear that helium must be the working cryogen for achieving "clean" vacuum with cryopumps.

4.2.2 *Defining Cryogenic Fluids or Liquids*

Cryogenic liquids, also known as cryogens are gases at normal temperatures and pressures. However, at low temperatures, they are in their liquid state. These liquids are extremely cold and have a boiling point less than −150 °C (−238 °F). Even the vapors and gases released from cryogenic liquids are very cold. They often condense the moisture in air, creating a highly visible fog. Different cryogens become liquids under different conditions of temperature and pressure, but all have two properties in common: extremely cold and small amounts of liquid can expand into very large volumes of gas. Everyone who works with cryogenic liquids must be aware of their hazards and know how to work safely with them. Figure 4.11 shows an image of liquid nitrogen.

The discovery of superconducting materials with critical temperatures significantly above the boiling point of liquid nitrogen has provided new interest in reliable, low-cost methods of producing high-temperature cryogenic refrigeration. The term "high-temperature cryogenic" describes temperatures ranging from above the boiling point of liquid nitrogen, −195.79 °C (77.36 K; −320.42 °F), up to −50 °C (223.15 K; −58.00 °F), the generally defined upper limit of study referred to as cryogenics [6]. Cryogenicists use the Kelvin or Rankine temperature scales present in nature [1].

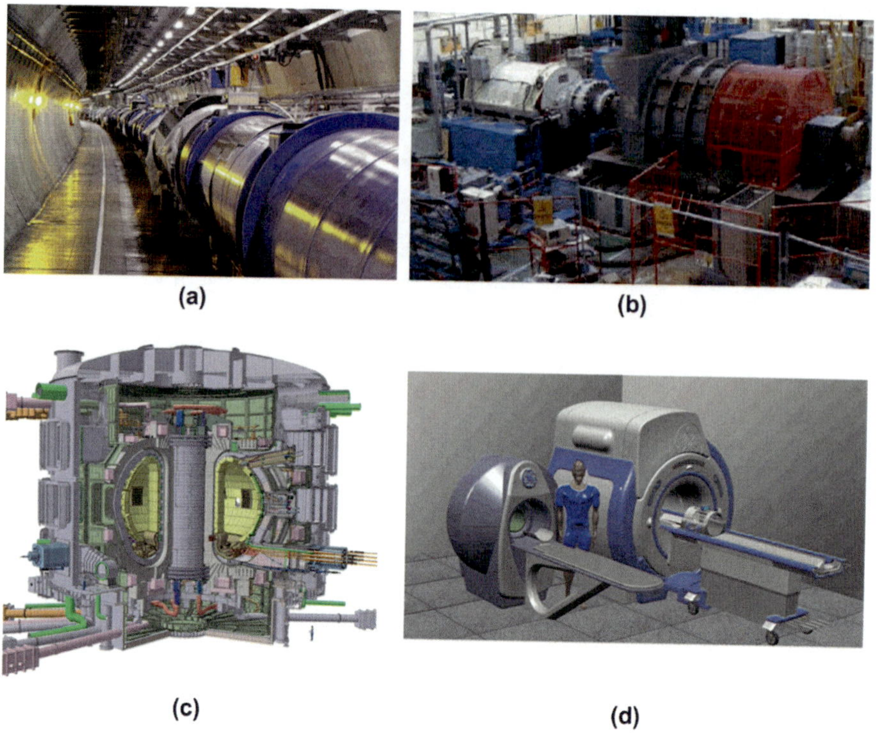

Fig. 4.9 Helium-cooled superconducting devices: (**a**) large hadron collider at CERN; (**b**) 5 MV HTS ship propulsion motor (AMS); (**c**) International Thermonuclear Experimental Reactor (ITER) experimental fusion reactor; and (**d**) whole-body magnetic resonance imaging (MRI) system (Bruker)

4.2.3 Type of Cryogenic Liquids

Each cryogenic liquid has its own specific properties, but most cryogenic liquids can be placed into one of three groups:

- *Inert gases:* Inert gases largely to any extent do not react chemically. They do not burn or support combustion. Examples of this group are nitrogen, helium, neon, argon, and krypton.
- *Flammable gases:* Some cryogenic liquids produce a gas that can burn in air. The most common examples are hydrogen, methane, carbon monoxide, and liquefied natural gas.
- *Oxygen:* Many materials considered as non-combustible can burn in the presence of liquid oxygen. Organic materials can react explosively with liquid oxygen. The hazards and handling precautions of liquid oxygen must therefore be considered separately from other cryogenic liquids.

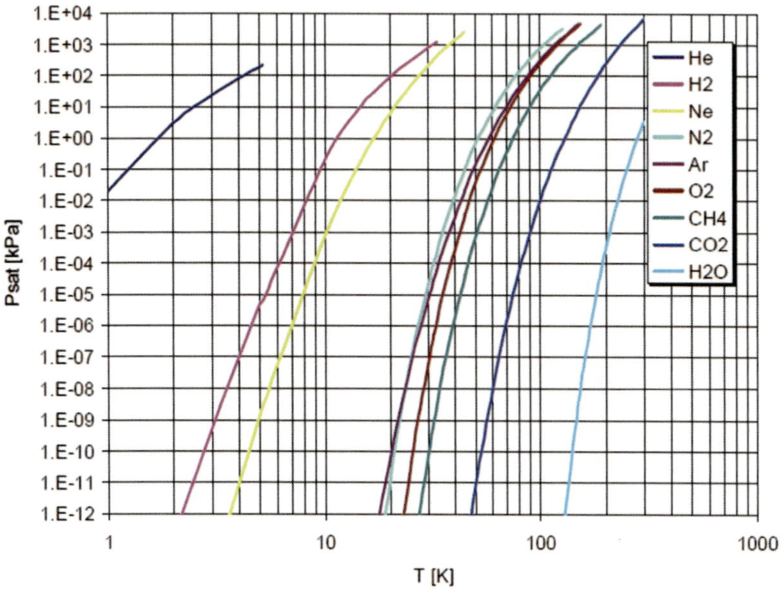

Fig. 4.10 Vapor pressure of common gases at cryogenic temperature

Fig. 4.11 Liquid nitrogen

It is generally agreed that cryogenic fluids are those whose boiling points at atmospheric pressure are about 120 K or lower, although liquid ethylene with its boiling point of 170 K is often included. A list of the cryogenic fluids, together with some selected properties, is given in Table 4.3. Detailed properties are available commercially on computer disc.

Perhaps the most important and widely used fluids are LNG (boiling point [bp] about 120 K), liquid oxygen (bp 90.2 K), and liquid nitrogen (bp 77.3 K).

The availability of cryogenic fluids forms an essential part of the infrastructure of a modem industrialized and civilized society. One of the major reasons for using

Table 4.3 Some properties of cryogens at their normal boiling points

	He^4	n-H_2	D_2	Ne	N_2	CO	F_2	Ar	O_2	CH_4	Kr	Xe	C_2H_4
Normal boiling point (K)	4.22	20.4	23.7	27.1	77.3	81.7	85.0	87.3	90.2	111.6	120.0	165.0	169.4
Liquid density (kg/m^3)	125	71.0	163	1205	809	792	1502	1393	1141	423	2400	3040	568
Liquid density/vapor density	7.4	53	71	126	175	181	267	241	255	236	270	297	272
Enthalpy of vaporization (KJ/kg)	20.42	446	301	86	199	216	157	161	213	512	108	96	482
Enthalpy of vaporization (KJ/kg-mol)	80.6	899	1211	2333	5565	6040	6659	6441	6798	8206	9042	12,604	13,534
Volume of liquid vaporization by energy input of 1 W-h (cm^3)	1410	114	74	35	22	21	14	16	15	17	14	13	13
Dynamic viscosity of liquid ($\mu Nsec/m^2$)	3.3	13.3	28.3	124	152	–	240	260	195	119	404	506	170
Surface tension (mN/m)	0.10	1.9	−3	4.8	8.9	9.6	14.8	12.5	13.2	13.2	5.5	18.3	16.5
Thermal conductivity· of liquid ($mWm^{-1} K^{-1}$)	18.7	100	−100	113	135	–	–	128	152	187	94	74	192
Volume of gas at 15 °C released from 1 volume of liquid	739	830	830	1412	168	806	905	824	842	613	689	520	475

Source: Hands [9]

[a]Pressure of 1.01325 bar

liquid cryogens is to allow transport and storage as liquid at atmospheric pressure, rather than as high-pressure gas in thick-walled vessels, although there is an energy penalty involved in refrigeration. However, the distillation of liquid air (air separation) enables the production of very high-purity oxygen and nitrogen. Plants producing up to several hundred tons per day and more of oxygen are commonplace, sometimes connected permanently to a chemical plant or steel works. Liquid nitrogen—formerly a by-product of the process—is now a product in its own right, being used principally as a convenient source of refrigeration, especially in the frozen food industry.

The other important by-product of air separation is liquid argon, which again can be produced at a very high purity. For welding, it is increasingly being stored as liquid at the factory rather than being delivered in high-pressure cylinders.

All cryogenic fluids except helium and hydrogen behave as "normal" fluids, their common distinguishing features in general being a low specific heat and enthalpy of vaporization. All gaseous cryogens are odorless, and all liquid cryogens are colorless apart from oxygen, which is pale blue, and fluorine, which is pale yellow. They are all diamagnetic except oxygen, which is quite strongly paramagnetic.

With the exception of oxygen, all the gases are asphyxiants, and even oxygen will not support human life in concentrations greater than about 60%. Fluorine and oxygen are powerful oxidizers even in liquid form. Some cryogens are flammable; hydrogen is especially delicate to handle.

Hydrogen is an unusual fluid in that the molecule exists in two forms known as ortho and para, with somewhat different properties. The ratio of ortho to para is determined by conventional thermodynamics and is dependent on temperature. There are also different forms of isotopes (deuterium and tritium) and these two isotopes are used in driving fusion energy production via either magnetic confinement fusion (MCF) [7] or inertial confinement fusion (ICF) [8].

An explanation of the behavior of the hydrogen molecule requires knowledge of quantum mechanics and is not discussed in this book. At low temperatures, equilibrium hydrogen (e-H^2) is entirely para. At room temperature, the ortho:para ratio is 3. The equilibrium state at room temperature is often known as normal hydrogen or n-hydrogen. The transition from the ortho to the para state involves a heat of conversion—which can be greater than the enthalpy of vaporization—so that the vaporization rates of hydrogen are often much larger than expected. It is for this reason that a catalyst is often included in a hydrogen liquefier to ensure that only para hydrogen is present in the liquid [9].

Helium is the one cryogenic fluid that can be claimed to be unique. Because of its low molecular weight and chemical inertness, quantum mechanical effects are important. There are two isotopic forms: the natural form He^4, which has a nucleus consisting of two protons and two neutrons; and the comparatively rare manufactured form He^3, with only one neutron. The two isotopes have markedly different properties due to their different nuclear spins. He^3 is not considered here. More details can be found in the book by Zohuri [1].

4.3 Way Forward

This chapter provides clear evidence that further research and development of cold and cryogenic composites for onboard vehicular hydrogen storage is needed. The stakeholders representing this area of manufacturing pay a lot of attention to DOE's future research activities within this technical research area. In addition, it is clear that innovations that are made in this area for automotive applications can have benefits in other aerospace applications where composite materials are exposed to cryogenic environments.

The need for additional standard testing specifications and procedures for low-temperature material testing should also be highlighted, since current standards are limited in both varieties of materials and temperature ranges. In addition, thermal expansion mismatch, fatigue, and impact performance are not currently measured.

Finally, it is clear that vacuum applications require low-vapor pressure materials. In space low vapor pressure is needed to prevent damage to other components in the system, but low vapor pressure is necessary to maintain thermal insulation properties for low-temperature onboard storage applications. Vacuum-compatible materials are readily available, but they are expensive for onboard applications. Also, outgassing of atmospheres (voids) is less of an issue in space, but is likely an issue in a vacuum-jacketed onboard storage system.

Not only should the development pathway be part of our research mission and focus, it also enables and accelerates the successful commercialization of hydrogen fuel cell technology through development of advanced hydrogen storage technologies able to cost effectively meet the performance requirements, in particular within cryogenics conditions.

To conclude, bear in mind that the cryogenics phenomenon has marked a new era in the growth of applied low temperature, with emphasis shifting from industrial gas liquidation and separation to the lower temperature involved in the use of liquid hydrogen for various applications. This included large-scale nuclear applications such as energy storage for renewable energy during peaks and off peak [4, 5], rocketry, a source of cost-effective motor fuel, and setting the stage for the exploration of superconducting devices as we now see in CERN, the European Organization for Nuclear Research.

Thus, the road ahead should include further research into the following areas:

- Cryogenic H_2 onboard storage
  - Temperature as a degree of freedom in H_2 storage
  - Overall DOE cry-compressed project history cross its organization that includes it operational national laboratories
  - 350-bar pressure test vehicles in park and drive modes
- Current project
  - 700-bar prototype (cryogenic) vessels
  - Refueling with a liquid hydrogen (LH_2) pump
  - Test vessel cycling facility

- System considerations
  - Vacuum jacketing
  - Vacuum, temperature, heat transfer
  - Material properties at low temperature.

Note that 700-bar cryogenic hydrogen gas (H_2) refueling offers volume, capacity, and safety advantages balanced by increasing technical demands.

These are issues that should be part of the road map and pathway ahead from a technical point of view. For example, there are thermodynamic limitations when we are dealing with liquid hydrogen (LH_2) and ambient hydrogen (H_2) storage that can be overcome with hydrogen pressure vessels operable across a broad range of temperatures, as indicated in Fig. 4.12 within cryogenic region of hydrogen gas.

LH_2 and H_2 have several thermodynamic limits, as follows:

1. Maximum density, minimum mass
2. Extended thermal endurance
3. Superior refuel thermodynamics
4. Thermal isolation
5. Low internal energy.

Furthermore, the thermodynamics of high-pressure cryogenic hydrogen gas refueling and storage can provide advantages for vehicles and drivers, as follows:

1. Minimum size and cost
2. Fuel economy and parking time

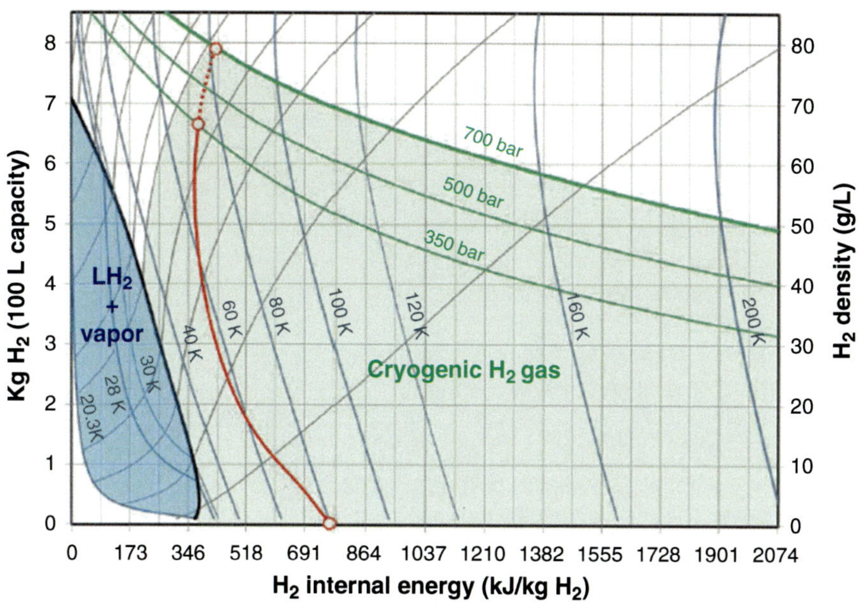

Fig. 4.12 Cryogenic hydrogen gas thermodynamic charts

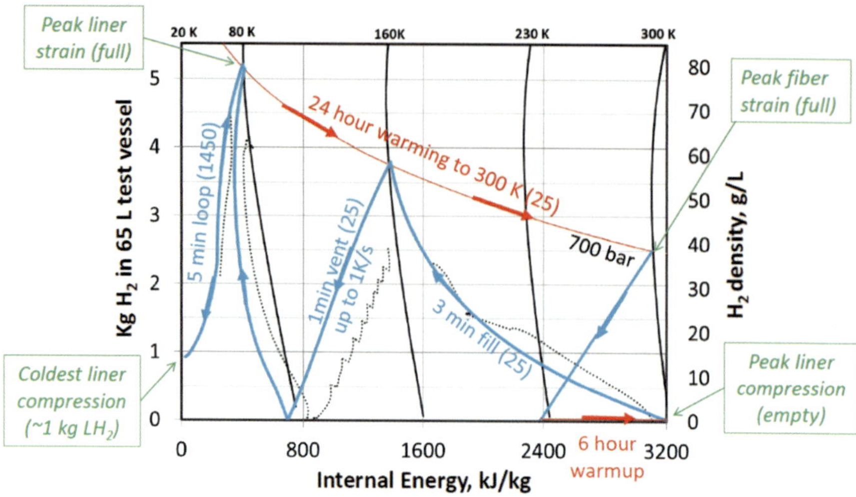

Fig. 4.13 Consumption chars for liquid hydrogen cycling

3. Low-energy rapid refueling at the fuel station
4. High on-road safety factor (5–10)
5. Low burst energy (3–5×).

Bear in mind that ideal cryogenic H_2 cycling covers the full pressure and temperature range, as illustrated in Fig. 4.13, here and emphasizes the maximum thermomechanical stress and time at pressure.

Last but not least, going forward we need to focus on gradients at moderate temperatures and dissimilar materials, as extreme cold (i.e., cryogenics regions) can maximize the degradation of thermomechanical properties.

In addition, low temperature material properties offer opportunities and challenges for cryogenic pressure vessels, a few of which are listed here:

Opportunities greatest at *coldest* temperatures (typically <100 K):

- Increased composite fatigue life
- Increased composite stiffness
- Increased metal strength, cycle life
- Declining thermal conductivity
- Asymptotic heat capacity
- Asymptotic thermal contraction coefficient.

Challenges due to temperature *change and variation:*

- Aluminum minimizes gradients, but high coefficient of thermal expansion (CTE)
- Stainless steel sustains gradients, but medium CTE
- Composites sustain highest gradients with small CTE
- Majority of thermal contraction typically occurs between 300 K and 200 K
- 10% of thermal contraction at temperatures <100 K.

The bottom line is that we need teamwork and collaboration among all of the players in this field—both technically and experimentally—by sharing our information and knowledge worldwide.

References

1. B. Zohuri, *Physics of Cryogenics, An Ultralow Temperature Phenomenon*, 1st edn. (Elsevier Publishing Company, New York, 2017)
2. *Oxford English Dictionary*, 2nd edn. (Oxford University Press, 1989)
3. J. Wilks, D.S. Betts, *An Introduction to Liquid Helium*, 2nd edn. (Oxford Science Publications, Clarendon Press, Oxford, 1987)
4. B. Zohuri, *Nuclear Energy for Hydrogen Generation Through Intermediate Heat Exchangers: A Renewable Source of Energy*, 1st edn. (Springer Publishing Company, New York, 2016)
5. A. Kanni Raj, *Cryogenics: Energy Storage in Nuclear Plants, Create Space Independent Publishing Platform* (November 20, 2015)
6. J.M. Nash, Vortex expansion devices for high temperature cryogenics, *Proceedings of the 26th Intersociety Energy Conversion Engineering Conference*, vol. 4 (1991), pp. 521–525
7. B. Zohuri, *Magnetic Confinement Fusion Driven Thermonuclear Energy*, 1st edn. (Springer Publishing Company, New York, 2017)
8. B. Zohuri, *Inertial Confinement Fusion Driven Thermonuclear Energy*, 1st edn. (Springer Publishing Company, New York, 2017)
9. B.A. Hands, *Cryogenic Engineering* (Academic Press, New York, 1986)
10. Office of Energy Efficiency & Renewable Energy. Advanced composite materials for cold and cryogenic hydrogen storage applications in fuel cell electric vehicles workshop. https://www.energy.gov/eere/fuelcells/downloads/advanced-composite-materials-cold-and-cryogenic-hydrogen-storage

Chapter 5
Hydrogen: Driving Renewable Energy

Hydrogen can be found in many organic compounds other than water. It is the most abundant element on Earth, but it does not occur naturally as a gas. It is always combined with other elements, such as with oxygen to make water. Once separated from another element, hydrogen can be burned as a fuel or converted into electricity. A fuel cell uses the chemical energy of hydrogen or another fuel to cleanly and efficiently produce electricity. If hydrogen is the fuel, electricity, water, and heat are the only products. Fuel cells are unique in terms of the variety of their potential applications; they can provide power for systems as large as a utility power station and as small as a laptop computer.

5.1 Introduction

Hydrogen can be considered the simplest element in existence. An atom of hydrogen consists of only one proton and one electron. It is also the most plentiful element in the universe and in the Earth's crust. Despite its simplicity and abundance, hydrogen does not occur naturally as a gas on Earth and must be manufactured—it is always combined with other elements in compound form such as water, coal, and petroleum. Water, for example, is a combination of hydrogen and oxygen (H_2O). This is because hydrogen gas is lighter than air and rises into the atmosphere as a result.

Hydrogen has the highest energy content of any common fuel by weight, but it has the lowest energy content by volume. It is the lightest element, and is a gas at normal temperature and pressure. Once separated from other elements, hydrogen can be burned as a fuel or converted into electricity.

Hydrogen is also found in many organic compounds, notably the *hydrocarbons* that make up many of our fuels, such as gasoline, natural gas, methanol, and propane. Two of the most common methods used for the production of hydrogen by separating it from hydrocarbons are *electrolysis* (or water splitting) and application of heat, known as *steam reforming*. Steam reforming is currently the least

© Springer International Publishing AG, part of Springer Nature 2019

B. Zohuri, *Hydrogen Energy*, https://doi.org/10.1007/978-3-319-93461-7_5

expensive method for producing hydrogen, and, currently, most hydrogen is made this way from natural gas. It is used in industries to separate hydrogen atoms from carbon atoms in methane; however, because methane is a fossil fuel, the process of steam reforming results in greenhouse gas emissions, which is linked to global warming. The other method for the production of hydrogen is electrolysis, which involves passing an electrical current through water to separate water into its basic elements: hydrogen and oxygen. Hydrogen is then collected at the negatively charged cathode and oxygen at the positive anode. Hydrogen produced by electrolysis is extremely pure, and results in no emissions since electricity from renewable energy sources can be used. Unfortunately, electrolysis is currently a very expensive process.

There are also several experimental methods of producing hydrogen such as photo-electrolysis and biomass gasification. Scientists have also discovered that some algae and bacteria produce hydrogen under certain conditions, using sunlight as their energy source.

5.2 Hydrogen as an Energy Carrier

Hydrogen can be considered a clean energy carrier similar to electricity. Hydrogen can be produced from various domestic resources such as renewable energy and nuclear energy. In the long-term, hydrogen will simultaneously reduce the United States dependence on foreign oil and the emission of greenhouse gases and other pollutants.

Hydrogen is also considered as a secondary source of energy, commonly referred to as an energy carrier. Energy carriers are used to move, store, and deliver energy in a form that can be easily used. Electricity is the most well-known example of an energy carrier.

Hydrogen as an important energy carrier in the future has a number of advantages. For example, a large volume of hydrogen can be easily stored in a number of different ways. Hydrogen is high in energy, yet an engine that burns pure hydrogen produces almost no pollution. It can be used for transportation, heating, and power generation in places where it is difficult to use electricity. In some instances, it is cheaper to ship hydrogen by pipeline than sending electricity over long distances by wire.

Currently, hydrogen is mainly used as a fuel in the United States National Aeronautics and Space Administration (NASA) space program, as illustrated in Fig. 5.1. Liquid hydrogen has been used to propel space shuttles and other rockets into orbit since the 1970s, while hydrogen fuel cells power the electrical systems of the shuttle, producing a clean byproduct—pure water, which the crew drinks [1].

In the future, hydrogen will join electricity as an important energy carrier, since it can be made safely from renewable energy sources and is virtually non-polluting. It will also be used as a fuel for “zero-emission” vehicles and to heat homes and offices, produce electricity, and fuel aircraft. Renewable energy sources, like the Sun and wind, cannot produce energy all the time but they could, for example, produce

Fig. 5.1 Shuttle launch

electric energy and hydrogen, which can be stored until needed. Hydrogen can also be transported (like electricity) to locations where it is needed.

Hydrogen has great potential as a way to reduce the United States' reliance on imported energy sources such as oil. However, before hydrogen can play a bigger energy role and become a widely used alternative to gasoline, many new facilities and systems must be built [2].

Fig. 1.20 in Chap. 1 illustrates the future hydrogen energy infrastructure required. The hydrogen is produced through a wind electrolysis system and is then compressed up to pipeline pressure and fed into a transmission pipeline. The pipeline transports the hydrogen to a compressed gas terminal where the hydrogen is loaded into compressed gas tube trailers. A truck delivers the tube trailers to a forecourt station where the hydrogen is further compressed, stored, and dispensed to fuel cell vehicles [2].

5.3 Hydrogen Fuel Cells

As stated at the beginning of this chapter, and discussed in Chap. 1, a fuel cell converts the chemical energy of hydrogen or another fuel to cleanly and efficiently produce electricity. If hydrogen is the fuel, electricity, water, and heat are the only products (see Fig. 1.19). Fuel cells are unique in terms of the variety of their potential applications; they can provide power for systems as large as a utility power station and as small as a laptop computer. Hydrogen-powered fuel cells are not only pollution-free, but a two- to three-fold increase in the efficiency can be experienced when compared to traditional combustion technologies.

Fuel cells can power almost any portable device that normally uses batteries, transportation such as vehicles, trucks, buses, and marine vessels, as well as provide auxiliary power to traditional transportation technologies. Hydrogen could play a particularly important role in the future by replacing the imported petroleum we currently use in our cars and trucks in the United States [3].

As already noted, the purpose of a fuel cell is to produce an electrical current that can be directed outside the cell to do work, such as powering an electric motor or a light bulb. Because of the way electricity behaves, this current returns to the fuel cell, completing an electrical circuit. The chemical reactions that produce this current are the key to how a fuel cell works (illustrated in Fig. 1.19). Oxygen enters the fuel cell at the cathode and, in some cell types (such as that in Fig. 1.19), combines with electrons returning from the electrical circuit and hydrogen ions that have traveled through the electrolyte from the anode. In other cell types the oxygen picks up electrons and then travels through the electrolyte to the anode, where it combines with hydrogen ions [4].

The electrolyte plays a key role: only the appropriate ions must be permitted to pass between the anode and cathode. If free electrons or other substances could travel through the electrolyte, they would disrupt the chemical reaction.

Whether they combine at anode or cathode, together hydrogen and oxygen form water, which drains from the cell. As long as a fuel cell is supplied with hydrogen and oxygen, it will generate electricity.

As fuel cells create electricity chemically, rather than by combustion, they are not subject to the thermodynamic laws that limit a conventional power plant and are thus more efficient in extracting energy from a fuel. Waste heat from some cells can also be harnessed, further boosting system efficiency [4].

Fuel cells are often compared to batteries as both convert the energy produced by a chemical reaction into usable electric power. However, the fuel cell will produce electricity as long as fuel (hydrogen) is supplied, never losing its charge. Fuel cell research aims to lower the cost and improve the performance and durability of fuel cell technologies.

Fuel cells offer a promising technology for use as a source of heat and electricity for buildings, and as an electrical power source for electric motors propelling vehicles. Fuel cells operate best on pure hydrogen. But fuels such as natural gas, methanol, or even gasoline can be reformed to produce the hydrogen required for fuel cells. Some fuel cells can even be fueled directly with methanol, without using a reformer.

Fuel cells can be used in a wide range of applications, including transportation, material handling, and stationary, portable, and emergency backup power applications. Fuel cells have several benefits over conventional combustion-based technologies currently used in many power plants and passenger vehicles. They can operate at higher efficiencies than combustion engines, and can convert the chemical energy in the fuel to electrical energy with efficiencies of up to 60%. Fuel cells have lower emissions than combustion engines. As hydrogen fuel cells emit only water, there are no carbon dioxide emissions and no air pollutants that create smog and cause health problems at the point of operation. Also, fuel cells are quiet during operation as they have fewer moving parts.

The United States Fuel Cell Technologies Office (FCTO) focuses on applied research, development, and innovation to advance hydrogen and fuel cells for transportation and diverse applications enabling energy security, resiliency, and a strong domestic economy in emerging technologies.

The United States Department of Energy's (DOE) hydrogen and fuel cell efforts are part of a broad portfolio of activities aimed at building a competitive and sustainable clean energy economy, reducing greenhouse gas emissions by 80% by 2050 [5], and eliminating dependence on imported fuel will require the use of diverse domestic energy sources and advanced fuels and technologies in all sectors of the economy. Achieving these goals requires a robust, comprehensive research and development (R&D) portfolio that balances short-term objectives with long-term needs and sustainability.

Fuel cells and hydrogen comprise key elements of the DOE portfolio. The DOE's efforts to enable the widespread commercialization of hydrogen and fuel cell technologies form an integrated program—the DOE Hydrogen and Fuel Cells Program, as reflected in the Hydrogen and Fuel Cells Program Plan [6]. This Program is coordinated across the DOE and includes activities in the offices of Energy Efficiency and Renewable Energy (EERE), Science, Nuclear Energy, and Fossil Energy.

As part of these R&D goals, the DOE works closely with its national laboratories, universities, and industry partners to overcome critical technical barriers to fuel cell development. Cost, performance, and durability are still key challenges in the fuel cell industry:

- *Cost*—Platinum represents one of the largest cost components of a fuel cell, so much of the R&D is focused on approaches that will increase activity and utilization of current platinum group metal (PGM) and PGM–alloy catalysts, as well as non-PGM catalyst approaches for long-term applications.
- *Performance*—To improve fuel cell performance, R&D is focused on developing ion-exchange membrane electrolytes with enhanced efficiency and durability at reduced cost; improving membrane electrode assemblies (MEAs) through integration of state-of-the-art MEA components; developing transport models and in situ and ex situ experiments to provide data for model validation; identifying degradation mechanisms and developing approaches to mitigate their effects; and maintaining core activities on components, subsystems, and systems specifically tailored for stationary and portable power applications.
- *Durability*—A key performance factor is durability, in terms of a fuel cell system lifetime that will meet application expectations. DOE durability targets for stationary and transportation fuel cells are 40,000 h and 5000 h, respectively, under realistic operating conditions. In the most demanding applications, realistic operating conditions include impurities in the fuel and air, starting and stopping, freezing and thawing, and humidity and load cycles that result in stresses on the

chemical and mechanical stability of the fuel cell system materials and components. R&D focuses on understanding the fuel cell degradation mechanisms and developing materials and strategies that will mitigate them.

More details of technical targets and goals can be found in the Fuel Cells section of the FCTO's *Multi-Year Research, Development, and Demonstration Plan*, where full details about technical targets, or individual target tables, can be found for the following [7]:

- Fuel cell systems, stacks, and components for light-duty transportation applications
  - Fuel cell systems and stacks
  - PEMFC components
  - Fuel cell system humidifiers and air compressions systems
- Fuel cell transit buses
- Fuel cell backup power systems
- Fuel cell systems for stationary (combined heat and power) applications
- Fuel cell systems for portable power and auxiliary power applications.

5.4 Fuel Cells

A fuel cell is a device that generates electricity by a chemical reaction. Every fuel cell has two electrodes, the anode (which is positively charged) and the cathode (which is negatively charged) [1]. The reactions that produce electricity take place at the two electrodes. Every fuel cell also has an electrolyte, which carries electrically charged particles from one electrode to the other, and a catalyst, which speeds the reactions at the electrodes [8]. Multiple fuel cells are usually assembled into a stack and generate direct current (DC).

A single fuel cell consists of an electrolyte sandwiched between two electrodes. Bipolar plates on either side of the cell help distribute gases and serve as current collectors. Hydrogen is the basic fuel for fuel cells, but fuel cells also require oxygen. The basic chemical reaction of fuel cell is as follows:

$$\mathbf{2H_2 + O_2 \rightarrow 2H_2O + 2e^-}$$

Depending on the application, a fuel cell stack may contain from a few to hundreds of individual fuel cells layered together. This "scalability" makes fuel cells ideal for a wide variety of applications, such as stationary power stations, portable devices, and transportation.

There are several kinds of fuel cells, and each operates a bit differently. But, in general terms, hydrogen atoms enter a fuel cell at the anode where a chemical reaction strips them of their electrons. The hydrogen atoms are now "ionized," and carry a positive electrical charge. The negatively charged electrons provide the current through wires to do work. If alternating current (AC) is needed, the DC output of the fuel cell must be routed through a conversion device called an inverter.

Oxygen enters the fuel cell at the cathode and in some cell types it combines with electrons returning from the electrical circuit and hydrogen ions that have traveled through the electrolyte from the anode. In other cell types the oxygen picks up electrons and then travels through the electrolyte to the anode, where it combines with hydrogen ions.

The type of fuel also depends on the electrolyte. Some cells need pure hydrogen, and therefore demand extra equipment such as a "reformer" to purify the fuel. Other cells can tolerate some impurities but might need higher temperatures to run efficiently. Liquid electrolytes circulate in some cells, which require pumps. The type of electrolyte also dictates a cell's operating temperature—"molten" carbonate cells run hot, just as the name implies.

Fuel cells are employed in stationary power generation, portable power supply and transportation. Small, stationary power generators provide 0.5–10 kW uninterrupted power supply to households, shopping malls, and data centers. Grid-scale fuel cell generation is also in development. Portable fuel cells are best suited for auxiliary power units (APUs), portable devices, personal computers, smartphones, and so on.

The application of fuel cells in transportation represents the future for the automotive and computation industries. Buses, light vehicles (cars), unmanned aerial vehicles (UAVs), and trains will soon be running on fuel cells.

Another great appeal of fuel cells is that they generate electricity with very little pollution—much of the hydrogen and oxygen used in generating electricity ultimately combines to form a harmless by-product, namely water. However, obtaining hydrogen is a challenge and can be energy intensive.

Despite its many advantages, the commercialization of fuel cell technology faces many technical and economic challenges. The durability and cost of fuel cell systems represent the biggest barriers. Fuel cells are still in the "technology development phase." Efforts are being made to reduce the cost and improve durability of fuel cells.

5.4.1 Different Types of Fuel Cells

Each type of fuel cell has advantages and drawbacks compared to the others, and none is cheap and efficient enough yet to widely replace traditional ways of generating power, such as coal-fired, hydroelectric, or even nuclear power plants.

While there are dozens of types of fuel cells, there are six principle kinds in various stages of commercial availability, or undergoing research, development, and demonstration (RD&D). These six fuel cell types are significantly different from each other in many respects; however, the key distinguishing feature is the electrolyte material.

The following sections and associated images describe the six main types of fuel cells. More detailed information can be found on the website provided by the

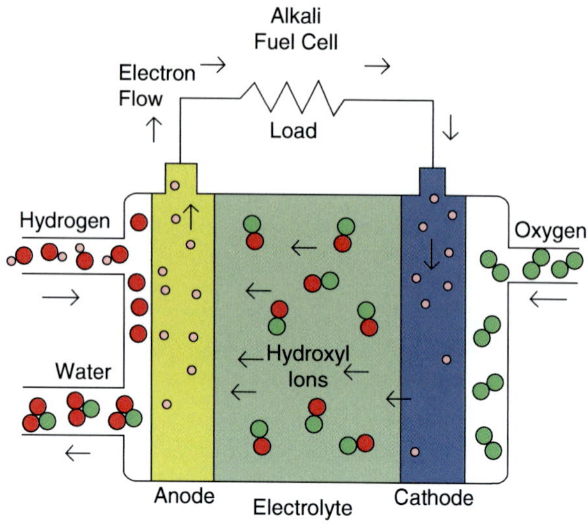

Fig. 5.2 Alkali fuel cell. (Courtesy of the Smithsonian Institution)

National Museum of American History, Smithsonian Institution at http://americanhistory.si.edu/fuelcells/basics.htm. Information and description of these fuel cells is courtesy of the Smithsonian Institution's website.

Alkali Fuel Cell

Alkali fuel cells (see Fig. 5.2) operate on compressed hydrogen and oxygen. They generally use a solution of potassium hydroxide (chemically, KOH) in water as their electrolyte. Efficiency is about 70%, and operating temperature is 150–200 °C (about 300–400 °F). Cell output ranges from 300 W to 5 kW. Alkali cells were used in Apollo spacecraft to provide both electricity and drinking water. However, they require pure hydrogen fuel, and their platinum electrode catalysts are expensive. And, like, any container filled with liquid, they can leak.

Phosphoric Acid Fuel Cell (PAFC)

Phosphoric acid fuel cells (PAFCs) (see Fig. 5.3) use phosphoric acid as the electrolyte. Efficiency ranges from 40% to 80%, and operating temperature is between 150 and 200 °C (about 300–400 °F). Existing phosphoric acid cells have outputs up to 200 kW, and 11 MW units have been tested. PAFCs tolerate a carbon monoxide concentration of about 1.5%, which broadens the choice of fuels they can use. If gasoline is used, the sulfur must be removed. Platinum electrode-catalysts are needed, and internal parts must be able to withstand the corrosive acid.

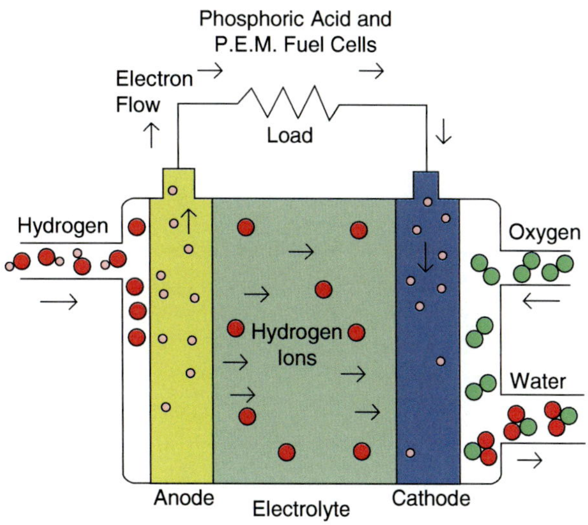

Fig. 5.3 Phosphoric acid fuel cell. (Courtesy of the Smithsonian Institution)

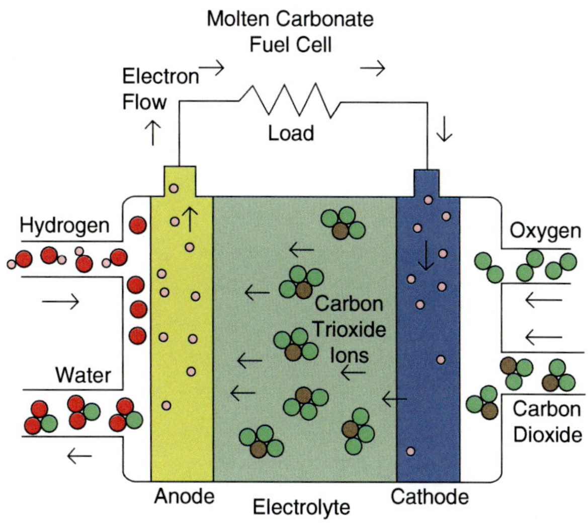

Fig. 5.4 Molten carbonate fuel cell. (Courtesy of the Smithsonian Institution)

Molten Carbonate Fuel Cell (MCFC)

The molten carbonate fuel cells (MCFCs) (see Fig. 5.4) use high-temperature compounds of salt (such as sodium or magnesium) carbonates (chemically, CO_3) as the electrolyte. Efficiency ranges from 60% to 80%, and operating temperature is about 650 °C (1200 °F). Units with output up to 2 MW have been constructed, and designs exist for units up to 100 MW. The high temperature limits damage from carbon monoxide "poisoning" of the cell and waste heat can be recycled to make additional electricity. Their nickel electrode-catalysts are inexpensive compared to the platinum used in other cells. But the high temperature also limits the materials and safe uses of MCFCs—they would probably be too hot for home use. Also, carbonate ions from the electrolyte are used up in the reactions, making it necessary to inject carbon dioxide to compensate for this.

Full-scale demonstration plants are now testing MCFCs. The electrolyte in an MCFC is an alkali carbonate (sodium, potassium, or lithium salts, i.e., Na_2CO_3, K_2CO_2, or Li_2CO_3) or a combination of alkali carbonates that is retained in a ceramic matrix of lithium aluminum oxide ($LiAlO_2$). An MCFC operates at 600–700 °C where the alkali carbonates form a highly conductive molten salt with carbonate ions (CO_3=) providing ionic conduction through the electrolyte matrix. Relatively inexpensive nickel (Ni) and nickel oxide (NiO) are adequate to promote reaction on the anode and cathode, respectively, at the high operating temperatures of an MCFC.

MCFCs offer greater fuel flexibility and higher fuel-to-electricity efficiencies than lower-temperature fuel cells, approaching 60%. The higher operating temperatures of MCFCs make them candidates for combined-cycle applications, in which the exhaust heat is used to generate additional electricity. When the waste heat is used for co-generation, total thermal efficiencies can approach 85%.

Proton Exchange Membrane (PEM)

Proton exchange membrane (PEM) fuel cells (see Fig. 5.5) work with a polymer electrolyte in the form of a thin, permeable sheet. Efficiency is about 40–50%, and operating temperature is about 80 °C (about 175 °F). Cell outputs generally range from 50 to 250 kW. The solid, flexible electrolyte will not leak or crack, and these cells operate at a low enough temperature to make them suitable for homes and cars. However, their fuels must be purified, and a platinum catalyst is used on both sides of the membrane, raising costs.

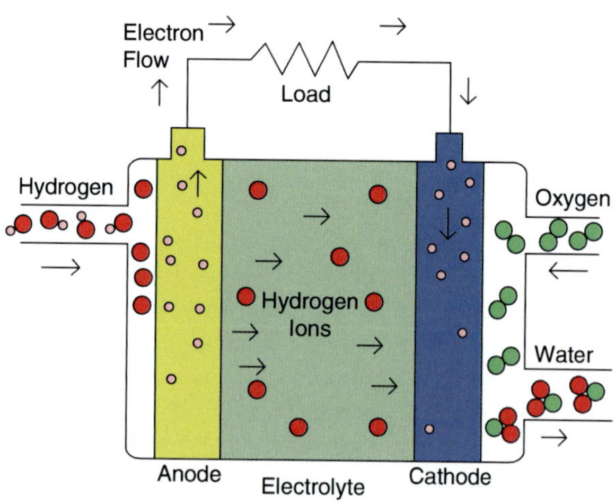

Fig. 5.5 Proton exchange membrane fuel cell. (Courtesy of the Smithsonian Institution)

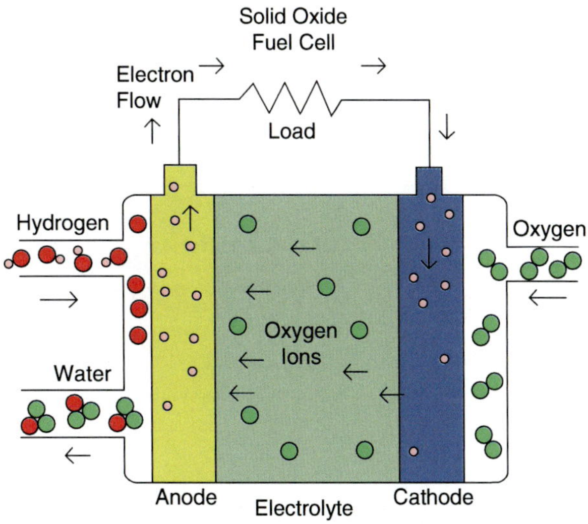

Fig. 5.6 Solid oxide fuel cell. (Courtesy of the Smithsonian Institution)

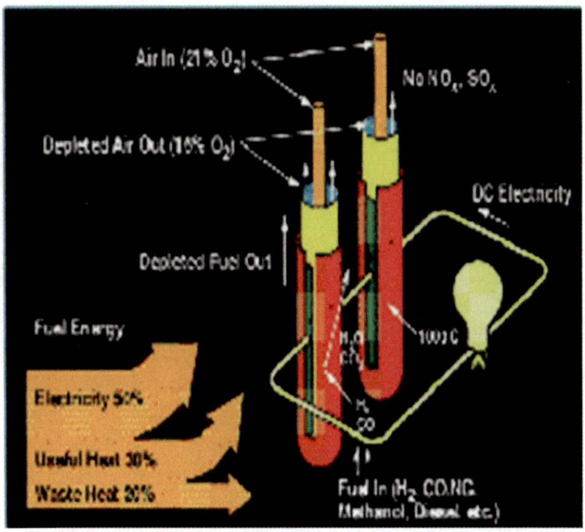

Fig. 5.7 Tubular solid oxide fuel cell. (Courtesy of Siemens Westinghouse Power Corporation)

Solid Oxide Fuel Cell (SOFC)

Solid oxide fuel cells (SOFCs) (see Figs. 5.6, 5.7, and 5.8) use a hard, ceramic compound of metal (such as calcium, zirconium, or Yttria-stabilized zirconia [Y_2O_3-stabilized ZrO_2]) oxides (chemically, O_2) as electrolyte. SOFCs approach 60% electrical efficiency in the simple cycle system, and 85% total thermal efficiency in co-generation applications (Singhal 1997). Operating temperatures are between 600 and 1000 °C (about 1100–1800 °F), and ionic conduction is accomplished by oxygen ions (O=). Cell output is generally up to 100 kW, but can range from 1 kW to 250 kW plants, with plans to reach the multi-megawatt range. Typically, the anode

Fig. 5.8 Planar solid oxide fuel cell. (Courtesy of Siemens Westinghouse Power Corporation)

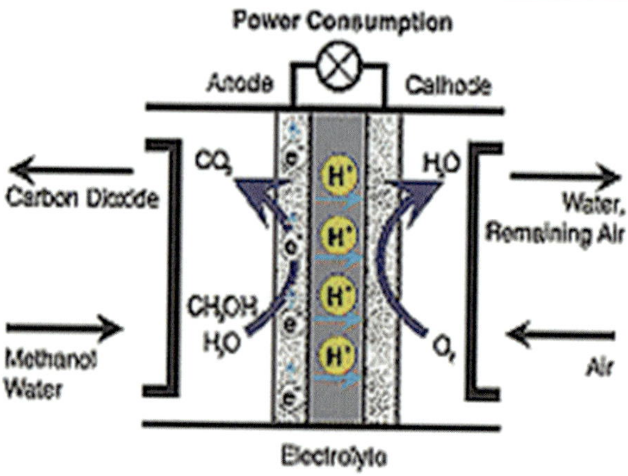

Fig. 5.9 Direct methanol fuel cell

of an SOFC is cobalt or nickel zirconia (Co-ZrO_2 or Ni-ZrO_2) and the cathode is strontium-doped lanthanum manganite (Sr-doped $LaMnO_3$). At such high temperatures, a reformer is not required to extract hydrogen from the fuel, and waste heat can be recycled to make additional electricity. High-temperature operation, up to 1000 °C, also allows more flexibility in the choice of fuels and can produce very good performance in combined-cycle applications. However, the high temperature limits applications of SOFC units and they tend to be rather large. While solid electrolytes cannot leak, they can crack.

The flat plate and monolithic designs are at a much earlier stage of development typified by sub-scale, single cell, and short stack development (kW scale). At this juncture, tubular SOFC designs are closer to commercialization.

Direct Methanol Fuel Cell (DMFC)

The direct methanol fuel cell (DMFC) (see Fig. 5.9) is similar to the PEM fuel cell in that it uses a polymer membrane as an electrolyte. However, a catalyst on the DMFC anode draws hydrogen from liquid methanol, eliminating the need for a fuel reformer. While potentially a very attractive solution to the issues of hydrogen storage and transportation (particularly for portable applications), the principal problem facing the commercial application of the DMFC today stems from its relatively low performance in comparison to hydrogen.

5.5 Fuel Cell Technologies

The Fuel Cell Technologies Program (FCT Program), situated within EERE, addresses key technical challenges for fuel cells and hydrogen production, delivery, and storage and the institutional barriers, such as hydrogen codes and standards, training, and public awareness that inhibit the widespread commercialization of hydrogen and fuel cell technologies. The FCT Program conducts applied research, technology development, and learning demonstrations, as well as safety research, systems analysis, early market deployments, and public outreach and education activities. These activities include cost-shared, public–private partnerships to address the high-risk, critical technology barriers preventing extensive use of hydrogen as an energy carrier. Public and private partners include automotive and power equipment manufacturers, energy and chemical companies, electric and natural gas utilities, building designers, standards development organizations, other federal agencies, state government agencies, universities, national laboratories, and other national and international stakeholder organizations. The FCT Program encourages the formation of collaborative partnerships to conduct RD&D and other activities, such as deployment, that support program goals.

The FCT Program addresses the development of hydrogen energy systems for transportation, stationary power, and portable power applications. Transportation applications include fuel cell vehicles (such as buses, automobiles, and heavy-duty vehicles), niche markets (such as lift trucks), and hydrogen refueling infrastructure. Stationary power applications include hydrogen used for backup emergency power, commercial/industrial power and heat generation, and residential electric power generation. Consumer electronics such as mobile phones, laptop computers, and recharging systems are among the portable power applications. The DOE funds RD&D efforts that will provide the basis for the near-, mid-, and long-term production, delivery, storage, and use of hydrogen derived from diverse energy sources, including renewable, fossil fuels, and nuclear energy as coordinated within the Program.

As stated earlier, fuel cell research aims to lower the cost and improve the performance and durability of fuel cell technologies. Research is performed on a variety of fuel cell types—PEMFCs, alkaline membrane fuel cells (AMFCs), and DMFCs—which are generally differentiated by the fuel used, as listed here:

1. *Catalysts:* Work in this area involves developing and optimizing advanced electrocatalysts and novel synthesis methods. Related projects concentrate on

extended-surface catalysts with reduced precious-metal loading and improved performance, durability, and activity compared with standard catalytic materials. Researchers are investigating fuel cells and electrolyzer catalysts under acidic and alkaline conditions, with the goal of "thrifting" platinum, iridium, and their alloys (in acidic-based systems) and silver, cobalt, nickel, and their oxides/alloys (in alkaline-based systems). Also under study are support materials for catalyst dispersion, with a focus on nitrogen-doped carbon supports and corrosion-resistant, non-carbon supports.

2. *Polymer Electrolytes:* AMFCs enable the use of non-precious-metal catalysts, but they are vulnerable to ambient carbon dioxide conditions. This vulnerability decreases, however, at higher operating temperatures. Researchers are developing novel chemistries to enable higher-temperature and higher-current-density operation via the use of perfluorinated alkaline membranes. Researchers are also exploring traditional PEMs with tethered heteropolyacid functionality to allow higher-temperature, lower-humidity operation and are investigating the stability of covalently tetherable captions.
3. *Electrode Design/High-Current-Density Operation:* This cross-cutting research area is focused on incorporating novel catalysts into high-performance devices and investigating the impact of low-precious-metal loading on high-current-density performance.
4. *Contaminants:* As fuel cell systems become more commercially competitive, and as automotive fuel cell R&D trends toward decreased catalyst loadings and thinner membranes, fuel cell operation becomes even more susceptible to contaminants. The National Renewable Energy Laboratory (NREL) also participates in the DOE's Fuel Cell Durability Working Group. Contaminants derived from fuel cell system component/structural materials, lubricants, greases, adhesives, sealants, and hoses have been shown to affect the performance and durability of fuel cell systems. Companies are currently performing research to identify and quantify these system-derived contaminants and to understand the effects of system contaminants on fuel cell performance and durability. The goal is to increase the understanding of fuel cell system contaminants and to help guide the implementation and, where necessary, development of system materials that will help enable fuel cell commercialization.

Again, more details can be found in official site of DOE office at Energy Efficiency & Renewable Energy as well as that of the NREL [9, 10].

5.6 Fuel Cell Backup Power Systems

The DOE technical targets for fuel cell backup power systems are given in Table 5.1. More information on these targets can be found in the Fuel Cells section of the FCTO's *Multi-Year Research, Development, and Demonstration Plan* [11].

Table 5.1 Technical targets: fuel cell backup power systems (1–10 kW) operating on direct hydrogen [10]

Characteristic	Units	2015 status[a]	2020 targets
Lifetime	Years	10	15
Durability[b]	Hours	8000	10,000
Energy efficiency[c]	%	50	60
Mean time between failures	Years	5	5
Ambient temperature range	°C	−10 to 40	−50 to 50
Noise	dB at 1 m	65	60
Start-up time[d]	Seconds	80	15
Availability	%	99.7	96.3
Equipment cost[e]	\$/kW	6100[f]	1000
Annual maintenance cost[e]	\$/kW	30	20
Annualized total cost of ownership[g]	\$/kW	500	200

[a]Unless otherwise stated, status based on input from DE-FOA-0000738
[b]Time until 10% voltage degradation when operated on a backup power duty cycle
[c]Ratio of direct current (DC) output energy from the power plant to the lower heating value of the input fuel (hydrogen, averaged over cycle
[d]Time indicated is start-up time for the fuel cell. The backup power system, including hybridized batteries, is expected to provide uninterruptible power
[e]Excludes tax credits and subsidies
[f]National Renewable Energy Laboratory (NREL), "Current Fuel Cell System Low Volume Price by Application"
[g]Annualized cost of ownership including cost of capital equipment, installation, operation and maintenance, fuel, and fuel storage. Based on a 5 kW system with 10-year lifetime

5.7 Fuel Cell Systems for Stationary Combined Heat and Power Applications

Tables 5.2 and 5.3 list the DOE technical targets for stationary fuel cell applications [12]. These targets have been developed with input from developers of stationary fuel cell power systems.

More information on these targets can be found in the Fuel Cells section of the FCTO's *Multi-Year Research, Development, and Demonstration Plan* [11].

5.8 Fuel Cell Systems for Portable Power and Auxiliary Power Applications

Tables 5.4 and 5.5 list the DOE technical targets for fuel cell systems for portable power and auxiliary power applications [12].

More information on these targets can be found in the Fuel Cells section of the FCTO's *Multi-Year Research, Development, and Demonstration Plan* [13].

Table 5.2 Technical targets: 1–25 kW residential and light commercial combined heat and power and distributed generation fuel cell systems operating on natural gas[a]

Characteristic	Units	2015 status	2020 targets
Electrical efficiency at rated power[b]	% (LHV)	34–40	>45[c]
CHP energy efficiency[d]	% (LHV)	80–90	90
Equipment cost[e], 5-kW_{avg} system[f]	$/kW	2300–2800[g]	1500
Transient response (10–90% rated power)	Min	5	2
Start-up time from 20 °C ambient temperature	Min	10	20
Degradation with cycling[h]	%/1000 h	<2%	0.3%
Operating lifetime[i]	h	12,000–70,000	60,000
System availability[j]	%	97	99

AC alternating current, *CHP* combined heat and power, *LHV* lower heating value, *SOFC* solid oxide fuel cell

[a]Pipeline natural gas delivered at typical residential distribution line pressures

[b]Regulated AC net/LHV of fuel

[c]Higher electrical efficiencies (e.g., 60% using SOFC) are preferred for non-CHP applications

[d]Ratio of regulated AC net output energy plus recovered thermal energy to the LHV of the input fuel. For inclusion in CHP energy-efficiency calculation, heat must be available at a temperature sufficiently high to be useful in space and water heating applications. Provision of heat at 80 °C or higher is recommended

[e]Complete system, including all necessary components to convert natural gas to electricity suitable for grid connection, and heat exchangers and other equipment for heat rejection to conventional water heater, and/or hydronic or forced air heating system. Includes all applicable taxes, and markups, based on projection to high-volume production (50,000 units per year)

[f]kW_{avg} is the average output (AC) electric power delivered over the life of system while unit is running

[g]Battelle preliminary 2015 cost assessment of stationary CHP systems, range represents different technologies (SOFC vs PEMFC) at manufacturing volumes of 50,000 units per year

[h]Durability testing should include effects of transient operation, startup, and shutdown

[i]Time until >20% net power degradation

[j]Percentage of time the system is available for operation under realistic operating conditions and load profile. Unavailable time includes time for scheduled maintenance

5.9 Hydrogen Storage

The FCTO is developing onboard automotive hydrogen storage systems that allow for a driving range of more than 300 miles while meeting cost, safety, and performance requirements.

Small amounts of hydrogen (up to a few MWh) can be stored in pressurized vessels at 100–300 bar or liquefied at 20.3 K (−423 F). Alternatively, solid metal hydrides or nano-tubes can store hydrogen with a very high density. Very large amounts of hydrogen can be stored in manmade underground salt caverns of up to 500,000 m^3 at 200 bar (2900 psi), corresponding to a storage capacity of 167 GWh hydrogen (100 GWh electricity). In this way, longer periods of flaws or of excess wind/photovoltaic energy production can be leveled. Even balancing seasonal variations might be possible.

Table 5.3 Technical targets[a]: 100 kW–3 MW combined heat and power and distributed generation fuel cell systems operating on natural gas[b]

Characteristic	Units	2015 status[c]	2020 targets
Electrical efficiency at rated power[d]	% (LHV)	42–47	> 50[e]
CHP energy efficiency[f]	% (LHV)	70–90	90
Equipment cost, natural gas	$/kW	1200[g]–4500[h]	1000[i]
Installed cost, natural gas	$/kW	2400[g]–5500[h]	1500[i]
Equipment cost, biogas	$/kW	3200–6500[j]	1400[i]
Installed cost, biogas	$/kW	4900–8000[j]	2100[i]
Number of planned/forced outages over lifetime	–	50	40
Operating lifetime[k]	h	40,000–80,000	80,000
System availability[l]	%	95	99

AC alternating current, *CHP* combined heat and power, *DOE* Department of Energy, *LHV* lower heating value, *SOFC* solid oxide fuel cell, *LT-PEMFC* Low Temperature-Proton-exchange membrane fuel cells

[a]Includes fuel processor, stack, and ancillaries

[b]Pipeline natural gas delivered at typical residential distribution line pressures

[c]Status varies by technology

[d]Ratio of regulated AC net output energy to the LHV of the input fuel

[e]Higher electrical efficiencies (e.g., 60% using SOFC) are preferred for non-CHP applications

[f]Ratio of regulated AC net output energy plus recovered thermal energy to the LHV of the input fuel. For inclusion in CHP energy-efficiency calculation, heat must be available at a temperature sufficiently high to be useful in space and water heating applications. Provision of heat at 80 °C or higher is recommended

[g]M. Wei, 100 kW LLT-PEMFC, projection at volume of 1000 systems/year

[h]DOE Hydrogen and Fuel Cells Program Record 11,014, "Medium-scale CHP Fuel Cell System Targets,"

[i]Includes projected cost advantage of high-volume production (totaling 100 MW per year)

[j]Assumed $2500/kW higher cost to operate on biogas than on hydrogen (DOE Hydrogen and Fuel Cells Program Record 11014, "Medium-scale CHP Fuel Cell System Targets")

[k]Time until > 10% net power degradation

[l]Percentage of time the system is available for operation under realistic operating conditions and load profile. Unavailable time includes time for scheduled maintenance

5.9.1 Why Study Hydrogen Storage?

Hydrogen storage is a key enabling technology for the advancement of hydrogen and fuel cell technologies in applications including stationary power, portable power, and transportation. Hydrogen has the highest energy per mass of any fuel; however, its low ambient temperature density results in a low energy per unit volume, requiring the development of advanced storage methods that have potential for higher energy density.

5.9.2 How Hydrogen Storage Works

Hydrogen can be stored physically as either a gas or a liquid (see Fig. 5.10). Storage of hydrogen as a gas typically requires high-pressure tanks (350–700 bar

Table 5.4 Technical targets: portable power fuel cell systems (5–50 W/100–200 W)[a]

Characteristic	Units	2015 status	Ultimate targets
Specific power[b]	W/kg	23[h]/25[i]	45/50
Power density[b]	W/L	24[h]/30[i]	55/70
Specific energy[b,c]	Wh/kg	121[j]/450[j]	650/640
Energy density[b,c]	Wh/L	200[i]/300[i,j]	650/900
Cost[d]	$/W	15[i]/15[i]	7/5
Durability[e,f]	Hours	1500[i]/2000[i]	5000/5000
Mean time between failures[f,g]	Hours	500[i]/500[i]	5000/5000

[a]These targets are technology neutral and make no assumption about the type of fuel cell technology or type of fuel used. In addition to meeting these targets, portable power fuel cells are expected to operate safely, providing power without exposing users to hazardous or unpleasant emissions, high temperatures, or objectionable levels of noise. Portable power fuel cells are also expected to be compatible with the requirements of portable electronic devices, including operation under a range of ambient temperature, humidity, and pressure conditions, and exposure to freezing conditions, vibration, and dust. They should be capable of repeatedly turning off and on and should have turndown capabilities required to match the dynamic power needs of the device. For widespread adoption, portable power fuel cell systems should minimize life-cycle environmental impact through the use of reusable fuel cartridges, recyclable components, and low-impact manufacturing techniques

[b]This is based on rated net power of the total fuel cell system, including fuel tank, fuel, and any hybridization batteries. In the case of fuel cells embedded in other devices, only device components required for power generation, power conditioning, and energy storage are included. Fuel capacity is not specified, but the same quantity of fuel must be used in calculation of specific power, power density, specific energy, and energy density

[c]Efficiency of 35% is recommended to enable high specific energy and energy density

[d]Cost includes material and labor costs required to manufacture the fuel cell system and any required auxiliaries (e.g., refueling devices). Cost is defined at production rates of 25,000 and 10,000 units per year for 5–50 W and 100–200 W units, respectively

[e]Durability is defined as the time until the system rated power degrades by 20%, though for some applications higher or lower levels of power degradation may be acceptable

[f]Testing should be performed using an operating cycle that is realistic and appropriate for the target application, including effects from transient operation, startup and shutdown, and off-line degradation

[g]Mean time between failures (MTBF) includes failures of any system components that render the system inoperable without maintenance

[h]Status calculated based on commercial products from myFC (myfcpower.com/pages/jaq)

[i]Department of Energy Hydrogen and Fuel Cells Program Record 11009

[j]Status calculated based on commercial products from ultra-cell (ultracell-llc.com)

[5000–10,000 psi] tank pressure). Storage of hydrogen as a liquid requires cryogenic temperatures because the boiling point of hydrogen at 1 atmosphere pressure is −252.8 °C. Hydrogen can also be stored on the surfaces of solids (by adsorption) or within solids (by absorption).

Table 5.5 Technical targets: fuel cell auxiliary power units (1–10 kW) operating on ultra-low-sulfur diesel fuel

Characteristic	Units	2015 status	2020 targets
Electrical efficiency at rated power[a]	% (LHV)	29[b]	40
Power density	W/L	16[b]	40
Specific power	W/kg	18[b]	45
Factory cost, system[c]	$/kWe	2100[d]	1000
Transient response (10–90% rated power)	Min	5[e]	2
Start-up time from 20 °C	Min	70[b]	30
Start-up time from standby conditions[f]	Min	–	5
Degradation with cycling[g]	%/1000 h	2.6[e]	1
Operating lifetime[g,h]	h	3000[e]	20,000
System availability[i]	%	97[e]	99

APU auxiliary power unit, *DOE* Department of Energy, *SOFC* solid oxide fuel cell
[a]Regulated DC net/LHV of fuel
[b]DESTA-Demonstration of 1st European SOFC Truck APU, Programmed Review Days 2015
[c]Cost includes materials and labor costs to produce system. Cost defined at 50,000 unit/year production of a 5-kW system. Today's low-volume cost is expected to be higher than quoted status. Allowable cost is expected to be higher than the target for systems with rated power below 5 kW, and lower than the target for systems with rated power above 5 kW
[d]Modeled cost of a 5 kW SOFC APU system produced at 50,000 units/year. F. Eubanks et al., "Stationary and Emerging Market Fuel Cell System Cost Analysis-Auxiliary Power Units," 2015 Annual Merit Review, slide 20
[e]DOE Hydrogen Program Record 11001, "Revised APU Targets"
[f]Standby conditions may be at or above ambient temperature depending on operating protocol
[g]Durability testing should include, at minimum, daily cycles to stand-by condition, and weekly cycles to full off condition (ambient temperature). The system should be able to meet durability criteria during and after exposure to vibration associated with transportation and highway operation, and during operation in a range of ambient temperature from −40 °C to 50 °C, a range of ambient relative humidity from 5% to 100%, and in dust levels up to 2 mg/m^3
[h]Time until > 20% net power degradation
[i]Percentage of time the system is available for operation under realistic operating conditions and load profile. Scheduled maintenance does not count against system availability

5.9.3 Research and Development Goals

The FCTO conducts R&D activities to advance hydrogen storage systems technology and develop novel hydrogen storage materials. The goal is to provide adequate hydrogen storage to meet the DOE hydrogen storage targets for onboard light-duty vehicle, material-handling equipment, and portable power applications. By 2020, the FCTO aims to develop and verify onboard automotive hydrogen storage systems achieving targets that will allow hydrogen-fueled vehicle platforms to meet customer performance expectations for range, passenger and cargo space, refueling time, and overall vehicle performance. Specific system targets include the following:

- 1.5 kWh/kg system (4.5 wt.% hydrogen)
- 1.0 kWh/L system (0.030 kg hydrogen/L)
- US$10/kWh (US$333/kg stored hydrogen capacity).

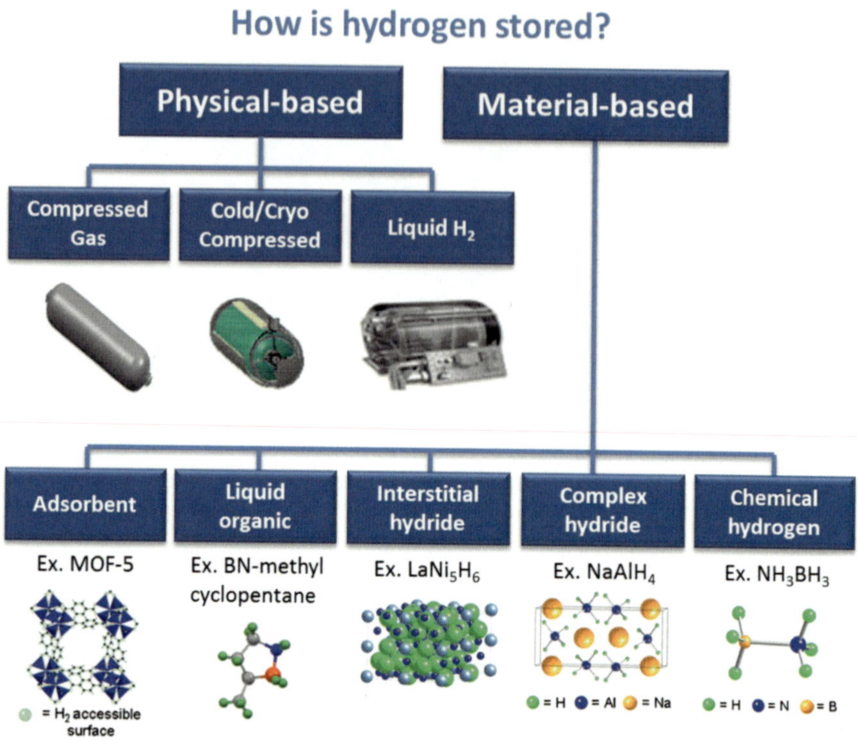

Fig. 5.10 Hydrogen storage. (Courtesy of the US Department of Energy)

The collaborative Hydrogen Storage Engineering Center of Excellence (HSECoE) conducts analysis activities to determine the current status of materials-based storage system technologies. The Hydrogen Materials—Advanced Research Consortium (HyMARC) conducts foundational research to understand the interaction of hydrogen with materials in relation to the formation and release of hydrogen from hydrogen storage materials.

5.9.4 Hydrogen Storage Challenges

High-density hydrogen storage is a challenge for stationary and portable applications and remains a significant challenge for transportation applications. Presently available storage options typically require large-volume systems that store hydrogen in gaseous form. This is less of an issue for stationary applications, where the footprint of compressed gas tanks may be less critical.

However, fuel cell-powered vehicles require enough hydrogen to provide a driving range of more than 300 miles with the ability to quickly and easily refuel

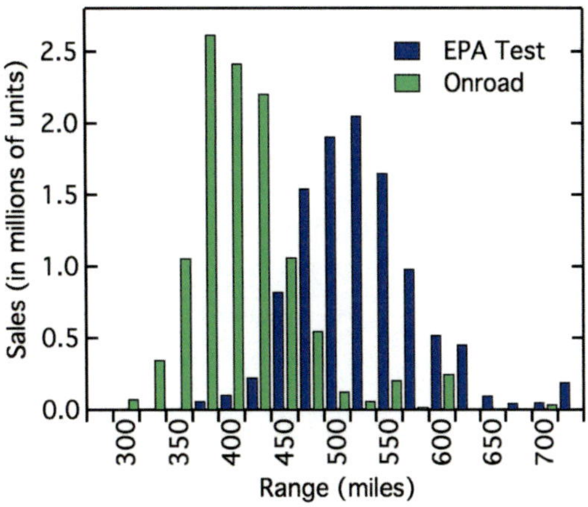

Fig. 5.11 2010 US light-duty vehicle sales distribution by driving range. (Courtesy of the US Department of Energy)

the vehicle. While some light-duty hydrogen fuel cell electric vehicles (FCEVs) that are capable of this range have entered the market, these vehicles will rely on compressed gas onboard storage using large-volume, high-pressure composite vessels (see Fig. 5.11).

The large storage volumes required may have less impact for larger vehicles, but providing sufficient hydrogen storage across all light-duty platforms remains a challenge. The importance of the 300-mile range goal can be appreciated by looking at the sales distribution by range chart (Fig. 5.11), which shows that most vehicles sold today are capable of exceeding this minimum.

On a mass basis, hydrogen has nearly three times the energy content of gasoline—120 MJ/kg for hydrogen versus 44 MJ/kg for gasoline. On a volume basis, however, this is reversed; liquid hydrogen has a density of 8 MJ/L, whereas gasoline has a density of 32 MJ/L, as shown in Fig. 5.12, which compares specific energy (i.e., energy per mass or gravimetric density) and energy density (i.e., energy per volume or volumetric density) for several fuels based on lower heating values. Onboard hydrogen storage capacities of 5–13 kg hydrogen will be required to meet the driving range for the full range of light-duty vehicle platforms.

To overcome these challenges, the FCTO is pursuing two strategic pathways, targeting both near-term and long-term solutions. The near-term pathway focuses on compressed gas storage, using advanced pressure vessels made of fiber reinforced composites that are capable of reaching 700 bar pressure, with a major emphasis on system cost reduction. The long-term pathway focuses on both:

1. Cold or cryo-compressed hydrogen storage, where increased hydrogen density and insulated pressure vessels may allow for DOE targets to be met; and
2. Materials-based hydrogen storage technologies, including sorbents, chemical hydrogen storage materials, and metal hydrides, with properties having potential to meet DOE hydrogen storage targets.

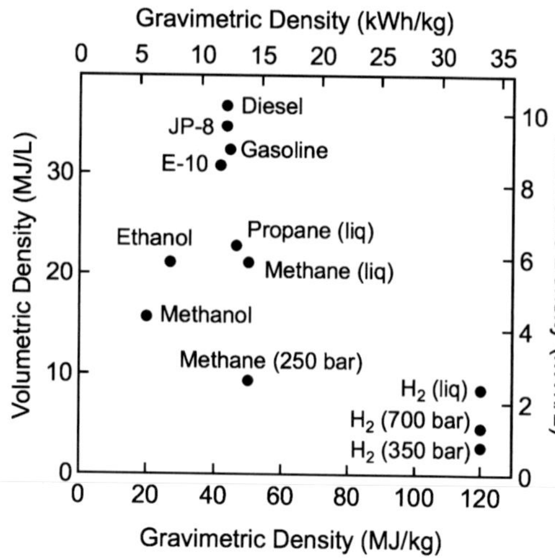

Fig. 5.12 Comparison of specific energy versus volumetric density. (Courtesy of the US Department of Energy)

5.10 Hydrogen Energy Storage

Hydrogen is the most versatile means of energy storage—it can be produced and stored in all scales and used as a fuel, chemical material, or natural gas substitute.

Electricity can be converted into hydrogen by electrolysis and the hydrogen can be then stored and eventually re-electrified. Currently, round-trip efficiency is as low as 30–40% but could increase up to 50% if more efficient technologies are developed. Despite this low efficiency, the interest in hydrogen energy storage (HES) is growing due to the much higher storage capacity than batteries (small scale) or pumped hydro and compressed air energy storage (CAES) (large scale).

Hydrogen is an energy-rich gas, which is one of the reasons that it is used as a rocket fuel. It can be produced from a variety of feedstock—and from electricity—and stored in many different ways, from a few grams in handheld cartridges to thousands of tons in an underground cavern. This gives hydrogen a unique potential to store renewable energy on both a small and very large scale. Especially for longer-term storage (weeks to months), hydrogen is the only viable alternative in sight currently. It is also versatile to use: it can be converted back to power, used as fuel for cars, used as a material for many industrial products (such as hardened fats), or even be converted to synthetic natural gas. Hydrogen makes all these markets accessible for renewable power.

HES systems have been the topic of numerous studies and analyses. These systems typically involve the production of hydrogen from electricity by electrolysis. Most electrolysis units involve alkaline or PEMFC conversion processes [14, 15]. As early as 1999, Ogden provided an overview of hydrogen infrastructure components, which included storage systems [16], and Yang reviewed general similarities and differences between hydrogen and electricity as energy carriers [17]. Many studies of future hydrogen scenarios have been developed [18, 19], and this complementarily between hydrogen and electricity has been the focus of high-renewable scenarios developed by Barton and Gammon for the United Kingdom [20], and more recently by Jacobson et al. [21] for California (2014). Several studies have compared hydrogen storage systems with other storage systems on the basis of cost, performance, and other attributes relevant to market viability and policy development [22–25].

In addition to numerous analytical studies, multiple grid-connected and remote demonstration projects have been executed during the past decade with approximately 80 hydrogen fueling stations currently based on electrolysis, 35 of which are located in North America [28]. Recently, interest has focused on power-to-gas applications, with several projects, especially in Germany, converting electrolytic hydrogen to synthetic methane (CH_4) by methanation. Methanation involves combining electrolytic hydrogen with carbon dioxide (CO_2) by a thermo-catalytic or biologic process. The concept of power-to-gas (a phrase derived from the German "*Strom zu Gas*") is to produce "green gas" with hydrogen from renewables and carbon dioxide from bioenergy or other sources, which allows for a significant increase in the overall utilization of renewable energy assets [26]. Power-to-gas and biogas projects in Austria, the Netherlands, Denmark, Sweden, Germany, and elsewhere were reviewed by Iskov and Rasmussen [27]. In 2013, Gahleitner [28] reviewed 41 international power-to-gas projects and concluded with recommendations to improve overall system performance, develop codes and standards, and determine optimum system configurations. Also in 2013, Grond et al. [29] reviewed technologies for power-to-gas systems and concluded that these systems can provide community energy storage, time shifting/load leveling, and transmission and distribution management services.

HES units can not only increase the utilization of renewable energy resources but also have the potential to provide services to the grid. These services can be on the transmission or distribution level and enable access to additional revenue streams for HES systems. Several studies have been performed to assess the ability and value for electrolyzers, acting as demand response devices, to provide grid services [40–42]. In this respect, electrolytic hydrogen can play a role within the larger architecture of a smart grid and/or "smart gas" system by providing increased flexibility and resiliency. As is the case with other energy storage options, there are challenges to characterizing the value of these grid services to equipment owners, utilities, and electricity market operators.

HES, is more than "electricity in, electricity out." To understand this topic better, the reader is referred to the report "Hydrogen Energy Storage, Grid and Transportation Services" published by the NREL in February 2015.

5.10.1 Hydrogen Production

Alkaline electrolysis is a mature technology for large systems, whereas PEM electrolyzers are more flexible and can be used for small decentralized solutions. The conversion efficiency for both technologies is about 65–70% (lower heating

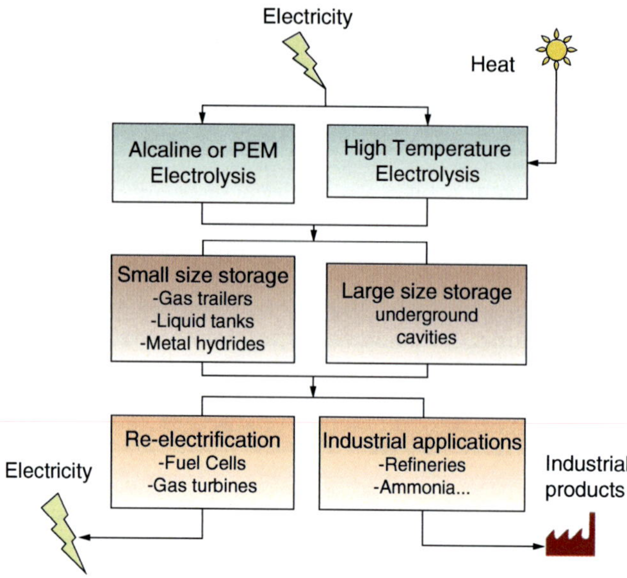

Fig. 5.13 Hydrogen production process. (Courtesy of European Institute for Energy Research (EIFER))

value). High-temperature electrolyzers are currently under development and could represent a very efficient alternative to PEM and alkaline systems, with efficiencies up to 90% (see Fig. 5.13).

5.10.2 *Hydrogen Re-Electrification*

Hydrogen can be re-electrified in fuel cells with efficiencies up to 50%, or alternatively burned in combined cycle gas power plants (efficiencies as high as 60%).

Because of the limited round-trip efficiency, direct uses of green hydrogen are under development, for example as feedstock for the chemical and the petrochemical industry, as fuel for future fuel cell cars, or blended with natural gas of up to 5–15% in natural gas pipelines. Electrolytic hydrogen can also be used for the production of synthetic liquid fuels from biomass, thereby significantly increasing the efficiency of the biomass utilization.

Deployment of hydrogen as an integrated solution by several European and American companies to supply electric power to small isolated sites or islands is via pipes. Demonstration projects have been performed since 2000 in Europe and the USA and commercial products are available. Large-scale hydrogen storage in salt caverns is the standard technology. To date, there are two full-sized hydrogen caverns in operation in Texas, USA, and a third is under construction; three older caverns are operating at Teesside in the UK.

5.11 Pipelines and Underground Hydrogen Storage

As part of the application of hydrogen as a source of renewable energy source during peak demand for electricity, one approach for power-to-gas applications is to inject hydrogen directly into natural gas pipelines rather than to undertake the additional step of methanation. This pathway was researched thoroughly in the European Union's NaturalHy project [30] and is discussed by Melaina et al. [31] in the context of the United States' natural gas pipeline systems. In general, few changes to existing natural gas transmission or distribution pipeline networks are required if the hydrogen blend level is very low. Although industry codes and standards have become more stringent and society's tolerance for risk has decreased, hydrogen was a major constituent of town gas used for heating and lighting in homes, commercial buildings, and industry for nearly a century until 1950 [32–34]. Dodds and Hawkes [35] reviewed issues related to hydrogen blending potential in the United Kingdom's natural gas system and advised that early blend levels be limited to 2–3% hydrogen by volume (2014). Standards in Germany suggest up to 5%, with potential to increase to 6–20% [36]. As is evidenced by these studies, there is continued interest in pipeline material research for enabling power-to-gas applications. Power-to-gas projects today have a bias toward methanation, partly because of the lack of standards and pipeline-specific analysis required to approve direct injection of hydrogen. However, if suitable gas quality standards exist to facilitate direct hydrogen blending, it will likely lower the development cost for these systems. Furthermore, methanation processes are not expected to achieve 100% conversion of the input hydrogen feedstock, so the development of gas quality standards for lower levels of direct hydrogen blending is also expected to facilitate the growth of the methanation technologies.

In addition to injection into the natural gas system, underground geologic formations can be used to store large amounts of natural gas or hydrogen. This concept has several successful demonstrations and continues to attract interest in North America and Europe [37]. Salt caverns, which are currently used to store natural gas seasonally, are perhaps the best example of very large-scale hydrogen storage [38]. For example, Ozarslan [39] recently evaluated a large-scale solar hydrogen storage system that used salt caverns (2012).

5.12 Materials-Based Hydrogen Storage

The FCTO's applied materials-based hydrogen storage technology RD&D activities focus on developing materials and systems that have the potential to meet DOE 2020 light-duty vehicle system targets with an overarching goal of meeting ultimate full-fleet, light-duty vehicle system targets.

Materials-based research is currently being pursued on metal hydride, chemical hydrogen storage, and sorbent materials:

- Metal hydride materials research focuses on improving the volumetric and gravimetric capacities, hydrogen adsorption/desorption kinetics, cycle life, and reaction thermodynamics of potential material candidates.
- Chemical hydrogen storage materials research focuses on improving volumetric and gravimetric capacity, improving transient performance, reducing release of volatile impurities, and developing efficient regeneration processes for the spent storage material.
- Sorbent materials research focuses on increasing effective adsorption temperature through increase of the dihydrogen binding energies and improving volumetric and gravimetric storage capacities through optimizing the material's pore size, increasing pore volume and surface area, and investigating effects of material densification.

A key component for advancing storage materials is the use of reliable material property measurement techniques. It is imperative to understand how the hydrogen storage properties of a material can be significantly influenced by not only individual sample characteristics—including chemical composition and distribution and microscopic and macroscopic material structure—but also pressure, temperature, and sample size. To help researchers better understand the proper measurement techniques, the FCTO commissioned a best practices manual that gives a detailed overview of the recommended best practices in measuring the hydrogen storage properties of materials.

5.12.1 Technical Targets and Status

Materials-based research offers a long-term solution to the challenge of onboard automotive storage, as well as opportunities for stationary and portable power applications, with the potential to significantly reduce the required storage pressure, increase gravimetric and volumetric capacity, and reduce cost. From 2005 through 2010, the DOE Hydrogen Storage program supported three collaborative efforts—the Metal Hydride Center of Excellence, the Hydrogen Sorption Center of Excellence, and the Chemical Hydrogen Storage Center of Excellence—as well as independent projects that investigated more than 400 materials for potential use in hydrogen storage applications. Analysis activities in the HSECoE have determined the current status of systems using these materials. The HSECoE has also developed spider charts showing three modeled systems for each material class and how they compare against all of DOE's 2020 targets.

Table 5.6 presents the projected performance and cost of materials-based automotive systems compared with the 2020 and ultimate DOE targets.

Figure 5.14 shows hydrogen gravimetric capacity as a function of hydrogen release temperature for many of the unique hydrogen storage materials investigated by the FCTO.

Table 5.6 Projected performance and cost of materials-based automotive hydrogen storage systems[a]

Storage system targets	Gravimetric density [kWh/kg system (kg H_2/kg system)]	Volumetric density [kWh/L system (kg H_2/L system)]	Cost [US $/kWh (US $/kg H_2)]
2020	1.5 (0.045)	1.0 (0.030)	10 (333)
Ultimate	2.2 (0.065)	1.7 (0.050)	8 (266)
Current status (from HSECoE)			
MH: $NaAlH_4$	0.4 (0.012)	0.4 (0.012)	43 (1430)
Sorbent: MOF-5, 100 bar, 80 K	1.3 (0.038)	0.7 (0.021)	15 (490)
CH storage: off-board regenerable[a, b]	1.5 (0.046)	1.3 (0.040)	17 (550)

CH chemical hydrogen, *HSECoE* Hydrogen Storage Engineering Center of Excellence, *MH* metal hydride, *MOF* Metal-Organic Framework

[a]Assumes a storage capacity of 5.6 kg of usable hydrogen

[b]MH reflects status at the end of phase I; CH and sorbent reflect status at the end of phase II

5.13 Industrial Application of Hydrogen Energy

When it comes to industrial applications, hydrogen (H_2) is everywhere. Hydrogen has been deployed as an industrial gas for over 100 years and large volumes are used across the widest range of applications every day. Hydrogen is also set to play a defining role in the much-publicized third, "green" industrial revolution. It is the most commonly occurring element in nature and—unlike fossil fuels such as crude oil or natural gas—will never run out. Like electricity, hydrogen is an energy carrier, not a source of energy. It must therefore be produced. Yet hydrogen offers several key benefits that increase its potential to replace fossil fuels. Stored hydrogen, for example, can be used directly as a fuel or to generate electricity.

Hydrogen will open up regenerative, sustainable mobility choices in our everyday lives. Hydrogen-powered vehicles have a long-distance range and can be rapidly fueled. Decades of research, development, and testing have shown that hydrogen technology is a workable, economically viable alternative suited to mass deployment. A series of illustrations presented here shows different applications of hydrogen in industry: a fuel cell bus at a hydrogen station (Fig. 5.15), hydrogen drive vehicles in an urban city (Fig. 5.16), a conceptual application of hydrogen driving a bicycle (Fig. 5.17), and different applications of hydrogen in various industries (Fig. 5.18).

The compressor unit is the key component of a hydrogen fueling station because fueling is carried out using compressed gaseous H_2 at pressures from 35 to 70 MPa. Apart from the initial state—gaseous or liquid—the technology used for fueling also depends on a range of other factors, such as, for example, the throughput and the type of vehicle to be fueled.

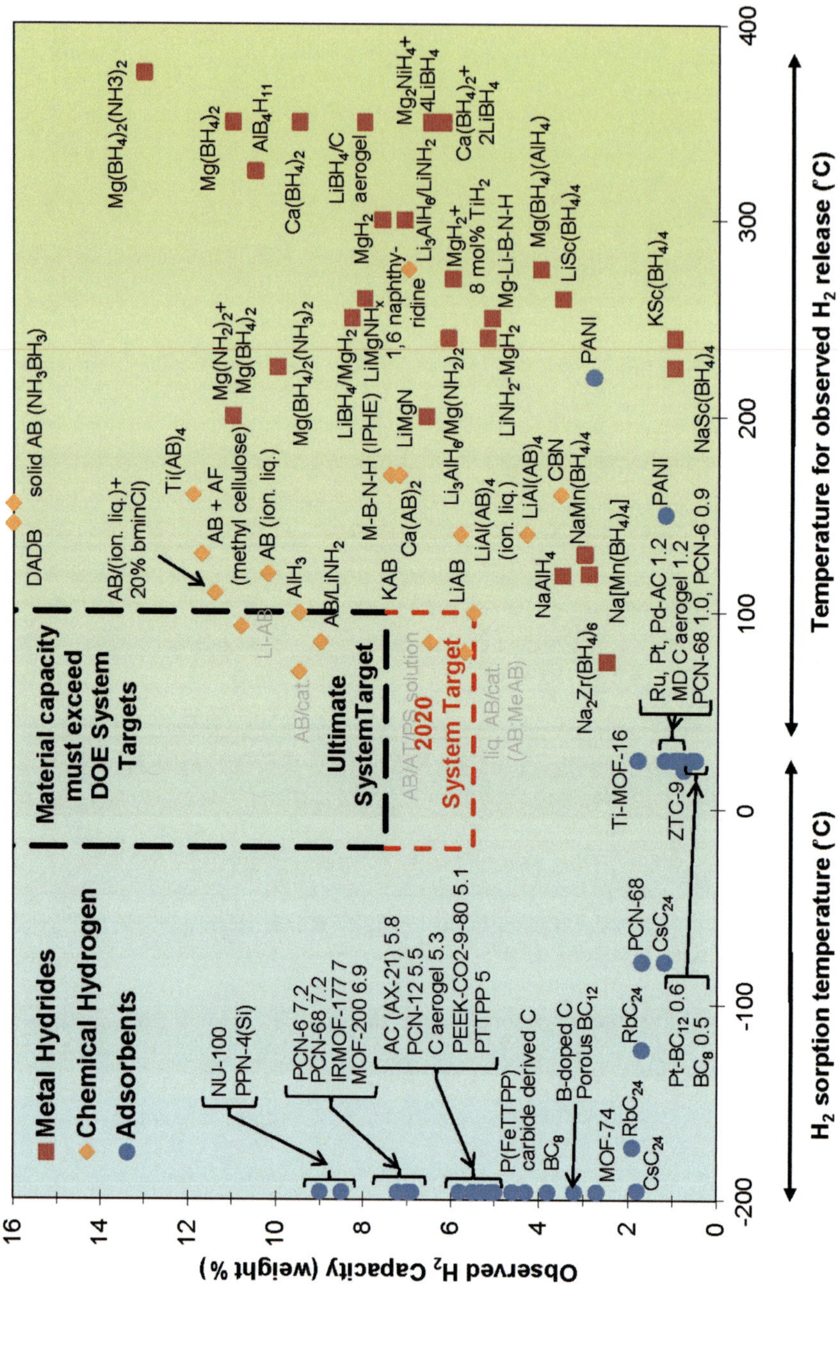

Fig. 5.14 Hydrogen gravimetric capacity as a function of hydrogen release temperature for many of the unique hydrogen storage materials. (Courtesy of the US Department of Energy)

Fig. 5.15 Bus with fuel cell drive functionality at a hydrogen fuel station. (Courtesy of Linde Group)

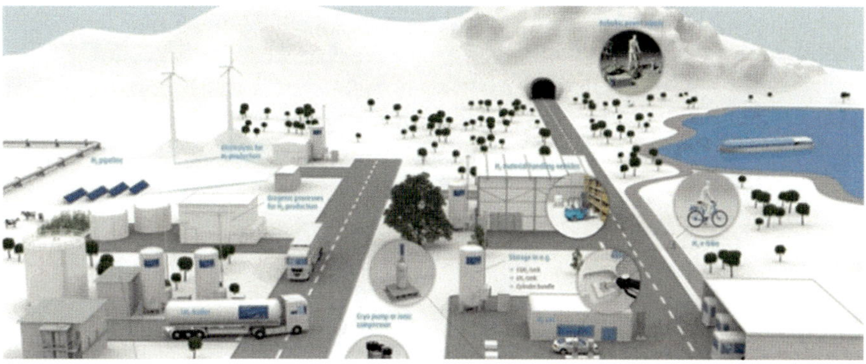

Fig. 5.16 Vehicle driven by hydrogen in a metropolitan area. (Courtesy of Linde Group)

Fig. 5.17 Bicycle driven by hydrogen. (Courtesy of Linde Group)

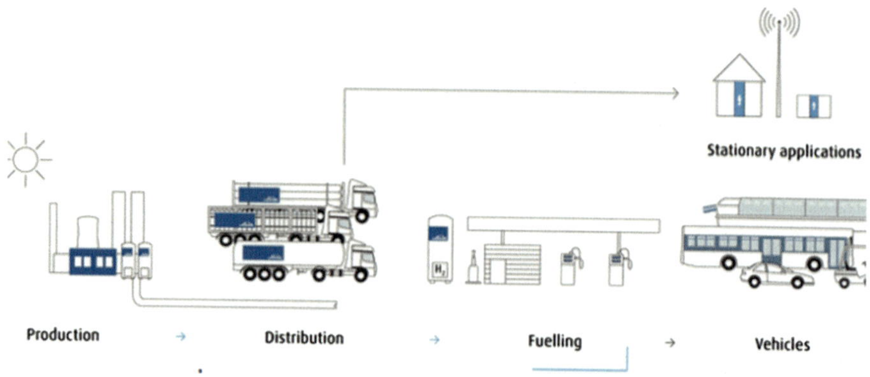

Fig. 5.18 Various industry applications of hydrogen. (Courtesy of Linde Group)

With the ionic compressor and the cryo pump, Linde has two cutting-edge, self-developed and patented technologies in its portfolio which are reliable, require little maintenance, and have high energy efficiency. Both systems can be tailored to meet individual requirements. Linde is also the first company worldwide that can produce small series of hydrogen fueling technologies.

5.14 Electrical Energy Storage

Electrical energy storage (EES) is one of the key technologies in the areas covered by the International Electrotechnical Commission (IEC). EES techniques have shown unique capabilities in coping with some critical characteristics of electricity, for example hourly variations in demand and price. In the near future EES will become indispensable in emerging IEC-relevant markets in the use of more renewable energy, to achieve CO_2 reduction, and for smart grids [43].

Historically, EES has played three main roles. First, EES reduces electricity costs by storing electricity obtained at off-peak times, when its price is lower, for use at peak times instead of electricity bought then at higher prices. Secondly, in order to improve the reliability of the power supply, EES systems support users when power network failures occur due to natural disasters, for example. Their third role is to maintain and improve power quality, frequency, and voltage [43].

Regarding emerging market needs, in on-grid areas, EES is expected to solve problems—such as excessive power fluctuation and undependable power supply—associated with the use of large amounts of renewable energy. In the off-grid domain, electric vehicles with batteries are the most promising technology to replace fossil fuels by electricity from mostly renewable sources.

The smart grid has no universally accepted definition, but in general it refers to modernizing the electricity grid. It comprises everything related to the electrical system between any point of electricity production and any point of consumption.

Through the addition of smart grid technologies the grid becomes more flexible and interactive and can provide real-time feedback. For instance, in a smart grid, information regarding the price of electricity and the situation of the power system can be exchanged between electricity production and consumption to realize a more efficient and reliable power supply. EES is one of the key elements in developing a smart grid [43].

5.14.1 Characteristics of Electricity

Two characteristics of electricity lead to issues in its use, and at the same time generate the market needs for EES. First, electricity is consumed at the same time as it is generated. The proper amount of electricity must always be provided to meet the varying demand. An imbalance between supply and demand will damage the stability and quality (voltage and frequency) of the power supply even when it does not lead to totally unsatisfied demand [43].

The second characteristic is that the places where electricity is generated are usually located far from the locations where it is consumed. Generators and consumers are connected through power grids and form a power system. Due to the locations and quantities of power supply and demand, much power flow may happen to be concentrated into a specific transmission line and this may cause congestion. Since power lines are always needed, if a failure on a line occurs (because of congestion or any other reason) the supply of electricity will be interrupted; also, because lines are always needed, supplying electricity to mobile applications is difficult. The following sections outline the issues caused by these characteristics and the consequent roles of EES.

5.14.2 Electricity and the Roles of Electrical Energy Storage

The high generation cost of electricity during the peak-demand period is a fundamental concern. Power demand varies from time to time (see Fig. 5.19), and the price of electricity changes accordingly. The price for electricity at peak-demand periods is higher and at off-peak periods lower. This is caused by differences in the cost of generation in each period.

During peak periods, when electricity consumption is higher than average, power suppliers must complement the base-load power plants (such as coal-fired and nuclear) with less cost-effective but more flexible forms of generation, such as oil and gas-fired generators. During the off-peak period, when less electricity is consumed, costly types of generation can be stopped. This is a chance for owners of EES systems to benefit financially. From the utilities' viewpoint there is a huge potential to reduce total generation costs by eliminating the costlier methods through storage

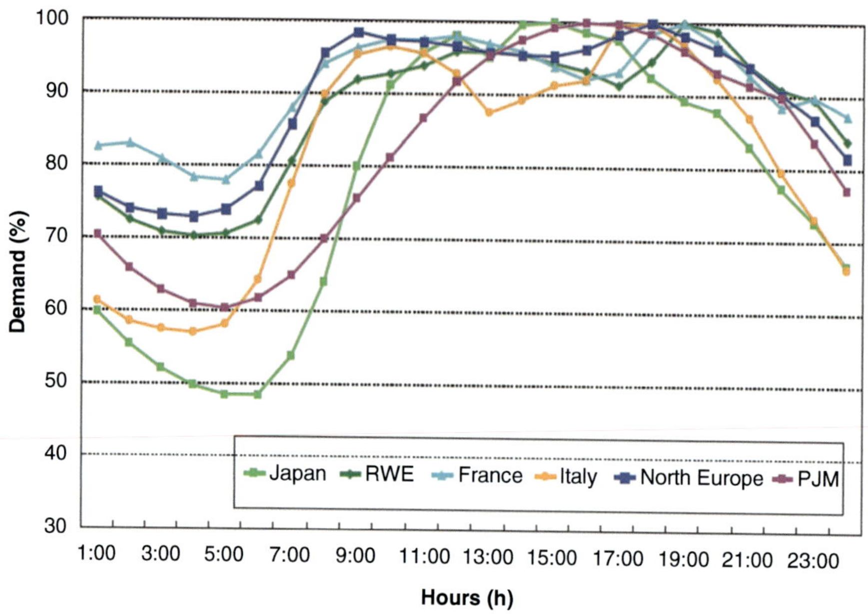

Fig. 5.19 Comparison of daily load curves. (Courtesy of IEEJ—The Institute of Energy Economics, Japan, 2005)

of electricity generated by low-cost power plants during the night being reinserted into the power grid during peak periods.

With high photovoltaic and wind penetration in some regions, cost-free surplus energy is sometimes available. This surplus can be stored in EES and used to reduce generation costs. Conversely, from the consumers' point of view, EES can lower electricity costs as it can store electricity bought at low off-peak prices and they can use it during peak periods in the place of expensive power. Consumers who charge batteries during off-peak hours may also sell the electricity to utilities or to other consumers during peak hours.

A fundamental characteristic of electricity leads to the utilities' second issue: maintaining a continuous and flexible power supply for consumers. If the proper amount of electricity cannot be provided when consumers need it, the power quality will deteriorate and, at worst, may lead to service interruption. To meet changing power consumption levels, appropriate amounts of electricity should be generated continuously, relying on an accurate forecast of the variations in demand.

Power generators therefore need two essential functions in addition to the basic generating function. First, generating plants are required to be equipped with a "kilowatt function," to generate sufficient power (kW) when necessary. Secondly, some generating facilities must possess a frequency control function, fine-tuning the output so as to follow minute-by-minute and second-by-second fluctuations in demand, using the extra power from the kilowatt function if necessary. Renewable energy facilities such as solar and wind do not possess both a kilowatt function and a

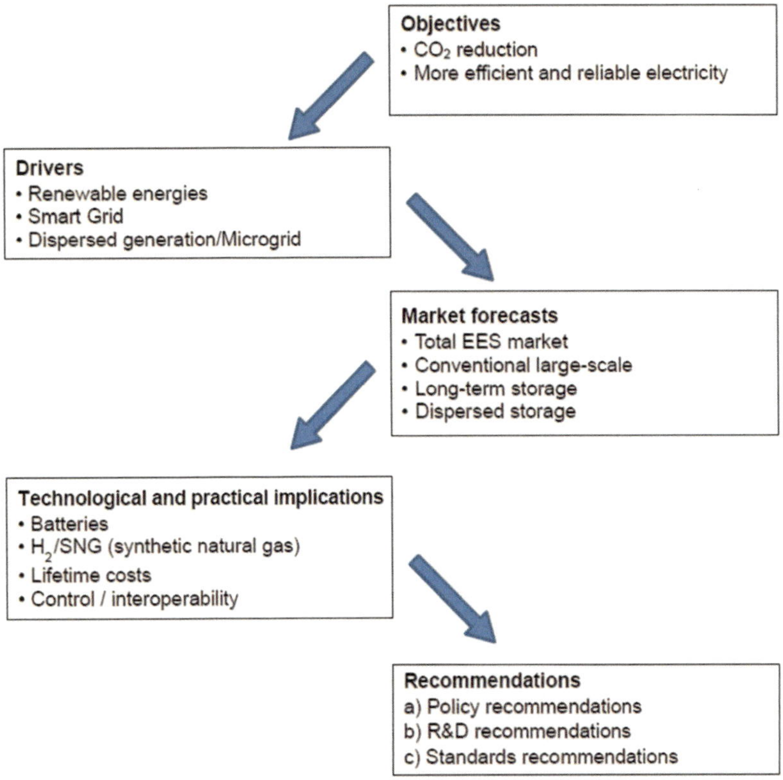

Fig. 5.20 Flowchart of the logical progression of electrical energy storage (EES) objectives. R&D research and development. (Courtesy of the International Electrotechnical Commission)

frequency control function unless they are suitably modified. Such a modification may be a negative power margin (i.e., decreasing power) or a phase-shift inverter.

EES is expected to be able to compensate for such difficulties with a kilowatt function and a frequency control function. Pumped hydro has been widely used to provide a large amount of power when generated electricity is in short supply. Stationary batteries have also been utilized to support renewable energy output with their quick response capability.

Figure 5.20 provides an overall illustration of ESS to given an overall perspective. Note that in the electricity market, global and continuing goals are carbon dioxide reduction and more efficient and reliable electricity supply and use.

Corresponding to these goals, three major drivers determining the future of EES have been identified: the foreseeable increase in renewable energy generation, the design and rollout of smart grids, and the future spread of dispersed generation and dispersed management of electrical energy—referred to here for simplicity as "microgrids." These drivers are only partly independent of each other: renewables clearly encourage, and simultaneously need, microgrids, and the increase in both renewables and dispersed sources demands a smarter grid. However, these three drivers illuminate different aspects of what will affect the future of EES systems.

The results of these drivers on future demand for EES may be divided into four market segments: the total EES market, conventional large-scale systems (e.g., pumped hydro storage [PHS]), long-term storage (e.g., H_2), and dispersed storage. How these markets are expected to develop has direct implications for which technologies will be most needed, which technology will need what type of further development, what considerations will influence rollout and penetration, and what implementation problems may be expected.

For further information refer to IEC white paper on EES [43].

5.15 Strategic Asset Management of Power Networks

Electricity networks around the world are facing a once-in-a-lifetime level of profound challenges, ranging from the massive uptake of distributed generation devices, such as rooftop solar generation, through to significant changes in the control and communications equipment used in the network itself. Power networks in developed nations are struggling with an equipment base nearing the end of its lifetime, whilst those in developing nations wrestle with trying to identify best-practice examples on which to model their operations. Compounding these challenges, there is ever-increasing regulatory and funding pressure being placed on electricity network businesses to justify their management actions and expenditure decisions [44].

There is great variation around the world on how electricity network companies approach what are arguably their number one challenge—the design, maintenance, and operation of a large network of electrical equipment. Network companies often take quite different approaches in testing equipment, calculating the lifetime and financial costs of various equipment maintenance options, and even reporting on the performance of their system. The variety here is hardly intentional—it stems from a lack of internationally accepted global standards or guidelines on how to practice asset management in the electricity network sector.

This current lack of international standards or guidelines on asset management for electrical networks will have a significant impact on the reliability and future viability of the electricity sector.

Whilst standards such as the ISO 55000 series provide general guidance on best-practice asset management procedures, they do not provide the industry-specific guidance that is needed given the operational methods and challenges of the electricity transmission and distribution industry.

The current situation means that:

- Network businesses around the world use different metrics to measure and report on the performance of their network. Without a commonly accepted definition of ways to calculate, for example, failure rates, it is very difficult to benchmark across organizations or jurisdictions.

- There is a lack of consensus on what are best practice methods for everything from testing the health of a particular item of equipment to prioritizing various asset management options. This makes stakeholder communication difficult, and means many electricity network businesses waste time and resources developing their own methods to address a particular problem. This situation is particularly exacerbated in developing nations or in the context of relatively small organizations, who could benefit greatly by simply adopting best practice methods developed by others.
- Without worldwide standards on measuring and reporting on electricity network asset management procedures and performance, broader stakeholder engagement is very difficult. When a network business cannot benchmark its performance against peers, or demonstrate that it is following industry recognized best practice, stakeholders such as regulators or funding bodies can struggle to trust the network business's management decisions or appreciate the full depth of challenges ahead.

Electricity networks in many developed nations face the very significant challenge of an aging asset base. In many nations, electricity network rollout proceeded apace throughout the 1940s to 1980s but has slowed in recent years. Many significant items of equipment are now operating close to, or even beyond, their expected retirement age.

In many developed nations, the age of the asset base and the current slow rate of replacement mean it would take hundreds of years to renew all assets. This has significant reliability implications.

The aging equipment problem is not just one of equipment wear—it also constitutes a human resources issue, as in many cases the people with the skills and expertise to complete maintenance, or the experience needed to make asset management decisions regarding this older equipment, have retired from the industry. With an equipment fleet nearing the end of its life and a shortage of parts or people to maintain it, there are very significant implications for the reliability of electricity networks in many developed nations.

Whilst aging equipment may not represent such a challenge in developing nations, or in others with more recently installed networks, simply understanding the optimal path forward amidst a plethora of technologies, management options, and an often challenging regulatory or funding environment can be very difficult.

We need to consider the elaboration of detailed international standards or guidelines to introduce a common language across the electricity network business industry regarding current system performance. Metrics such as System Average Interruption Duration Index (SAIDI) and System Average Interruption Frequency Index (SAIFI) are vital to the benchmarking of electricity network performance, yet such metrics are calculated differently around the world.

5.16 Orchestrating Infrastructure for Sustainable Smart Cities

Cities are facing unprecedented challenges. The pace of urbanization is increasing exponentially. Every day, urban areas in the United States grow by almost 150,000 people, either due to migration or births. Between 2011 and 2050, the world's urban population is projected to rise by 72% (i.e., from 3.6 billion to 6.3 billion) and the population share in urban areas from 52% in 2011 to 67% in 2050. In addition, due to climate change and other environmental pressures, cities are increasingly required to become "smart" and take substantial measures to meet stringent targets imposed by commitments and legal obligations [45].

Furthermore, the increased mobility of our societies has created intense competition between cities to attract skilled residents, companies, and organizations. To promote a thriving culture, cities must achieve economic, social, and environmental sustainability. This will only be made possible by improving a city's efficiency, and this requires the integration of infrastructure and services. While the availability of smart solutions for cities has risen rapidly, the transformations will require radical changes in the way cities are run today.

Thus, developing smart cities is not only a process whereby technology providers offer technical solutions and city authorities procure them. Building up smart cities also requires the development of the right environment for smart solutions to be effectively adopted and used.

The development of a smart city requires participation, input, ideas, and expertise from a wide range of stakeholders. Public governance is naturally critical, but participation from the private sector and citizens of the community are equally important. It also requires a proper balance of interests to achieve the objectives of both the city and the community at large.

The IEC [45] proposes a number of answers on the *what, who*, and *how* of smart city development in their executive summary. It calls for a wider collaboration between international standardization bodies that will ultimately lead to more integrated, efficient, cheaper, and environmentally friendly solutions.

> "*Needs of cities differ strongly but… the main three pillars of development remain the same.*" [45]

There is no single trend, solution, or specific approach for smart cities. Regional trends illustrate that there are divergent urban growth patterns among major regions with different levels of economic development. Still, significant disparities in the level of urbanization can also be observed across different countries within the same region. Nevertheless, all cities aiming to develop into smart cities have to be built on three sustainability pillars:

- *Economic Sustainability*

  Cities need to provide citizens with the capacity to develop their economic potential and attract business and capital. With the global financial crisis, the economic sustainability of cities has taken center stage. The crisis has unearthed considerable weaknesses in the financial models and planning strategies of public authorities in the provision of services and in their infrastructure investments. Their financial sustainability now depends also on new financial models, as well as more efficient and better-integrated services and infrastructures.

- *Social Sustainability*

  A city's attractiveness for people, business, and capital is closely related to the quality of life, business opportunities, and security and stability, which are guaranteed by social inclusiveness.
- *Environmental Sustainability*

  Cities face a number of environmental sustainability challenges, generated by the city itself or caused by weather or geological events. To reduce the impact of the city on the environment resource it is important to promote the efficient and intelligent deployment of technology and to integrate infrastructures. This process can also be developed in such a manner as to increase the resilience of the city to environmental shocks. These three pillars have one common denominator, namely the need to achieve more and better with less: efficiency. Efficiency must also be achieved in a manner that brings benefits and opportunities to citizens, making the city more dynamic and participatory.

5.16.1 Smart Technology Solutions Create Value

Rather than being an expense, smart technology integration can create considerable opportunities for added value in any city. Technology integration helps cities to improve efficiency, enhance their economic potential, reduce costs, open the door to new business and services, and improve the living conditions of its citizens. A key condition for value creation through integration is the compatibility of technologies; which is best achieved through common and consensus-based standards that ensure interoperability [45].

Presently, however, smart city projects concentrate mainly on vertical integration within existing independent infrastructure and services silos, for example, energy, transport, water, or health. A truly "smart" city requires horizontal integration as well as creating a system of systems capable of achieving considerable increases in efficiency and generating new opportunities for the city and its citizens.

5.16.2 New Approaches Needed to Smart City Solution

Cities are faced with a complex challenge, as the traditional processes of planning, procuring, and financing are not adequate for their needs. Smart cities can only exist if fundamental reforms are undertaken. Thus, there is a need for new approach and it is necessary to design, implement, and finance smart city solutions [45].

5.16.3 *Stakeholders are Key Drivers to Smart City Solution*

A smart city cannot be imposed by decree, as the city is shaped by a large number of individual decisions and social and technological changes cannot be fully accounted for [45]. The present advances in telecommunications, information and communication technologies (ICT), and affordable energy efficiency and energy production tools are changing the relationship between citizens and city services. Citizens are increasingly becoming providers of city services and not only users. A good plan requires participation, input, and ideas from a wide range of stakeholders within the city. This means that city planning needs to allow for bottom-up processes of modernization. The stakeholders are:

- Political leaders, managers, and operators of the local government (city).
- The service operators—public or private: water, electricity, gas, communication, transport, waste, education, and so on.
- End users and producers: inhabitants and local business representatives.
- Investors: private banks, venture capitalists, pension funds, international banks.
- Solution providers: technology providers, financiers, and investors.

Giving each of these groups a true stake in smart city development is important to achieve the necessary consensus for the changes. Their concerns need to be carefully considered and acknowledged, and ultimately the direction and next steps have to be collectively approved. In the absence of proper consultation, the authorities will sooner or later face considerable additional obstacles to make their vision a reality.

5.16.4 *Without Integration Rising to the Level of Systems There Cannot be a Smart City*

The transformation of a city into a smart form presents its stakeholders a wide range of challenges, including benefits and consequences when such a transformation is undertaken. A promising approach to support city planners, but also standards developing organizations (SDOs), is to model a city as a collection of activity domains in an integrated virtual organization (the city), where various groups of stakeholders (local governments, public and private corporations, academia, healthcare institutions, cultural associations, religious congregations, and financial firms) participate in operating and sustaining the city as a whole. Modeling the interrelations allows identification of pain points, gaps, and overlaps in standardization and clarification of the technical needs for integration [45].

While the technologies to develop smart cities are mostly already readily available and improving, their deployment is hampered by technical, social, and administrative challenges. Horizontal integration of infrastructures through technology is essential to reap the benefits of innovation and the potential and necessary efficiency.

Thus, interoperability is essential; without it, city planning is marred by unexpected inefficiencies leading to suboptimal outcomes and higher costs. The planning requirements for city authorities are very complex, as there are thousands of organizations and companies working in parallel to bring on the tools, systems, and products that offer potentially affordable/sustainable solutions.

To ensure that smart integrated systems are put in place in practice, internationally agreed standards that include technical specifications and classifications in order to support interoperability (i.e., devices and systems working together) are essential conditions. These include technical specifications and classifications in order to support interoperability. These are metrics against which benefits can be assessed as well as best practice documents that detail controls.

5.16.5 Horizontal and Vertical Integration a Key to Interoperability

Electric grids, gas/heat/water distribution systems, public and private transportation systems, and commercial buildings/hospitals/homes play a key role in shaping a city's livability and sustainability. To increase their performance and efficiency, these critical city systems need to be integrated [45].

The successful development of a smart city will require the combining of a bottom-up systems approach with a top-down service development and a data-centric approach. Technology integration includes everything from vertical integration from sensors, to low-cost communication, real-time analysis and control, and horizontal integration of historically isolated systems up to citizen-based services. Combined, this creates a system of systems.

Today's smart city projects are mainly focused on improving the integration of historical verticals, that is, parts of existing utilities, for example, improving energy efficiency or reducing water leakage. The next step is horizontal integration. Data from the different sectors can be combined to better manage the city and reduce risks. Thus, horizontal as well as vertical integration is key to creating value and interoperability.

5.16.6 Interoperability is the Key to Open Markets and to Competitive Solutions

Interoperability is the key to managing systems of systems and opening markets to competitive solutions. While we are now experiencing the Internet of Things (IoT) revolution (driven by the appearance of smart devices, such as wireless sensors, radio-frequency identification [RFID] tags and internet protocol [IP]-enabled devices), different producers are generating technologies using their own communication specifications and data protocols [45].

IoT market forecasts show that IoT is already making an impact on the global economy. While estimates of the economic impact during the next 5–10 years vary slightly (the International Data Corporation [IDC] estimates US$1.7 trillion in 2020 [46], Gartner sees a benefit of US$2 trillion by that time [47], and McKinsey predicts growth of US$4–11 trillion by 2025 [48]), there seems to be a consensus that the impact of IoT technologies is substantial and growing.

Future interoperability can only be guaranteed through the existence of international standards ensuring that components from different suppliers and technologies can interact seamlessly. Continued best practice sharing and development of common standards to ensure that data can flow freely between systems is essential, while maintaining the need to protect confidentiality and individual privacy.

Common terminology and procedures have to be developed in order to also ensure that organizations and businesses can efficiently communicate and collaborate, which can also be guaranteed through standards.

In addition, the multiplicity of technologies within a city now demands a top-down approach to standardization. This requires new coordination approaches between SDOs, in which all the parts of the city are jointly considered by the several technical committees involved by the different organizations. This methodology is essential as systems-level standards will enable the implementation and interoperability of smart city solutions.

5.16.7 *Guiding Principles and Strategic Orientation for the International Electrotechnical Commission and Messages to Other Standards Developing Organizations*

Electricity is core to any urban infrastructure system and the key enabler of cities development. As a result, the IEC has a specific role to play in the development of a smart city's set of standards. The IEC will call for, take initiative, invite, and strongly contribute to a more global and collaborative approach including not only international standardization organizations, but also all stakeholders of the smart city landscape (city planners, city operators, etc.) and specifically the citizens [45].

Technology and system integration are critical to ensure interoperability and the IEC will support active collaboration between the relevant actors as described in the following guiding principles:

- The IEC will continue to foster technology integration (electrotechnical, electronics, digital and information technology [IT]), and make sure that digital technology is fully integrated in all IEC products in a connect and share data perspective.

- The IEC shall make sure digital and IT technology suppliers are actively contributing in its work. Data aspects shall become a key issue in IEC, including IoT, data analytics, data utilization, data privacy, and cyber security.
- The system approach shall be accelerated as a top IEC priority, taking into account flexibility, interoperability, and scalability. Value creation for users (citizens and city infrastructure and service planners and operators) will remain the main driver of standardization work.
- Smart development requires solutions to be adapted to the specific needs of the city and its citizens, and standards have to be developed with this purpose in mind, removing technology barriers that prevent technology integration.

In conclusion, smart cities are necessary not only to reduce emissions, but to handle the rapid urbanization growth that the world is experiencing. Inefficiencies in urban areas bring large negative environmental and social impacts. City infrastructure is the backbone of the cities, delivering the necessary services to the population and creating the conditions for citizens to develop their professional, social, and cultural activities. Infrastructure is also essential in guaranteeing the city's resilience to environmental risks.

Until now city infrastructures have been built independently and operated separately in parallel silos (water supply, electricity, transport). Furthermore, the citizen has mainly been a consumer of services with little direct influence on the system. In a smart city, this needs to change. First of all, efficiency requires that infrastructures are appropriately interlinked horizontally. Secondly, citizens are becoming producers and service providers. In the area of energy, individuals are starting to produce energy from renewables and, thanks to the data revolution, also to deliver information and services in a number of areas. With smart systems, goods owned by citizens can be active in improving efficiency. For example, smart meters and electric cars can interact with the grid, data produced by the smart applications of the citizens can contribute to traffic control, improve emergency response, and so on. Citizens can also use the technologies to sell new services.

This change in cities needs to be accompanied by enabling conditions, which means reforming the ways cities are governed and financed—administrative reforms and new financial systems.

However, the glue allowing infrastructures to link and operate efficiently are standards. Standards are necessary to ensure interoperability of technologies and the transfer of best practices. But standards are not yet adapted to the level of technology integration we are requiring. Standard bodies still operate in sectorial parallel silos, developing standards which are not easy to understand by non-specialists, particularly city managers. Standards are facilitators for city planners, and they need to incorporate standards in planning and procurement. There is thus a need to reform the way standards are produced, ensuring those are adapted to the needs of the city planners and other service operators within the city.

Close collaboration is needed between standard bodies themselves and collaboration with outside organizations, in particular the city planners.

References

1. B. Zohuri, *Physics of cryogenics: an ultra-low temperature phenomena* (Elsevier Publishing Company, USA), p. 2017
2. AzoCleantech. Hydrogen energy – the perfect energy source for the future? https://www.azocleantech.com/article.aspx?ArticleID=29
3. B. Zohuri, *Nuclear Energy for Hydrogen Generation Through Intermediate Heat Exchangers: A Renewable Source of Energy*, 1st edn. (Springer Publisher, USA, 2016)
4. Smithsonian Institute. Fuel cell basics. http://americanhistory.si.edu/fuelcells/basics.htm. Accessed 2018
5. http://change.gov/agenda/energy_and_environment_agenda/
6. http://www1.eere.energy.gov/hydrogenandfuelcells/program_plans.html
7. https://energy.gov/eere/fuelcells/fuel-cells
8. X. Li, (2006) *Principles of Fuel Cells*, 1st ed. (Taylor & Francis Group, 2006), ISBN-13: 978–1591690221
9. https://energy.gov/eere/fuelcells/fuel-cell-technologies-office
10. https://www.nrel.gov/hydrogen/publications.html
11. https://energy.gov/eere/fuelcells/doe-technical-targets-fuel-cell-backup-power-systems
12. https://energy.gov/eere/fuelcells/doe-technical-targets-fuel-cell-systems-stationary-combined-heat-and-power
13. https://energy.gov/eere/fuelcells/doe-technical-targets-fuel-cell-systems-portable-power-and-auxiliary-power
14. F. Barbir, PEM electrolysis for production of hydrogen from renewable energy sources. Sol. Energy **78**(5), 661–669 (2005). https://doi.org/10.1016/j.solener.2004.09.003
15. K. Zeng, D. Zhang, Recent progress in alkaline water electrolysis for hydrogen production and applications. Prog Energy Combust Sci. **36**(3), 307–326 (2010)
16. M. Joan, Annual Review of Energy and the Environment. Ogden, Center for Energy and Environmental Studies, Princeton University, Princeton, New Jersey. **24**, 227–279 (Volume publication date November 1999). https://doi.org/10.1146/annurev.energy.24.1.227
17. C. Yang, Hydrogen and electricity: Parallels, interactions, and convergence. Available at www.sciencedirect.com. (2008)
18. D.L. Greene, P.N. Leiby, B.D. James, J. Perez, M. Melendez, A. Milbrandt, S. Unnasch, M. Hooks, Analysis of the transition to hydrogen fuel cell vehicles & the potential hydrogen energy infrastructure requirements. McQueen, S., ed. ORNL/TM-2008/30. (Oak Ridge National Laboratory, Oak Ridge, 2008). http://cta.ornl.gov/cta/Publications/Reports/ORNL_TM_2008_30.pdf. Accessed 4 Aug 2014
19. National Research Council, *Transitions to Alternative Transportation Technologies: A Focus on Hydrogen. National Research Council of the National Academies, Committee on Assessment of Resource Needs for Fuel Cell and Hydrogen Technologies* (The National Academies Press, Washington, DC, 2008)
20. J. Bartona, R. Gammonb, The production of hydrogen fuel from renewable sources and its role in grid operations. Journal of Power Sources **195**(24), 8222–8235 (2010)
21. M.Z. Jacobson, M.A. Delucchi, A.R. Ingraffea, R.W. Howarth, G. Bazouin, B. Bridgeland, K. Burkart, M. Chang, N. Chowdhury, R. Cook, G. Escher, M. Galka, L. Han, C. Heavey, A. Hernandez, D.F. Jacobson, D.S. Jacobson, B. Miranda, G. Novotny, M. Pellat, P. Quach, A. Romano, D. Stewart, L. Vogel, S. Wang, H. Wang, L. Willman, T. Yeskoo, A roadmap for repowering California for all purposes with wind, water, and sunlight. Energy **73**, 875–889 (2014)
22. S.M. Schoenung, W.V. Hassenzahl, *Long- vs. Short-Term Energy Storage Technologies Analysis: A Life-Cycle Cost Study—A Study for the DOE Energy Storage Systems Program. SAND2003–2783* (Sandia National Laboratories, Albuquerque, 2003)
23. D. Steward, G. Saur, M. Penev, T. Ramsden, *Lifecycle Cost Analysis of Hydrogen Versus Other Technologies For Electrical Energy Storage. NREL/TP-560-46719* (National Renewable

Energy Laboratory, Golden, 2009). http://www.nrel.gov/docs/fy10osti/46719.pdf. Accessed 4 Aug 2014

24. P.W. Parfomak, *Energy Storage for Power Grids and Electric Transportation: A Technology Assessment.* Tech. Rep. R42455. (Congressional Research Service, Washington, DC, 2012). http://fas.org/sgp/crs/misc/R42455.pdf. Accessed 4 Aug 2014:
25. A. Oberhofer, *Energy storage technologies & their role in renewable integration* (Global Energy Network Institute, San Diego, 2012). http://www.geni.org/globalenergy/research/energy-storage-technologies/Energy-Storage-Technologies.pdf. Accessed 4 Aug 2014
26. M. Sterner, (2010) 100% renewable energy supply for cities and nations. Presented at the Sustainable Energy Week, March 22–26, 2010. http://www.klimabuendnis.org/fileadmin/inhalte/dokumente/EUSEW2010_1.IWES_M.Sterner.pdf. Accessed 4 Aug 2014
27. H. Iskov, N.B. Rasmussen, *Global Screening of Projects and Technologies for Power-To-Gas and Bio-SNG: A Reference Report* (Danish Gas Technology Centre, Horsholm, 2013). https://www.energinet.dk/SiteCollectionDocuments/Engelske%20dokumenter/Forskning/global_screening_08112013_final.pdf. Accessed 4 Aug 2014
28. G. Gahleitner, Hydrogen from renewable electricity: an international review of power-to-gas pilot plants for stationary applications. Int J Hydrog Energy **38**(5), 2039–2061 (2013)
29. L. Grond, P. Schulze, J. Holstein, Systems analyses power to gas: final report—deliverable 1: technology review. GCS 13.R.23579. Groningen, the Netherlands: DNV KEMA Energy & Sustainability (now DNV GL). (2013). http://www.dnv.com/binaries/dnv%20kema%20%282013%29%20-%20systems%20analyses%20power%20to%20gas%20-%20technology%20review_tcm4–567461.pdf. Accessed 4 Aug 2014
30. O. Florisson, NATURALHY: an overview. Presented at the NATURALHY Final Public Presentation, November 19, Nederlandse Gasunie at Groningen, the Netherlands (2009)
31. M.W. Melaina, O. Antonia, M. Penev, *Blending Hydrogen into Natural Gas Pipeline Networks: A Review of Key Issues.* NREL/TP-5600-51995. (National Renewable Energy Laboratory, Golden, 2013). http://www.nrel.gov/docs/fy13osti/51995.pdf. Accessed 6 Nov 2014
32. C.J. Castaneda, *Invisible fuel: manufactured and natural gas in America 1800–2000* (Twayne Publishers, New York, 1999)
33. J.A. Tarr, History of manufactured gas, in *Encyclopedia of Energy*, vol. **3**, (Elsevier, Inc., 2004), pp. 733–743
34. Melaina, M.. Market transformation lessons for hydrogen from the early history of the manufactured gas industry. in *Hydrogen Energy and Vehicle Systems*, ed. by S.E. Grasman, (CRC Press, USA, 2012)
35. P.E. Dodds, A. Hawkes, *The role of hydrogen and fuel cells in providing affordable, secure low-carbon heat* (H2FC SUPERGEN, London, 2014)
36. R. Winkler-Goldstein, A. Rastetter, Power to gas: the final breakthrough for the hydrogen economy? Green **3**(1), 69–78 (2013). https://doi.org/10.1515/green-2013-0001
37. HyUnder, Assessment of the potential, the actors and relevant business cases for large scale and long term storage of renewable electricity by hydrogen underground storage in Europe. (2014). http://www.hyunder.eu/deliverables-and-publications.html. Accessed 30 Sep 2014
38. A.S. Lord, P.H. Kobos, G.T. Klise, D.J. Borns, *A Life Cycle Cost Analysis Framework for Geologic Storage of Hydrogen: A User's Tool.* SAND20116221. (Sandia National Laboratories, Albuquerque, 2011). http://prod.sandia.gov/techlib/access-control.cgi/2011/116221.pdf. Accessed 4 Aug 2014
39. A. Ozarslan, Large-scale hydrogen energy storage in salt caverns. Int J Hydrog Energy **37**(19), 14265–14277 (2012)
40. Hydrogenics, Hydrogenics successfully completes utility-scale grid stabilization trial with Ontario's independent electricity system operator. Press release. June 16, 2011. (2011). http://www.hydrogenics.com/about-the-company/news-updates/2011/06/16/hydrogenics-successfully-completes-utility-scale-grid-stabilization-trial-with-ontario's-independent-electricity-system-operator. Accessed 4 Aug 2014

41. J. Eichman, K. Harrison, M. Peters, *Novel Electrolyzer Applications: Providing More Than Just Hydrogen. NREL/TP-5400-61758* (National Renewable Energy Laboratory, Golden, 2014)
42. J. Judson-McQueeney, T. Leyden, C. Walker, Clean Energy Group: resilient power project webinar—energy storage: new markets and business models. (2013). http://www.cesa.org/assets/Uploads/RPP-Webinar-Presentations-Energy-Storage-New-Markets-and-Business-Models.pdf. Accessed 4 Aug 2014
43. International Electrotechnical Commission. White paper on Electrical Energy Storage. http://www.iec.ch/whitepaper/energystorage/
44. International Electrotechnical Commission. White paper on Strategic Asset Management of Power Networks. http://www.iec.ch/whitepaper/pdf/iecWP-assetmanagement-LR-en.pdf
45. International Electrotechnical Commission. White paper on Orchestrating Infrastructure for Sustainable Smart Cities. http://www.iec.ch/whitepaper/pdf/iecWP-smartcities-LR-en.pdf
46. International Data Corporation (IDC). Explosive internet of things spending to reach $1.7 trillion in 2020, according to IDC, 2 June 2015. http://www.idc.com/getdoc.jsp?containerId=prUS25658015. Accessed 22 Aug 2016
47. Gartner Inc., *The Internet of Things Is A Revolution Waiting to Happen*, 30 April 2015. http://www.gartner.com/smarterwithgartner/the-internet-of-things-is-a-revolution-waiting to happen. Accessed 22 Aug 2016
48. McKinsey Global Institute. Unlocking the potential of the internet of things, June 2015. .http://www.mckinsey.com/business-functions/business-technology/our-insights/the-internet-of-thingsthe-value-of-digitizing-the-physical-world. Accessed 22 Aug 2016

Chapter 6
Nuclear Hydrogen Production Plants

Hydrogen is an environmentally friendly energy carrier that, unlike electricity, can be stored in large quantities. It can be converted into electricity in fuel cells, with only heat and water as by-products. It is also compatible witdecreases sharply with temperature. At ah combustion turbines and reciprocating engines to produce power with near-zero emission of pollutants, as discussed in Chap. 2 and 3. Therefore, hydrogen could play a major role in energy systems and serve all sectors of the economy, substituting for fossil fuels and helping mitigate global warming. The quest for better and cheaper production of this clean substance for consumption is an important task for engineers and scientists, who are working toward zero emissions and a decarbonized environment for the present and future generations.

6.1 Introduction

Hydrogen promises to be an important energy carrier in the future. However, scarcely any hydrogen exists in nature—it must be produced using huge thermal energy. A promising method for hydrogen production is the use of nuclear power as a provider of high-temperature heat, while fusion power is a candidate nuclear power due to it being clean, powerful, renewable, and environmentally benign. Therefore, the design of the fusion blanket, the main component for the extraction of heat power, is the key issue for hydrogen production. An innovative concept of a high-temperature lithium lead blanket with "multilayer flow channel inserts" is discussed in this chapter. The maximum outlet temperature of 1000 °C of lithium lead coolant could be achieved while assuring the temperature of the structural material well below the engineering limit of 550 °C for reduced activation ferritic martensitic (RAFM) steel, which is the benchmark structural material for in-vessel components of fusion reactors. The theoretical analyses and numerical calculations have been performed to validate the feasibility of this concept. Technology issues, such as

© Springer International Publishing AG, part of Springer Nature 2019

B. Zohuri, *Hydrogen Energy*, https://doi.org/10.1007/978-3-319-93461-7_6

tritium permeation and material corrosion related to high-temperature conditions, need to be clarified for improvement of this innovative blanket concept.

Nuclear thermal energy is not only a feasible but attractive heat source for the production of hydrogen in the large quantities and with the thermal efficiency required to meet future demand. Hydrogen can be produced by three conventional processes: steam reforming, water electrolysis, and thermochemical water-splitting. However, high-efficiency hydrogen production is not always available.

Thermochemical water-splitting using a sulfur–iodine (S–I) cycle, which consists of three chemical reactions, offers the prospect of high heat-to-hydrogen efficiency of ~ 50% when the temperature reaches 850 °C. High-temperature steam electrolysis (HTSE) can also achieve similar efficiency when combined with high-efficiency generation with an advanced gas turbine.

As already stated, the S–I cycle is a three-step thermochemical cycle used to produce hydrogen. The net reactant of the three chemical reactions is water and the net products are hydrogen and oxygen. All other chemicals are recycled. The S–I process requires an efficient source of heat. Figure 6.1 provides a simplified diagram of the S–I cycle.

Some high-temperature DEMOnstration Power Station (DEMO) (i.e., the fusion demonstration reactor being considered in Europe) blanket concepts suitable for providing high-temperature heat for hydrogen production have been studied by engineers and scientists [1]. The recommended blanket concepts for hydrogen production need to have advanced technology development to meet the high-temperature requirements.

The high-temperature liquid (HTL) blanket has a high-temperature capability and relatively mature technology base to satisfy hydrogen thermal processes. It has been developed to obtain a temperature of 1000 °C with RAFM steel as the structural material by using MFCI technology in lithium lead (LiPb) flow channels. The

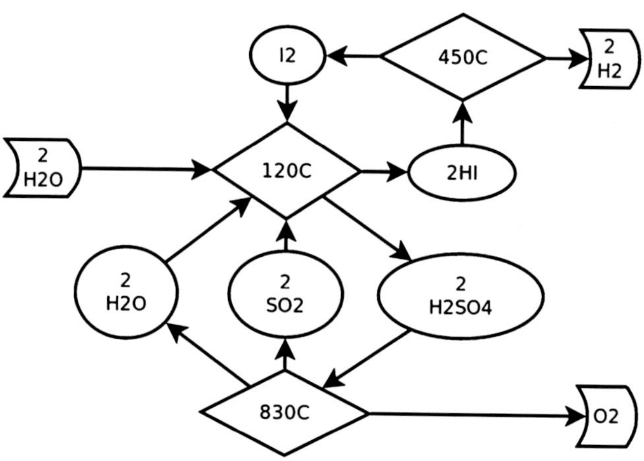

Fig. 6.1 Simplified diagram of the sulfur–iodine (S-I) cycle

potential combination of the HTL blanket concept coupled with the thermochemical water-splitting S–I cycle is recommended in China [2].

For an integrated fusion reactor hydrogen production system such as a DEMO reactor, some important technical issues have been specified and need to be developed in the next stage, such as material characterization and selections for the hydrogen production system and intermediate heat exchanger (IHX), inter-contamination issues between the hydrogen production system and fusion reactor, and the area of subsystems such as the power conversion system, tritium control system, and so on.

In this chapter, I explain how nuclear power plants are helping to produce hydrogen for consumption and as a renewable source of energy to meet the demand for of electricity supply in terms of near-future and long-term strategies.

6.2 Nuclear Hydrogen Production Technology

Nuclear energy, in addition to its application for producing electricity, can also be used to generate hydrogen for direct use by energy consumers. Generating hydrogen using nuclear energy has important potential advantages over other processes. For example, it requires no fossil fuels, results in lower greenhouse gas emissions and other pollutants, and can lend itself to large-scale production.

As a greenhouse gas-free alternative primary energy source for centralized hydrogen production, methods to use nuclear energy to produce hydrogen from water by electrolysis, HTSE, high-temperature thermochemical, and hybrid processes are being explored.

Energy efficiency is important in providing hydrogen economically and in a climate-friendly manner. High operating temperatures are needed for more efficient thermochemical and electrochemical hydrogen production using nuclear energy. Therefore, high-temperature reactors, such as the gas-cooled, molten salt-cooled, and liquid metal-cooled reactor technologies, are the candidates for use in hydrogen production. Several candidate technologies, spanning the range from well-developed to conceptual, are compared in this analysis.

Research into producing hydrogen based on nuclear energy is proceeding. Hydrogen production processes necessitate high temperatures that can be reached in the Generation-IV (GEN-IV) nuclear reactors. Technological studies are now underway in order to define and qualify components that will enable us to retrieve and transfer heat produced by these reactors in the future. Hydrogen combustion turbine power could be one of the solutions to our future energy needs, particularly for on-peak demand for electricity, but until recently the problem with hydrogen power was its production for use as an energy source. Although hydrogen is the most common element in the known universe, actually capturing it for energy use is a process which itself usually requires some form of fuel or energy.

6.3 Considerations Regarding Nuclear Reactor Types and Processes

In this section, the initial considerations regarding nuclear reactor types and processes are discussed, giving the facts regarding upcoming Generation-IV (GEN-IV) technologies which are rapidly enhancing, especially with advanced small modular reactors (SMRs), using an innovative highly efficient open-air combined Brayton cycle as suggested by this author and others [3–8].

As long as it can provide electricity and process heat, any type of nuclear reactor can be used for the production of hydrogen. However, the reactor coolant and its maximum temperature are essential criteria for determining which reactor type is more appropriate for different production processes. Power size is also an important factor, as large reactors are more suitable for co-generation of electricity and hydrogen production, whereas small-sized plants are more suitable as single-purpose plants (e.g., for hydrogen production only).

As part of a study for the Next Generation Nuclear Plant (NGNP) Project under the United States Department of Energy (DOE), the Idaho National Laboratory (INL) is tasked with evaluating the integration of high-temperature gas-cooled reactor (HTGR) technology with conventional chemical processes, such as HTSE, where the main products are hydrogen and oxygen [9]. Such an integration between HTGR technology for NGNP and HTSE is illustrated in Fig. 6.2.

Fig. 6.2 Integration of a high-temperature gas-cooled reactor (HTGR) with high-temperature steam electrolysis (HTSE). (Courtesy of the US Department of Energy)

In 2009, an independent review team considered three hydrogen production technologies to be combined with a NGNP [1]. Those technologies included the S–I process, the hybrid sulfur (HyS) process, and the HTSE process. The review team recommended the HTSE process as the first choice for the NGNP Project, with HyS as the second option. The purpose of this independent review team was to present the process modeling and economic results from producing hydrogen from HTSE combined with a high-temperature gas reactor. These results are used in other process models developed under the NGNP Program where HTGR-integrated hydrogen may be integrated with industrial processes.

6.4 Hydrogen Production Driven by Nuclear Energy

Nuclear-generated hydrogen has important potential advantages over other sources that will be considered for a growing hydrogen share in a future world energy economy. Still, there are technical uncertainties in nuclear hydrogen processes that need to be addressed through a vigorous research and development (R&D) effort. Safety issues as well as hydrogen storage and distribution are important areas of research to be undertaken to support a successful hydrogen economy in the future.

The hydrogen economy is gaining higher visibility and stronger political support in several parts of the world. In recent years, the scope of the International Atomic Energy Agency (IAEA) program has been widened to include other more promising applications such as nuclear hydrogen production and higher-temperature process heat applications. The OECD (Organization for Economic Co-operation and Development) Nuclear Energy Agency, Euratom, and the Generation IV International Forum have also shown interest in the non-electric applications of nuclear power based on future-generation advanced and innovative nuclear reactors.

Nuclear power already produces electricity as a major energy carrier. It is well-placed, though it is beyond the capability of most current plants, to produce hydrogen if this becomes a major energy carrier also.

The evolution of nuclear energy's role in hydrogen production over an estimated three decades is predicted to be:

1. Electrolysis of water, using off-peak capacity;
2. Use of nuclear heat to assist steam reforming of natural gas;
3. High-temperature electrolysis of steam, using heat and electricity from nuclear reactors; and
4. High-temperature thermo-chemical production using nuclear heat.

The first three are essentially co-generation or known as combined heat and power (CHP). The the top three 2,3,4 on above this statement are described in detail in the paper in this series Nuclear Process Heat for Industry. See also the 2013 I.E. technical report: *Hydrogen Production Using Nuclear Energy* [10].

The projection of the European Commission High Level Group shown in Fig. 6.3 offers a realistic scenario of the hydrogen market and the application areas. The

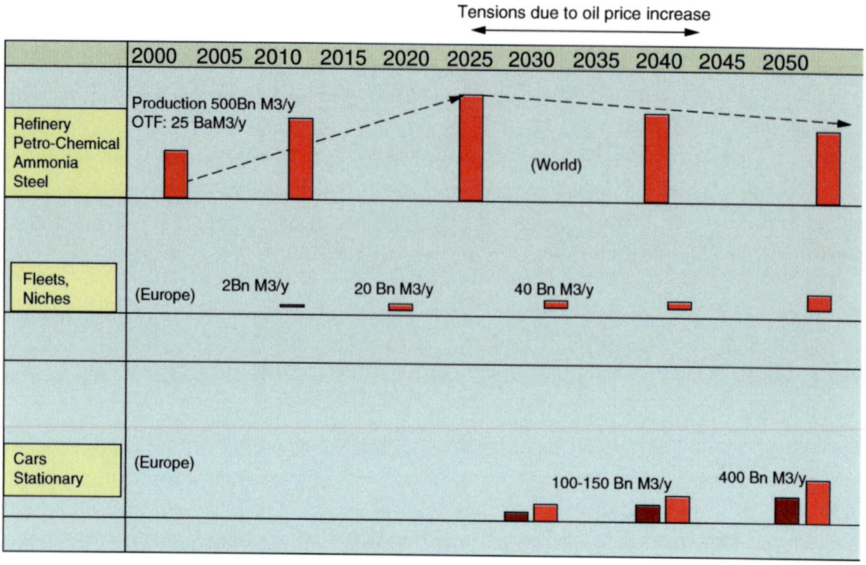

Fig. 6.3 Hydrogen production roadmap

figure clearly shows that the largest near-term markets will be the petrochemical industries, requiring massive amounts of hydrogen (H_2) for the conversion of heavy oils, tar sands, and other low-grade hydrocarbons, as well as the fertilizer and steel industries.

Hydrogen's main application as source of fuel for transportation use comes as fuel cells. Conceptually speaking, a fuel cell can be considered as a refuelable battery, making electricity as a direct product of a chemical reaction. Except where the normal battery has all the active chemical ingredients built in at the factory, fuel cells are supplied with fuel form an external source and oxygen from air.

These fuel cells catalyze by converting the oxidation of hydrogen straight to electricity energy at a relatively cryogenic (low temperature) mode in a very efficient way. A Japanese researcher in the field of hydrogen as a renewable energy claimed that the theoretical efficiency of converting chemicals to electrical energy to drive the cars is about 60% or more. However, in practice, about half that amount of efficiency has been demonstrated, except for in the high-temperature solid oxide fuel cells, which have peaked at 46%.

Definition of Co-Generation

Co-generation or combined heat and power (CHP) is the use of a heat engine or power station to generate electricity and useful heat at the same time. Trigeneration or combined cooling, heat and power (CCHP) refers to the simultaneous generation of electricity and useful heating and cooling from the combustion of a fuel or a solar heat collector. Figure 6.4 shows the layout of a typical trigeneration power plant.

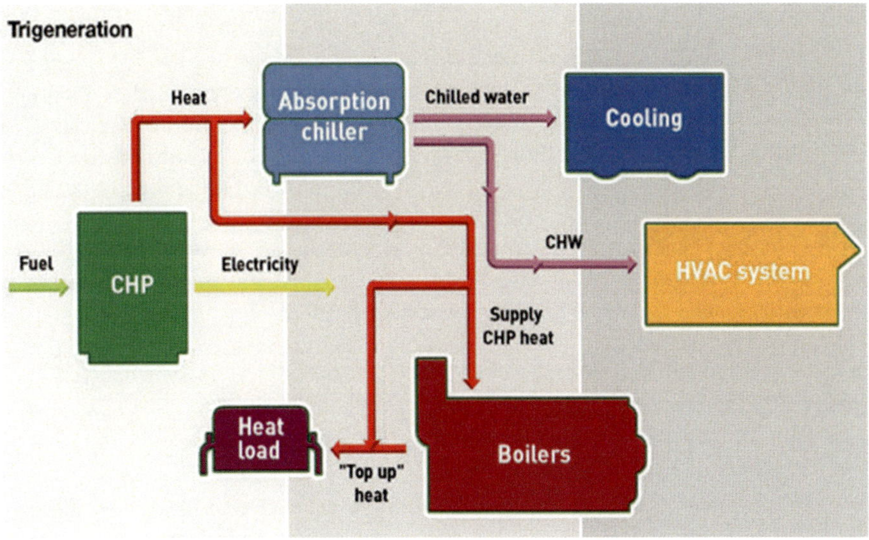

Fig. 6.4 Layout of a typical trigeneration power plant

Although this type of combustion is a carbon-free process and can be distributed to consumers, on-board storage is a challenging problem for manufacturers of vehicles where hydrogen is going to be used as the fuel source. It is impossible to store hydrogen as simply and in the same compact manner as having a tank of gasoline or liquefied natural gas (LNG) onboard. Therefore, to overcome this problem one technical approach is the suggestion of storing the hydrogen cryogenically at very high pressure, or chemically as hydrides. The latter seems more promising as a potential approach than the former, although from a refueling point of view it is not as straightforward as would be ideal for usage purposes in fuel cell vehicles (FCVs).

Pressurized storage of hydrogen is the main technology available now and this means that at 345 times atmospheric pressure (34.5 MPa, 5000 psi), ten times the storage volume is required than for an equivalent amount of petrol/gasoline. By 2010, however, 680 atmospheres (70 MPa) was practical, and the weight penalty of a steel tank was reduced using carbon fiber. The tank had previously been about 50 times heavier than the hydrogen it stored—now it is about 20 times as heavy, and the new target is ten times as heavy.

Another promising potential for a storage system currently under research is utilization of sodium borohydride ($NaBO_2$) as the energy carrier, with energy density. The $NaBO_2$ is catalyzed to yield hydrogen, leaving a borate $NaBO_2$ to be reprocessed.

Fuel cells are currently being used in electric forklift trucks and this usage is expected to increase steadily. The fuel cells are apparently not very cost effective, costing about three times as much as batteries, but last twice as long (10,000 h) and have less downtime. The hydrogen needs to be produced at lower cost to increase its production feasibility at existing nuclear power plants and align with the NGNP in the near future. The first fuel cell electric cars running on hydrogen were expected to be on the fleet market soon after 2010, but this target was met in 2015. Fuel cell buses have clocked up over 2 million kilometers and a fleet of 20 has been used in Vancouver, Canada. Another project involves three Mercedes Citaro buses in each of 11 cities worldwide. Japan has a goal of 5 million FCVs on the road by 2020.

Current fuel cell design consists of bipolar electrode plates in a frame with electrolyte between, of the most common proton exchange membrane (PEM) type. Fuel cells using hydrogen can also be used for stand-alone small-scale stationary generating plants—where higher-temperature operation (e.g., of solid oxide fuel cells) and hydrogen storage may be less of a problem or where it is reticulated like natural gas. Co-generation fuel cell units for domestic power and heat are being deployed in Japan under a subsidy scheme which terminated in 2012, by which time unit costs were expected to drop from US$50,000 to US$6000; the units will need to last for a decade [11].

However, at present, fuel cells are much more expensive to make than internal combustion engines (burning petrol/gasoline or natural gas). In the early 2000s, PEM units cost over US$1000/kW, compared with US$100/kW for a conventional internal combustion engine. The target cost for a PEM fuel cell stack is below €100/kW, which will require a reduction in the amount of palladium catalyst used.

Proton Exchange Membrane (PEM) Units

Proton exchange membrane (PEM) fuel cells work with a polymer electrolyte in the form of a thin, permeable sheet. This membrane is small and light, and works at low temperatures (about 80 °C [175 °F]). Other electrolytes require temperatures as high as 1000 °C.

PEM technology was invented at General Electric (GE) in the early 1960s, through the work of Thomas Grubb and Leonard Niedrach. GE announced an initial success in mid-1960 when the company developed a small fuel cell for a program with the United States Navy's Bureau of Ships (Electronics Division) and the United States Army Signal Corps. The unit was fueled by hydrogen generated by mixing water and lithium hydride. This fuel mixture was contained in disposable canisters that could easily be supplied to personnel in the field. The cell was compact and portable, but its platinum catalysts were expensive.

A study of the feasibility of hydrogen production at existing nuclear power plants has been conducted by most nuclear-powered countries and at the forefront of this is the United States, with most nuclear power being produced using the Generation III (GEN-III) production line supporting the network of electricity as part of the grid system within the nation. The INL is leading R&D efforts on this matter (Idaho National Laboratory INL/EXT-09-16326) [12].

Although the INL's report concentrated on existing nuclear power and the current design of such plants based on GEN-III configurations and specifications, it had no recommendation for producing hydrogen based on the current designs of GEN-IV or NGNP efforts. Currently, electrolysis equipment that is sized to produce 1 kg of hydrogen per second does not exist due to lack of demand for such plant, but demand for this is increasing significantly due to recent efforts related to the quest for a renewable source of energy for the base load to be able to meet electricity demand during the peak power use.

As discussed so far, the cost of electricity is an important consideration in the economic feasibility of any hydrogen production facility. Prior to the release of this report, much of the discussion on keeping electrical costs low centered on the use of production facilities during utility off-peak hours when costs are generally lower. However, the INL's study shows that off-peak use only results in higher breakeven hydrogen pricing because the high capital cost plant is idle for many hours. Indeed, there are times when the on-peak electrical cost is lower than some of the off-peak cost.

A need for such a feasibility study based on existing nuclear power plant should extend beyond GEN-III in transition to GEN-IV and should be included as part of the overall effort of R&D regarding NGNPs as well.

Recent experience with combined-cycle natural gas power plants has demonstrated efficiencies approaching 60%. A combined-cycle molten salt nuclear Brayton–Rankine cycle appears to achieve efficiencies 5–10% higher than a straight high-temperature Rankine cycle with peak coolant temperatures near 1000 K [13–15].

The key to achieving these efficiencies is a multi-turbine, multi-heat exchanger system that adds as much heat as possible near the peak output temperature of the molten salt coolant. Since the nuclear heating system does not consume the oxygen in the Brayton working fluid, it might be possible to build a hybrid system that burns the oxygen in the air after it has been heated by the nuclear heat exchangers.

This would provide a topping capability for a hybrid nuclear plant that could use natural gas to expand its power output to meet peak demand above the base-load capability of the plant. If all that is required is to initiate injection of natural gas into the flow path of the nuclear Brayton cycle, the response time for adding significant power will be very short. The details of how this might be done are intriguing. Ignition of the injected gas should not be a problem as the working fluid would be at high temperature.

In any gas turbine system that drives a rotating shaft (turboprop, turbo-shaft, etc.), the turbine load is usually split between a gas generator turbine and a power turbine. The two turbines are not on the same shaft and usually rotate at different speeds. Normally it is desirable for the power turbine (connected to the drive shaft) to maintain rpm control separate from the level of torque delivered. The simplistic hybrid approach that has often been proposed would have the natural gas injected after the gas generator turbine and before the power turbine.

However, this will be a problem as the pressure driving the gas through the turbine will not be strong enough to overcome the resistance of the higher-temperature gas to flow through the power turbine nozzles. Some of the flow could be bled off prior to the power turbine, but that would lower the efficiency of the injected combustible mixture by about 50%.

The combustible gas will have to be injected prior to the gas generator turbine so that it can drive the pressure up and get the hotter working fluid through the power turbine nozzles. This essentially requires the Brayton power system to be designed for two conditions: the peak power with the combustion taking place will be the "design" condition and the base-load nuclear heating will be the "off-design" condition.

With support from INL under its National University of Educational Planning (NUEP) program, which involves universities in this effort—including the University of New Mexico, with which this author has an association—models are currently being built to address this combination of conditions: "design" and "off-design" to quantify the tradeoffs of both the high-temperature Brayton system and the bottoming Rankine cycle.

It is the classic "off-design" problem, where the machinery is designed to operate at a specific pressure temperature point and it is then asked to also operate at a different set of conditions. "Off-design" operation will affect system efficiency and performance. It is important to quantify the effects of this multiple operating point scenario.

The base-load nuclear only system will probably become the "off-design" point and if the effects of operating at this "off-design" point are severe, the concept of a hybrid system may not be viable.

The variables associated with hydrogen production, including storage and transpiration as discussed earlier, as well as variation of the cost of electricity due to on-peak and off-peak per season consumption and from year-to-year, makes the issue of production a challenging problem using existing hydrogen production technology. The cost variance has an impact on the total cost of production, transportation, and storage as well.

Production processes such as the electrolyzing process require a large amount of water and this raises its own challenging issue of water purification for the process. In addition, availability of such a water source varies from year-to-year—there are years that drought becomes an issue for certain states and regions in the United States.

Producing hydrogen via nuclear energy by combing a hydrogen production facility with a new nuclear power plant has its own set of rules, regulations, safety and licensing processes, and policy involved, and these have not been well-defined by regulators so far. Study is needed into what would be involved in building such a facility side-by-side with a nuclear power plant facility and utilities must provide generation resources capable of supplying their customer base through the variations of daily and seasonal energy demands. Although there are tremendous moves towards a means for a renewable source of energy and today's policies and regulation are asking for clean energy and encourage the use of renewable resources such as wind or solar, the demand on electricity per population growth in the future shows that these two resources are not enough to meet this demand; therefore, we need to

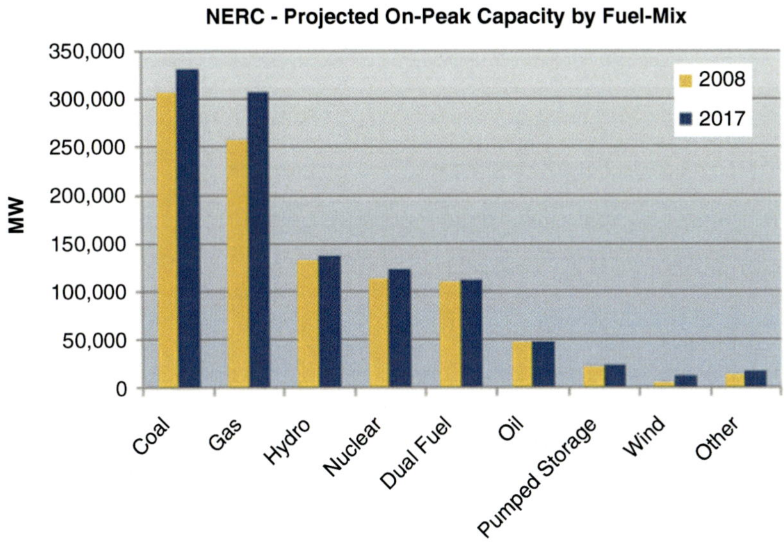

Fig. 6.5 North American Electric Reliability Corporation (NERC) current and projected generation capacity for electricity

think of an alternative solution. An assessment of the reliability of the bulk electric power system regarding future generation capabilities and trends according to the North American Electric Reliability Corporation (NERC) depicts fuel mix as a source of producing energy. To accomplish this requirement, owners of electrical utilities utilize various fuels, including nuclear, hydro, coal, gas, oil, wind, and solar, as the energy source for electric generation. This diversity provides the electric utility industry independence from any single source of fuel [12].

The capability of the electric utility industry to utilize a variety of energy sources to generate electric energy, and to move quickly to the most economical fuel resource, is an important factor in the planning for future-generation resources (see Fig. 6.5).

The projected growth in capacity in response to the need to meet an increase in electricity demand per NERC's projections is depicted in Fig. 6.6. Note that these graphs provide overall United States capacity and demand—more localized graphs might show a different situation.

The National Renewable Energy Laboratory (NREL) published a report in 2004 called *Hydrogen Demand*, in which several sizes of production units were classified based on the number of vehicles that are expected to utilize hydrogen as their source of fuel. Their findings were structured on 12,000 miles/year being the typical distance that will be driven by each hydrogen-fueled car and that each car will averaging 60 miles/kg of hydrogen consumption (i.e., 12,000 miles per year equates to about 32 miles per day, which is close to that projected for each electric vehicle [EV]). Therefore, each hydrogen fuel cell vehicle (HFCV) requires approximately 200 kg of hydrogen per year. This is calculated based on the fact that 1 kg of

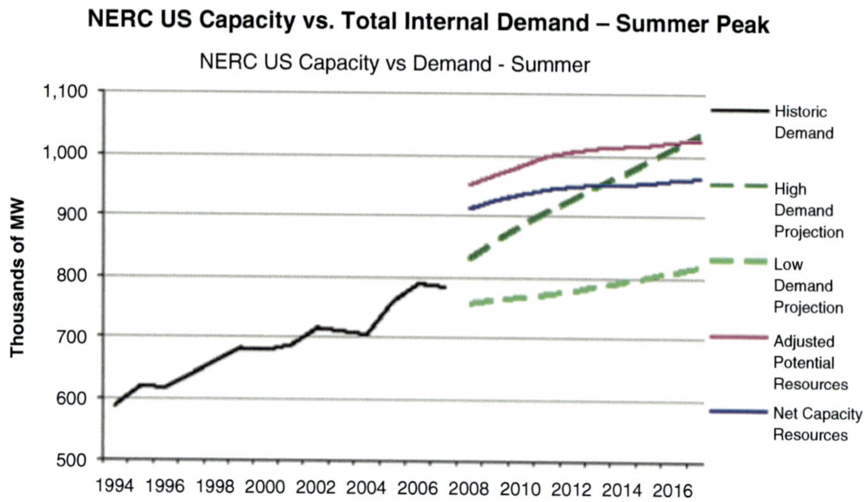

Fig. 6.6 United States electricity capacity versus total demand

hydrogen contains approximately the same energy as 1 gallon of gasoline, although of course each type of vehicle has a different consumption rate per mileage that they cover, so this assumption is averaged over all vehicles types and may not be a good foundation; however, it is a good point from which to start to have an idea of the demand for hydrogen as a fuel cell.

On the other hand, if the average HFCV achieves 45 miles/kg of hydrogen, it would mean that each of the five hydrogen production plant designs listed below would serve fewer vehicles [16]:

- The home size will serve the fuel needs of one to five cars with a hydrogen production rate of 200–1000 kg H_2/year.
- The small neighborhood size will serve the fuel needs of five to 50 cars with a hydrogen production rate of 1000–10,000 kg H_2/year.
- The neighborhood size will serve the fuel needs of 50–150 cars with a hydrogen production rate of 10,000–30,000 kg H_2/year.
- The small forecourt (refueling station) size, which could be a single hydrogen pump at an existing station, will serve the fuel needs of 150–500 cars with a hydrogen production rate of 30,000–100,000 kg H_2/year.
- A full hydrogen forecourt size will serve more than 500 cars per year with a hydrogen production rate greater than 100,000 kg H_2/year.

Manufacturers of electrolytic hydrogen generators are sizing up their production equipment to meet this demand and be able to deliver the quantity of hydrogen needed, but if production of HFCVs and hydrogen internal combustion engines (HICE) rises, then growth in hydrogen production would be justified accordingly. It is expected that this would be up to nearly 1 kg/s for production facilities, which is equivalent to 86,400 kg/day and would lead to 31,500 metric tons per year or 0.032

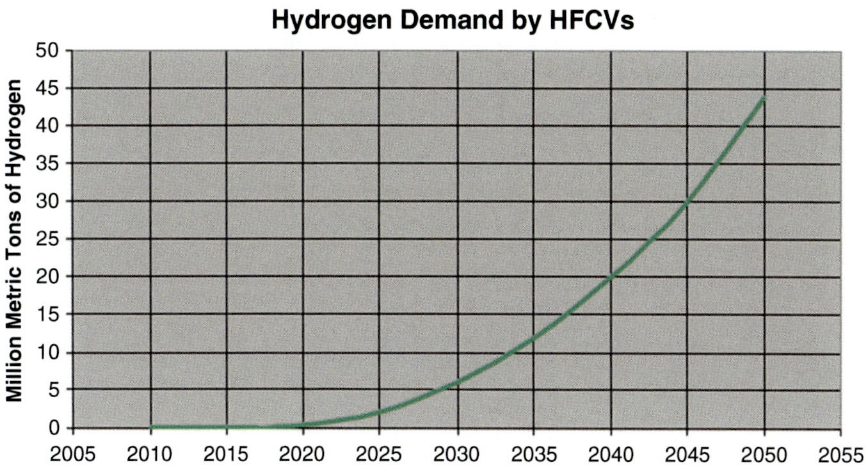

Fig. 6.7 Hydrogen demand in million metric tons per year [11]

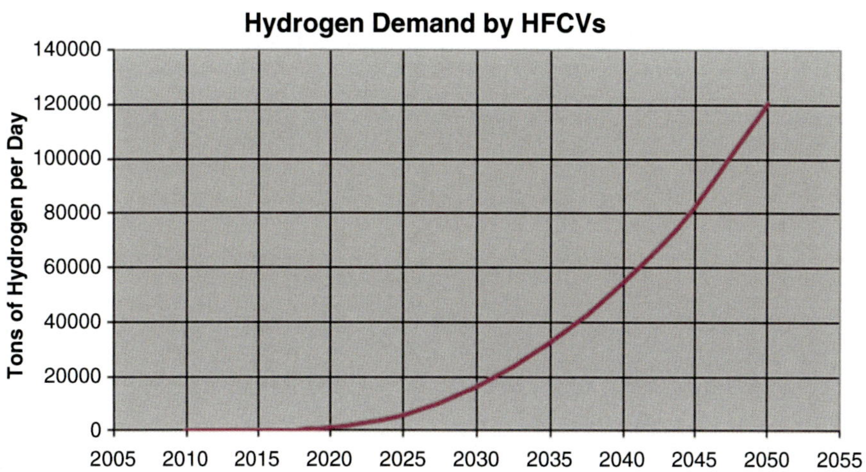

Fig. 6.8 Hydrogen demand in tons per day [12]

million metric tons of hydrogen. Thus, production facilities of this magnitude will require the ability to meet such projected demand for growth (depicted in Figs. 6.7 and 6.8).

Demand for hydrogen will be driven by the local user. Figure 6.9 shows the demand increase for various market sizes based on HFCV market penetration.

The INL report (INL/EXT-09-16326) [12], based on their analysis of existing gasoline stations in four major United States cities, has determined general trends in station sizes and geographic distribution in each city. Stations of different sizes are more or less uniformly distributed across the urban area. Although there is a slight tendency for the large stations to be found away from downtown areas and city

Market \ Penetration	1%	10%	30%	70%
Small Urban				
Population	100,000	100,000	100,000	100,000
Vehicles	116,000	116,000	116,000	116,000
H_2 fueled vehicles	1,160	11,600	34,800	81,200
H_2 fuel stations	12[a]	6[b]	17[b]	39[b]
H_2 demand (tpd)	1	8.3	2.5	58
Large Urban				
Population	1,000,000	1,000,000	1,000,000	1,000,000
Vehicles	890,000	890,000	890,000	890,000
H_2 fueled vehicles	8,900	89,900	267,000	623,000
H_2 fuel stations	86[a]	43[b]	128[b]	298[b]
H_2 demand (tpd)	9	83	250	580

a. 100 kg/d station (home size)
b. 1,500 kg/d station (small neighborhood)

Fig. 6.9 Key demand assumptions by market and penetration [17]

centers in three of the four cities, when normalized by average station size and total number of outlets in each city, relative station size distributions are nearly identical in each city.

This result is preserved during cluster analysis, which simulates reduced station networks that might resemble early hydrogen station networks. The relative station size distributions for both existing gasoline networks and simulated early hydrogen networks suggest that some 10% of the stations will be at least twice as large as the average station size, and some 30% of stations will be smaller than half the average station size [18].

The mix of home, neighborhood, and forecourt sizes will evolve as demand changes. The flexibility of a location able to adapt with the demand will also be important. Other issues potentially limiting the locations of sites and permission for hydrogen production facilities include the "determination of need" for hydrogen, as required by state environmental reviews, and water availability.

6.5 Constraints Associated with Hydrogen Production Driven by Nuclear Energy

Certain constraints are associated with producing hydrogen from operation of a nuclear power plant, and these need to be addressed and overcome in order to deal with the design and construction of large-scale hydrogen production facilities. These requirements relate to codes and standards and consideration of the regulatory, environmental, and licensing aspects of these facilities. The task becomes greater when considering co-locating a hydrogen facility with an existing or new nuclear reactor or even locating the hydrogen facility adjacent to a nuclear site.

6.5.1 Safety: Hydrogen Generation

The development and promulgation of codes and standards are essential if hydrogen is to become a significant energy carrier and fuel because these are critical to establishing a market-receptive environment for commercializing hydrogen-based products and systems. The Hydrogen, Fuel Cells, and Infrastructure Technologies (HFCIT) Program of the DOE and NREL, with the help of the leading standards and model code development organizations, other national laboratories, and key stakeholders in the United States, are coordinating a collaborative government–industry effort to prepare, review, and promulgate the hydrogen codes and standards needed to expedite hydrogen infrastructure development [19].

The DOE has undertaken a comprehensive program to support and facilitate the development of hydrogen codes and standards based on the research, development, and testing (RD&T) needed to establish the scientific and technical foundation for requirements embodied in the codes and standards. The overall structure of the program is shown in Fig. 6.10.

Over recent years, a coordinated national agenda for hydrogen and fuel cell codes and standards has emerged through DOE leadership and the support and

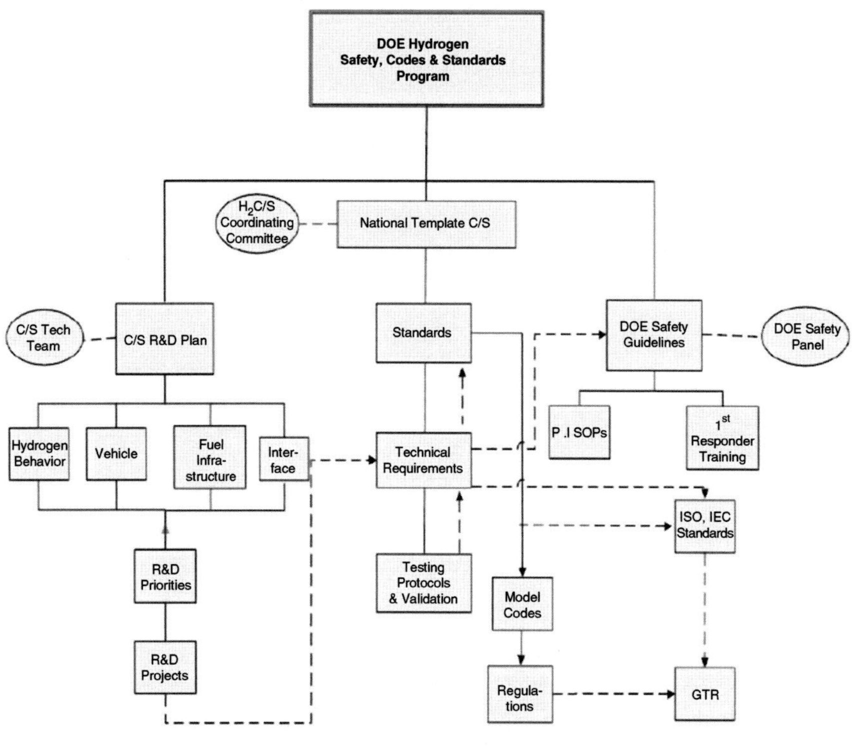

Fig. 6.10 US Department of Energy hydrogen safety, codes, and standards program [15]

collaboration of industry and key standards development organizations (SDOs) and model code organizations [19].

The R&D Roadmap provides a guide to the research, development, and demonstration activities needed to obtain data required for SDOs to develop performance-based codes and standards for a commercial hydrogen-fueled transportation sector in the United States.

Currently, no large-scale, cost-effective, environmentally attractive hydrogen production process is available for commercialization and nor has such a process been identified. The goal is to determine the potential for efficient, cost-effective, large-scale production of hydrogen utilizing high-temperature heat from an advanced nuclear power station. The benefits of this effort will include generation of a low-polluting transportable energy feedstock in a highly efficient method from an energy source that has little or no effect on greenhouse gas emissions and whose availability and sources are domestically controlled. This will help to ensure the energy supply for a future transportation/energy infrastructure that is not influenced and controlled by foreign governments.

Conventional nuclear plants readily generate electric power, but fossil fuels are firmly entrenched in the transportation sector. Hydrogen is an environmentally attractive transportation fuel that has the potential to displace fossil fuels. Hydrogen will be particularly advantageous when coupled with fuel cells. Fuel cells have higher efficiency than conventional battery/internal combustion engine combinations and do not produce nitrogen oxide during low-temperature operation. Contemporary hydrogen production is primarily based on fossil fuels, particularly on natural gas. When hydrogen is produced using energy derived from fossil fuels, there is little or no environmental advantage [20].

6.5.2 *Safety: Hydrogen Generation by Facility Location*

Public acceptance of nuclear energy is still greatly dependent on the risk of radiological consequences in the case of severe accidents. Such consequences were recently emphasized with the Fukushima–Daiichi accident in Japan in 2011. However, the nation's nuclear power plants are among the safest and most secure industrial facilities in the United States. Multiple layers of physical security, together with high levels of operational performance, protect plant workers, the public, and the environment.

However, despite the highly efficient prevention measures adopted for the current plants, some accident scenarios may, with a low probability, result in a severe accident, potentially leading to core melting, plant damage, and dispersal of radioactive materials out of the plant containment.

Even if the Japanese power station was not equipped with the newest devices for the prevention or mitigation of severe accidents, Fukushima, as well as the Three Mile Island accident in the United States in 1979, confirmed the key role of the containment barrier in the significant mitigation of radioactive releases.

Improvement of the nuclear designs and the set-up of adequate accident management strategies require confinement structures and emergency systems that must be properly dimensioned (configuration, choice of materials, and cooling circuits) to guarantee the integrity of the safety barriers and avoid release of radioactive gasses and aerosols to the outside environment.

United States nuclear plants are well-designed, operated by trained personnel, defended against attack, and prepared for emergencies. The following measures are in place:

1. *Emergency Preparedness:* Every nuclear power plant in the United States has a detailed plan for responding in the event of an emergency. Operators test that plan regularly, with the participation of local and state emergency response organizations.
2. *Operational Safety:* Stringent federal regulations, automated, redundant safety systems, and the industry's commitment to comprehensive safety procedures keep nuclear power plants and their communities safe.
3. *Personnel Training and Screening:* Operators receive rigorous training and must hold valid federal licenses. All nuclear power plant staff are subject to background and criminal history checks before they are granted access to the plant.
4. *Plant Security:* Each nuclear power plant has extensive security measures in place to protect the facility from intruders. Since September 11, 2001, the nuclear energy industry has substantially enhanced security at nuclear plants. A nuclear power plant protected by armed guards, physical barriers, and surveillance equipment from a top-level point of view is depicted in Fig. 6.11.

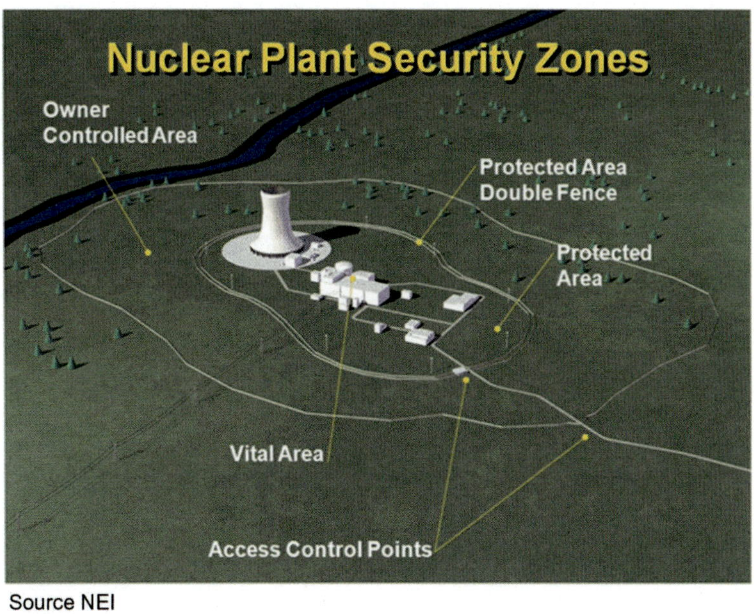

Fig. 6.11 Nuclear plant security zones. (Source: Nuclear Energy Institute (NEI))

5. *Nuclear Licensing:* All United States nuclear power plant facilities that are currently operational are reviewed and granted operating licenses under licensing regulation 10 CFR Part 50 by the Nuclear Regulatory Commission (NRC), and by requirement each particular nuclear facility is described in its Final Safety Analysis Report (FSAR). Included in the FSAR is a description of all facility activities that take place at the site, "including the products and materials most likely to be processed, stored, or transported" (such as listed in the four points above that are mentioned) [21]. Therefore, the licensees make changes to their facility accordingly and, pursuant to 10 CFR 50.59, each facility regulates itself. This includes all subsequent changes or modifications that the licensee wants to impose to its operations at the facility.
6. *Nuclear Liability Insurance:* The NRC requires all owners of nuclear power plants to maintain financial protection through primary and secondary liability insurance coverage mandated by the rules.
7. *Electrical System Up-Time and Stability:* Preventive maintenance is required in nuclear power plants, including long-term shutdown cooling requirements that consume power and have very restrictive voltage and frequency limitations. The electrical system associated with a nuclear power plant is not common in systems for oil-, gas-, and coal-generated electricity and requires stability. Safety-related issues during shutdown for preventive maintenance has a different set of rules than in traditional power plants. Even if the hydrogen production facility is powered from the grid, a shutdown or degraded operation could cause a grid disturbance that may adversely impact the nuclear plant load. The nature of hydrogen generation is that it can go very quickly from full production to zero. This type of load rejection is a feasible occurrence. Nuclear plants are currently designed and licensed for a loss of load event. However, depending on the power distribution provided to the hydrogen production facility (dedicated or off the grid) and the reliability and frequency of load disturbances, the licensee will likely need to review the electrical system stability. If it is determined that the frequency of a loss of external load event is increased, the licensee would need to evaluate the change under 10 CFR 50.59, as previously discussed, and request a license amendment from the NRC and approval, as required.
8. *Environment Review:* When issuing an amendment to a license it is necessary for the NRC to make a determination as to whether an environmental review is required. An environmental review may be conducted in the form of an environmental assessment or an environmental impact statement.

6.6 Efficient Generation of Hydrogen Fuels Utilizing Nuclear Power

Combustion processes using fossil fuel to provide power for transportation, electricity generation, heat for homes, and fuel for industry accounts for 86% of the world's energy use [22, 23]. The disadvantages and drawbacks of utilizing fossil fuel lie in supply limitations, pollution, and emission of carbon dioxide (CO_2), which is considered to be a factor involved in global warming and is now the subject of international treaties among countries around the world [24, 25].

Hydrogen is an environmentally attractive transportation fuel that has the potential to displace fossil fuels. Hydrogen will be particularly advantageous when coupled with fuel cells. Fuel cells have higher efficiency than conventional battery/internal combustion engine combinations and do not produce nitrogen oxides during low-temperature operation. Contemporary hydrogen production is primarily based on fossil fuels and most specifically on natural gas. When hydrogen is produced using energy derived from fossil fuels, there is little or no environmental advantage [26].

Currently no large-scale, cost-effective, environmentally attractive hydrogen production process is available for commercialization and nor has such a process been identified.

To overcome these combined drawbacks, the quest is on for a new source of renewable energy to replace fossil fuels, in particular with a less polluting process, primarily in parallel with nuclear energy coupled with an efficient HPP adjacent to it. This type of combination requires a nuclear power plant that can operate at very high temperatures coupled with efficient HPP via IHXs [20].

For example, the Japan Atomic Energy Agency (JAEA), using the concept of an SMR for their high-temperature test reactor (HTTR), has selected the option of coupling an HTGR to an SMR using an IHX for the steam reforming process.

This approach has been applied in the Japanese HTTR project. A flow diagram of the hydrogen production system based on an SMR and its potential coupling to the HTTR is shown in Fig. 6.12. The total system is subdivided by the dotted line into the existing nuclear part on the left-hand side and the—currently non-existent—chemical part on the right-hand side.

The requirements for a system with safe operation and high hydrogen production efficiency have initiated engineering design work on key components for the nuclear steam reforming process:

- A new-concept steam reformer heated by helium gas from the nuclear reactor has been designed to achieve high hydrogen production performance and competitiveness with an economical, fossil-fired HPP.
- A natural convection type of steam generator has been selected to achieve sufficient system controllability accommodating the large difference in thermal dynamics between the nuclear reactor and the steam reformer.
- An air-cooled radiator is connected to the steam generator to operate as a final heat sink during normal and anticipated operational occurrence conditions.

The separation of the primary circuit and the chemical process avoids the possibility of contamination in the steam reformer and reduces the permeation rates of hydrogen and tritium to negligible values. However, the heat fluxes in the steam reformer have values of around 40 kW/m^2 if the same conditions in the reforming process are fulfilled. The fabrication of the steam reformer and steam

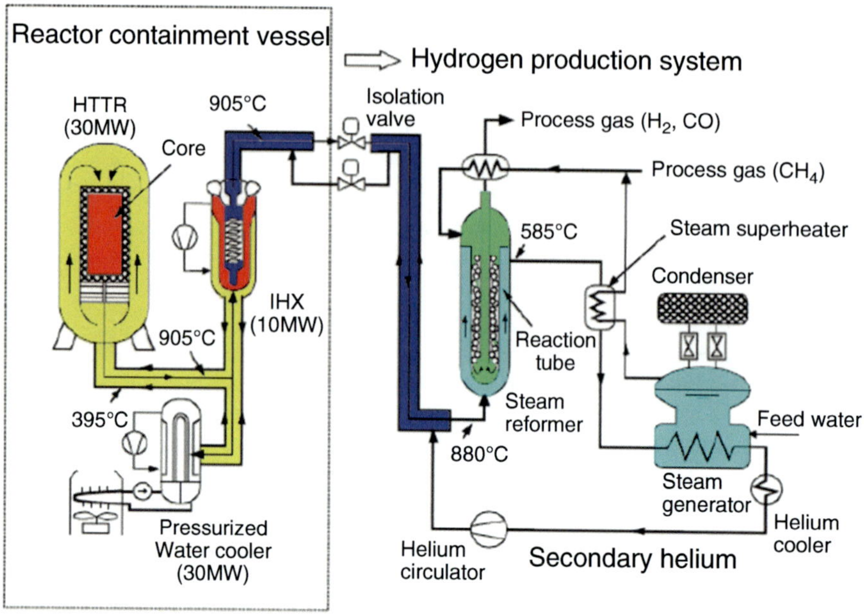

Fig. 6.12 High-temperature test reactor (HTTR) coupled to a hydrogen production plant based on a small modular reactor (SMR)

generator requires different standards than those of components that are directly integrated into the primary helium circuit [27].

For several years, steam reforming of methane has been considered the top candidate process to be connected to the HTTR for the world's first nuclear hydrogen production. The HTTR nuclear steam reforming system is therefore taken here as an example and described in more detail.

The HTTR steam reforming system has been designed to provide about 4200 Nm ^{3}/h of hydrogen production using a Ni-based catalyst with 10 MW of thermal energy. A heat utilization ratio (defined as the ratio of output hydrogen energy to total input thermal energy) of 73% is expected. This value is competitive with that of the conventional system, where the heat utilization ratio is about 80%.

The HTTR can provide high-temperature helium gas of 905 °C at the outlet of the IHX and, owing to further heat loss from the hot gas duct between the IHX and the steam reformer, secondary helium of 880 °C at the inlet of the steam reformer. The steam reformer component is shown in Fig. 6.13 [27].

With the recent worldwide increased interest in hydrogen as a clean fuel of the future, Europe has also embarked on comprehensive research, development, and demonstration activities, with the main objective of moving from a carbon-based economy toward a carbon dioxide emission-free energy structure. Due to the growing demand for hydrogen in the petrochemical, fertilizer, and refining industries, however, the near and medium terms will be characterized by coexistence between the energy carrier's hydrogen and hydrocarbons.

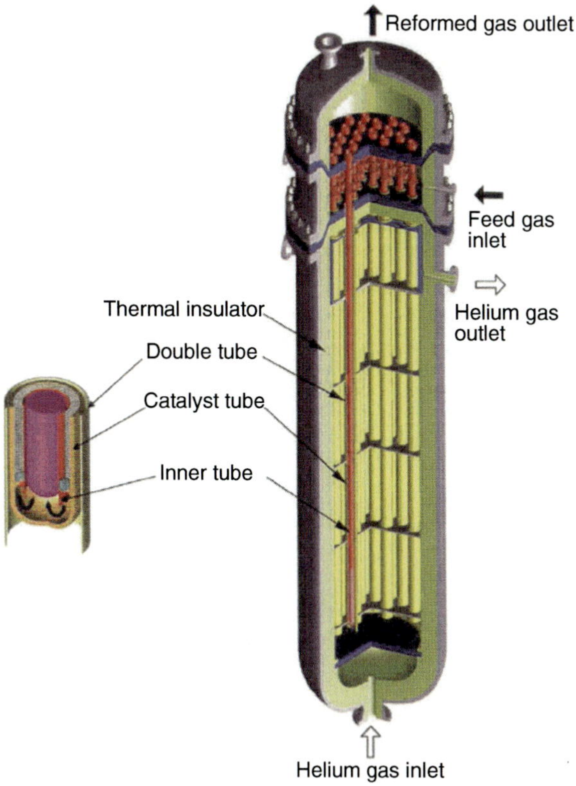

Fig. 6.13 Steam reformer component for connection to the high-temperature test reactor (HTTR) [22]

In Europe many countries are part of the Framework Program (FP), and closer collaboration is required among them as part of their strategies to provide a balanced choice of energy supply technologies while achieving the principal objectives of energy supply and continue further developments in a Network of Excellence (NOE) with long-term joint planning and Integrated Projects (IP).

For example, the coupling should involve a full-scale proven industrial process with high reliability of the nuclear heat supply to the process. A schedule for the European demonstrator high-temperature reactor (HTR)/very high-temperature reactor (VHTR) is suggested in Fig. 6.14.

Among the tasks of Framework Program-6 (FP-6) and the nuclear projects with a certain relationship to hydrogen, the most important was the IP and Reactor for Process Heat, Hydrogen, and Electricity Generation (RAPHAEL) which started in 2005 and was terminated in 2010. This IP consisted of 33 partners from ten European countries, with the objectives being, on the one hand, a study of advanced gas cooled fast reactor (GFR) technologies needed for industrial reference designs, but also concurring with and benefiting from the Japanese (HTR/VHTR) and Chinese (HTR-10) efforts toward such technologies with their current demonstrator projects.

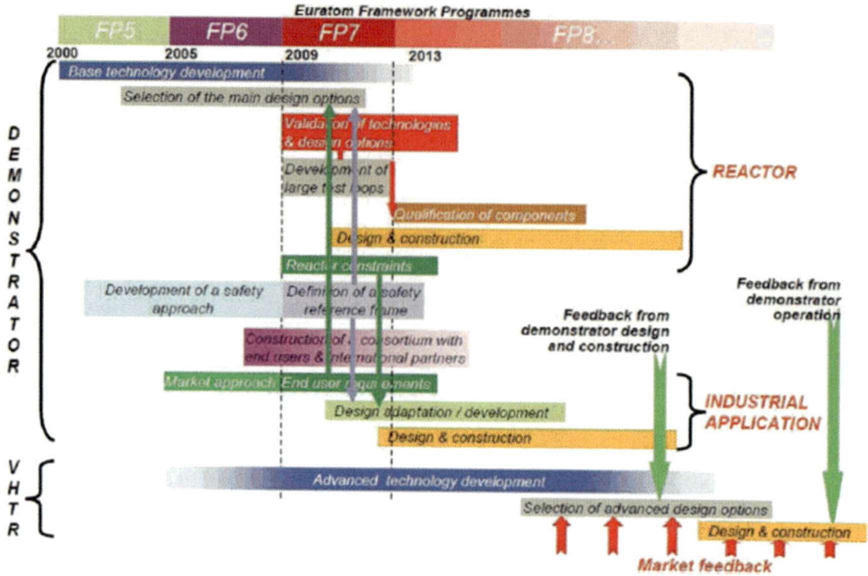

Fig. 6.14 Suggested schedule for the development of a demonstrator high-temperature reactor (HTR)/very high-temperature reactor (VHTR) for industrial process heat applications [22]

While RAPHAEL was fully concentrated on the development of a VHTR, five more activities were launched in the form of specific targeted research projects (STREPs) to deal with the other GEN-IV reactor systems:

- RAPHAEL: Reactor for Process Heat, Hydrogen, and Electricity Generation (VHTR), 2005–2010;
- GCFR: Gas Cooled Fast Reactor (GFR), 2005–2009;
- HPLWR: High Performance Light Water Reactor, Supercritical water reactor (SCWR), 2006–2010;
- ELSY: European Lead Cooled SYstem, Lead-cooled fast reactor (LFR), 2006–2010;
- EISOFAR: Road map for European Innovative Sodium Cooled Fast Reactor (SFR), 2007–2008;
- ALISIA: Assessment of liquid salts for innovative applications, Molten Salt Reactor (MSR), 2007.

Several years' joint research among the European industries dedicated to HTGR technologies in conjunction with hydrogen production using these power plants has concluded that building such a demonstration reactor with industrial process heat applications in mind and collaboration also requires strong partnerships with end-user industries. Figure 6.14 demonstrates the scheduling and timeline required for demonstration of HTR/VHTR types under their FP. The main objectives are to:

- Identify the main applications for nuclear process heat;
- Determine the viability of combining a nuclear heat source with conventional industrial processes and CHP applications;

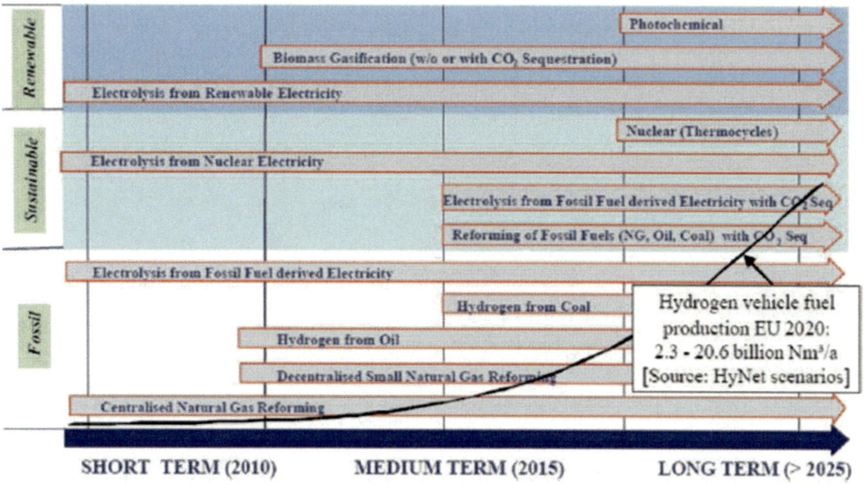

Fig. 6.15 Hydrogen Network (HyNet) timeline for hydrogen production technologies [22]

- Elaborate a program for the development of a coupled demonstrator between a VHTR and industrial processes that require heat supply; and
- Form a strategic alliance between nuclear industry and process industries.

An essential prerequisite for the success of this project is the significant involvement of private companies in the form of industrial participation to develop and deploy innovative energy supply systems. As part of their effort on hydrogen side of R&D, under a Hydrogen Network (HyNet) companies became active from 2001 to 2004 as a "thematic network" with their FP-5 with 12 European contractors and more than 70 interested partners. Figure 6.15 shows the road map working on the development of strategies for the introduction of a European hydrogen fuel infrastructure.

Hydrogen production is deemed a crucial element for the introduction of hydrogen into the energy sector in form of renewable energy. Research efforts need to be concentrated on further improvement of known reforming and gasification methods, also with regard to high-temperature primary energy systems such as GEN-IV nuclear reactors and solar–thermal concentrating systems; the development of carbon dioxide sequestration systems; gas separation technologies; and efficiency improvements in hydrogen liquefaction technologies and system integration with hydrogen production facilities [27].

6.7 Thermal Characteristics for Coupling a Hydrogen Product Plant to a High-Temperature Reactor/Very High-Temperature Reactor (HTR/VHTR)

As part of the DOE effort in collaboration with universities and national laboratories under NUEP toward R&D for employing the next-generation high-temperature nuclear reactors, huge movement by these universities and industries across the world along with national laboratories led by INL has taken place. The use of high-temperature nuclear reactors to produce hydrogen utilizing either thermochemical cycles or high-temperature electrolysis as a renewable source of energy is underway under the current budget of the DOE and NUEP.

Although the production of hydrogen using these types of processes in conjunction with next-generation nuclear reactors coupled with HPPs is at an early stage of R&D, there is a need to study such coupling for either of these processes to ensure that the HTRs have both efficient heat transfer and adequate separation to ensure that facility safety and security during off-normal events in both plants does not impact each other.

To prevent such events, there is a need to implement an intermediate heat transport loop such as IHX to separate the operations and safety functions between these two plants. The proposed investigation by the DOE to utilize such high-temperature nuclear power plants in order to produce hydrogen as source of renewable energy could be either single purpose or dual purpose [28].

Early plants, such as those proposed by the NGNP, may be dual-purpose facilities that demonstrate both hydrogen and efficient electrical generation. Later plants could be single-purpose facilities. At this stage of development, both single- and dual-purpose facilities need to be understood. Either way, both should be studied and understood regardless of which way the production of plants is going.

Seven possible configurations for such an intermediate loop have proposed by INL and they are all reported in the INL/EXT-05-00453 report [27], while other researchers at universities have also reported new design configurations [29].

The configuration proposed by Peterson is depicted here (see Fig. 6.16a) and is designed based on transportation of heat from the reactor to the chemical processing plant, where the IHX needs to couple these two plans. He suggested this design based on following configuration:

- Small-diameter channels ($D_0 = 3$ mm, $D_1 = 1$ mm, $L = 2$ m).
- Salt laminar flow regime (Re ~ 150).
- Linear heat rate of 200 °C/M.

Peterson proposed a capillary tube and shell heat exchanger showing the tube-bundle geometry formed by diffusion bonding of multiple bundles (approximately 2500) of 3.0 mm diameter tubes with hexagonally tapered ends to form inlet and outlet tube sheets [29].

These types of IHXs are coupled between two production plants if they are adjacent to each other with a facility for removing heat from the reactor side of the intermediate loop and a process heat exchanger (PHX), where the high quantity of heat can then be utilized by the HPP in order to produce the hydrogen via a process such as either thermochemical cycles or high-temperature electrolysis.

As mentioned earlier, in order to take advantage of HTR/VHTR of the next generation to produce hydrogen as a new source of renewable energy, acquisition of IHXs is required, and further studies and research should be conducted, both from a thermal hydraulic point of view and to test the integrity of materials that these IHXs are built from to ensure that they can withstand the high temperature that NGNPs

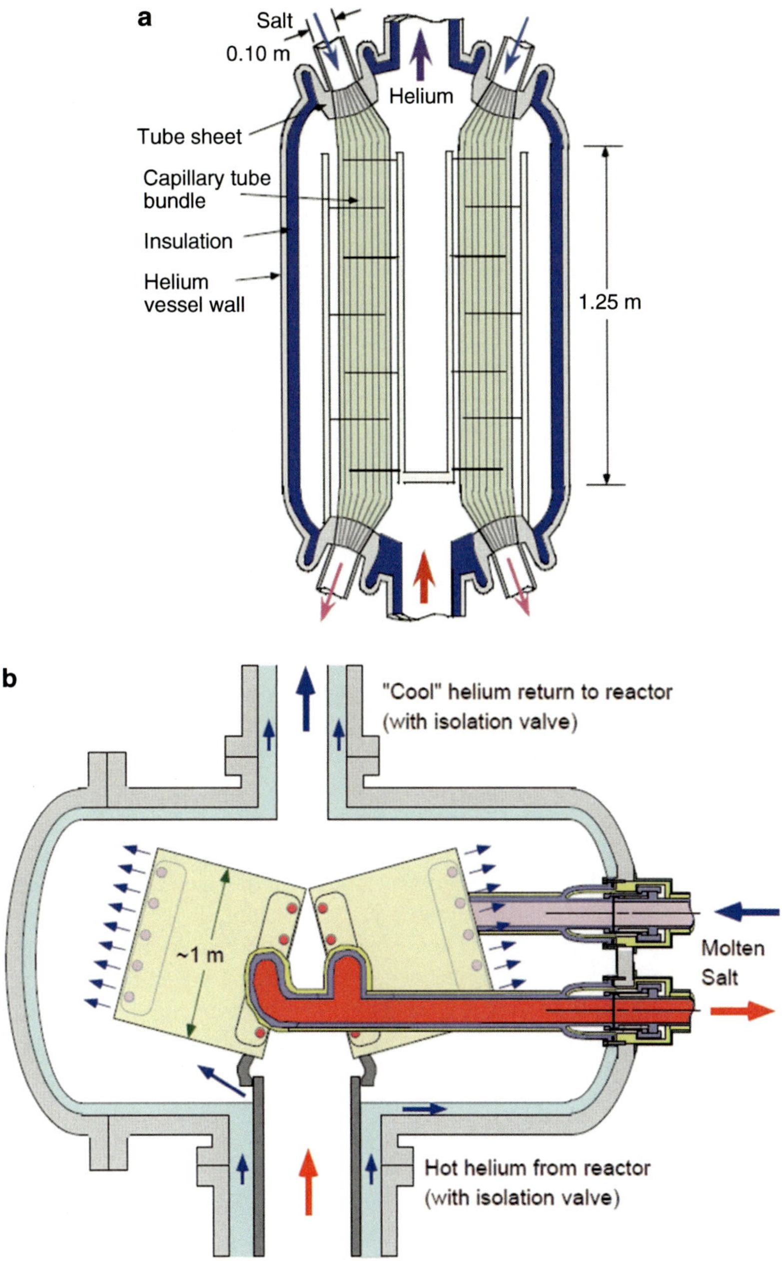

Fig. 6.16 (**a**) Intermediate heat exchanger (IHX) design proposed at University of California, Berkeley (UCB) (UCBTH-07-003) [24]. (**b**) Pre-conceptual design for a 50 MW(t) IHX in next-generation nuclear plant, based on a plate type, compact high-temperature composite design [30]

demand for their operational optimum designs. Another example of an NGNP is the HTGR, which is being envisioned to generate not just electricity but also hydrogen to charge fuel cells for cars, trucks, and other mobile energy uses. INL engineers studied various heat-transfer working fluids—including helium and liquid salts—in seven different configurations. In computer simulations, serial configurations diverted some energy from the heated fluid flowing to the electric plant and HPP.

For these IHXs to be able to act as an intermediate loop between a liquid salt or gas-cooled loop as a safety net or buffer to separate the reactor side from the hydrogen or chemical side of the plant combined within a facility, we need these intermediate loops, because by increasing the thermal inertia in the system with liquid salt in the loop in particular, the intermediate loop midgets for the reduction of the sensitivity to temperature transients' stresses.

Another IHX configuration by Peterson et al. (Report UCBTH-03-004) [31], in which they comparing the molten salt and high-pressure helium for the NGNP intermediate heat transfer fluid, is depicted in Fig. 6.16b.

They suggest a base design of a 50 MW(t) intermediate loop to be applied by the NGNP to transfer heat from the primary coolant to generate heat for the HPP using thermochemical and high-temperature electrolysis processes in order to achieve an efficiency higher than 50% where the high temperature is required for the baseline S–I thermochemical process. The functional requirements for NGNP include a 1000 °C core outlet temperature for this proposed design.

The high temperature imposes substantial technical challenges for the IHX and PHXs, as well as the intermediate heat transfer loop components. Peterson et al. have also considered the two candidate fluids for the NGNP intermediate loop for this approach to be intermediate-pressure of molten salt, and high-pressure helium (He) [31].

The best approach suggested for thermal analysis of the design of such heat exchangers is a method well known in Mechanical engineering and thermal analysis Number of Transfer Units (NTU)-effectiveness in a one-dimensional coordinate system by neglecting the longitudinal conduction in x-direction for the main flow and then expanding it to a more accurate two-dimensional model to include cross-flow conduction of fluids as well as spatially varying both longitudinal and latitudinal constraints. A finite elements method (FEM) can be employed to handle such computational analysis for each unit cell and can be built based on each region that captures the most important information of that particular region both in a steady state to start with and then expand in a transient mode for better accuracy of results. Utilization of dimensional analysis methods [25] is also recommended in order to deal with complex forms of fluid mechanics and fluid dynamics partial differential equations for thermal hydraulic and heat transfer analysis.

As mentioned earlier regarding INL/EXT-05-00453 [28], seven possible configurations for the high-temperature reactor primary coolant system and the intermediate heat transfer loop have been proposed and advantages/disadvantages of each are identified along with the working fluid (i.e., LiF-NaF-KF [Flinak] in the form of molar concentrations of 46.5%, 11.5%, and 42%, respectively as well as NaBF4-

NaF in molar concentrations of 92% and 8%) having been specified for each of these suggested configurations.

However, the recommendation of a specific design requires input from a variety of disciplines related to materials, thermal-hydraulics, economics, safety, and plant operability. This report also describes each of these intermediate heat transport loop configurations and summarizes the thermal hydraulic, structural, and efficiency calculations that have been processed and computed to characterize the advantages and issues associated with each configuration. The key issues that were taken into consideration include:

- Configuration options;
- System parameters, such as temperature and pressure;
- Structural issues; and
- Working fluid options and materials issues.

In order to perform this analysis, Davis et al. [26] identified key requirements by picking two top-level HTRs from NGNPs that fit into this calculation and a basic thermal hydraulic analysis for interfacing and coupling between the nuclear power and HPP via the intermediate heat transport loop is required.

Therefore, for the purpose of any heat transfer and thermal hydraulic analysis one needs to consider the following conditions:

1. Identify key requirements of the high-temperature reactor and the HPP that affect the choice of the intermediate heat transport loop;
2. Identify and justify assumptions that were used in the evaluation of such analysis;
3. Identify possible configurations of the intermediate heat transport loop along with the choice of fluids that is going to be used for heat transfer media;
4. Perform preliminary stress evaluations to determine allowable materials for the intermediate heat transport loop to deal with the high temperature of base plants;
5. Estimate the size and thermal hydraulic performance of various components in the intermediate heat transport loop, including the heat exchangers and loop piping;
6. Estimate the overall cycle efficiency of each configuration;
7. Determine the sensitivity of the cycle efficiency to various parameters; and
8. Compare and contrast the different options to help in the selection of the configuration and working fluid together.

One other important key requirement is the choice of nuclear reactor for the NGNP generation, based on its output of thermal power, and what coolant system is going to be used in that particular reactor. For example, the thermal power output among SMRs varies from 20 to 100 MW, such as that in production by NuScale Corporation [31]. This choice of thermal power output impacts the nominal rise in fluid temperature across the core, based on point design, and other variables such as pressure drop across the hot steam of the IHX can be assumed, which also drives the dependency on the pumping power associated with this pressure drop across the IHX as well.

All these constraints play a huge rule in determining the separation distance between the nuclear and HPPs which are part of the safety factors and also of licensing criteria. For example, for a Japanese HTTR this distance was calculated to be 300 m [32], while a similar distance for an NGNP, based on point design and other variables mentioned above, set this at between 60 and 120 m [27].

Such a separation distance has a direct impact and affects the diameters and insulation requirements of the hot and cold legs in the heat transport loop as well as the nominal temperature drop between the outlet of the NGNP and the maximum temperature delivered to the HPP, which is assumed to be 50 °C. This temperature drop imposes requirements on the effectiveness of the heat exchangers that connect the NGNP and production plant and the amount of heat loss that can be tolerated in the intermediate loop. In order to perform preliminary calculations, heat loss was assumed to cause the fluid temperature to drop 10 °C in the hot leg of the intermediate loop at nominal conditions [26].

As part of the IHX acquisition strategies one can select a compact heat exchanger rather than shell and tube design; therefore, the NTU-effectiveness design should be taken into consideration [34], while PHX, which is the heat exchanger that connects the heat transport loop to the HPP, can be chosen as the tube-in-shell type and logarithmic mean temperature difference (LMTD) analysis may possibly be utilized to design heat exchanger such as PHX.

These shell-and-tube type heat exchangers consist of tubes spirally wound into bundles and fitted into a shell. Because of the tube bundle geometry, a considerable amount of surface can be accommodated inside the shell. These heat exchangers are used for gas–liquid heat transfer applications, primarily when the operating temperature and/or pressure are very high. The challenging part of this type configuration is in preventive maintenance and the cleaning associated with this.

Sizing of these IHXs as part of the thermal hydraulic analysis is a function of the overall temperature difference between the outlet of the reactor core and the inlet on the cold side of the PHX and could be assumed to be tube-in-shell heat exchangers with the heat transport fluid flowing on the shell side. This configuration allows the tubes to contain the catalysts necessary for hydrogen production, and is judged to be the most convenient configuration. The tube side can be assumed to be at low pressure (less than 1 MPa). The hot and cold legs of the intermediate loop may be assumed to be separate pipes, as opposed to an annular configuration, for the purpose of the analysis [23].

For stress analysis, a simplified approach can be taken for different components in any desired configuration based on points of design in order to determine the thickness requirement so that the circumferential stress can be within allowed optimum vales and configurations. The creeping and rupture strength of materials also need to be taken into consideration as part of the stress analysis and it depends on the operating time at a given temperature.

Figure 6.17 shows that the rupture strength of Alloy 800 decreases sharply with temperature. At an operating time of 10^5 h (about 11 years), the rupture strength is 240 MPa at 500 °C, but decreases to 8 MPa at 900 °C. The rupture strength also depends on the time at temperature. At 900 °C, the rupture strength increases from

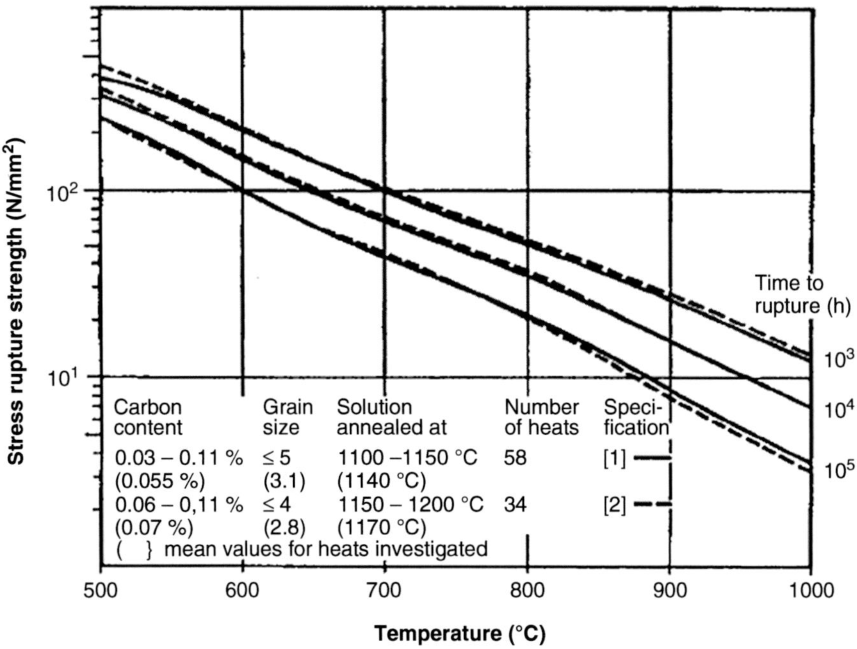

Fig. 6.17 Creep rupture strength of Alloy 800. (From Diehl and Bodman 1990 [45])

8 to 16 MPa when the operating time decreases from 10^5 to 10^4 h. The data presented in Fig. 6.17 suggest that the mechanical design of the heat transport loop will be a challenge because of the desired high temperature and the long lifetime, both of which act to reduce the rupture strength.

The creep rupture strengths of three candidate materials for the heat transport loop are shown in Fig. 6.3 for a temperature of 900 °C. These three materials are Alloy 800HT, which is a high-temperature variation of Alloy 800 (Special Metals 2004a), Alloy 617 (Special Metals 2004b), and Hastelloy X (Haynes International 2005). Alloy 617 has the highest rupture strength of these three materials at 900 °C. The allowable stress will eventually be specified by an applicable code but will be less than the strengths shown in Fig. 6.18 to account for safety factors. For this analysis, the allowable stress was assumed to be half of the creep rupture strength.

Crandall et al. (1972) [33] suggest using the following mathematical relationship to calculate the tangential stress σ for a thick wall:

$$\sigma = \frac{P_{\mathrm{i}}\left[(r_{\mathrm{O}}/r)^2 + 1\right] - P_{\mathrm{O}}\left[(r_{\mathrm{O}}/r_{\mathrm{i}})^2 + (r_{\mathrm{O}}/r)^2\right]}{(r_{\mathrm{O}}/r_{\mathrm{i}})^2 - 1} \tag{6.1}$$

where r is the radius, P is the pressure, and i and o refer to the inner and outer surfaces, respectively.

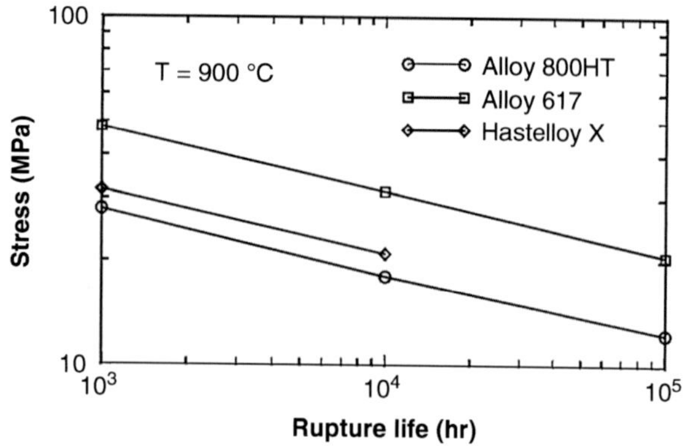

Fig. 6.18 Creeerall heat transfer coefficientp rupture strengths of candidate materials [23]

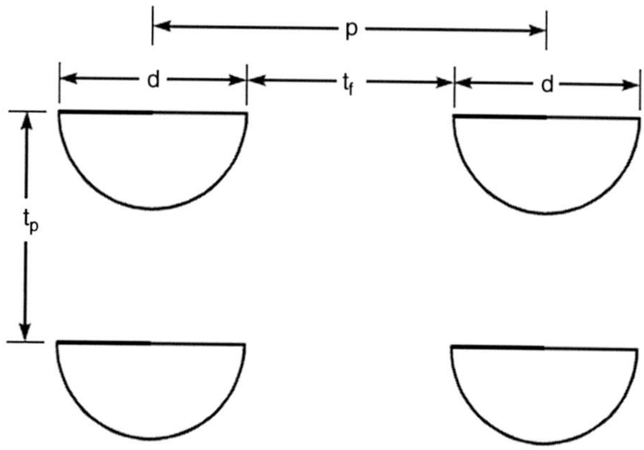

Fig. 6.19 Illustration of intermediate heat exchanger channels

If the external pressure exceeds the internal pressure, then the stress is a negative value, but the maximum magnitude always occurs at the inner surface. The radius ratio that causes the maximum stress to be less than or equal to the allowable stress, σ_D, can be calculated from Eq. 6.1. For those cases where the internal pressure exceeds the external pressure, the limiting ratio is:

$$\frac{r_o}{r_i} \geq \sqrt{\frac{\sigma_D + P_i}{\sigma_D + 2P_o - P_i}} \tag{6.2}$$

For cases where the external pressure exceeds the internal pressure, the maximum, absolute value of the stress will be less than or equal to the allowable stress when the radius ratio is:

$$\frac{r_o}{r_i} \geq \sqrt{\frac{\sigma_D - P_i}{\sigma_D - 2P_o - P_i}} \tag{6.3}$$

A simple stress analysis was also performed for the IHX assuming that it is a compact heat exchanger of the type designed by Heatric [34]. The design of the heat exchanger channels is defined by the channel diameter, d, pitch, p, and plate thickness, t_p, as illustrated in Fig. 6.19. Each plate contains either hot or cold fluid, but not both. Adjacent plates contain the other fluid. Following the method used by Dostal et al. (2004) [35], the minimum wall thickness between channels, t_f, can be approximated as follows:

$$t_{\mathrm{f}} \geq \frac{p}{\frac{\sigma_{\mathrm{D}}}{\Delta P}} + 1 \tag{6.4}$$

where σ_D is the allowable stress and ΔP is the differential pressure between the hot and cold streams. Expressing Eq. 6.4 in terms of the pitch:diameter ratio yields:

$$\frac{p}{d} \geq 1 + \frac{\Delta P}{\sigma_{\mathrm{D}}} \tag{6.5}$$

The required plate thickness can also be calculated based on the method explained and reported by Dostal et al. (2004) [35]. The plate is assumed to be a thick-walled cylinder, with an inner radius of $d/2$ and an outer radius of tp. Equations 6.2 and 6.5 can be used to calculate the thickness:diameter and pitch:diameter ratios for the IHX as a function of allowable stress and various pressures of the hot and cold streams. The allowable stress is assumed to be 10 MPa, which is approximately half of the rupture strength of Alloy 617 at 900 °C [26].

As part of thermal hydraulics and heat analysis, component sizing comes to play when we are looking at the nominal temperature drop between the outlet of NGNP and the maximum temperature delivered to the HPP per design point assumption as a fixed value rather than being variable. This temperature drop has a constraint on the effectiveness of the heat exchangers that bridge between the NGNP and the HPP and the amount of heat loss than would be tolerable in the intermediate loop. The distribution of the temperature drop between the heat exchangers and heat loss can be taken as a variable value. For example, if the heat loss can be reduced while the temperature drop across the heat exchanger can be increased, as a result a smaller or compact heat exchanger can be implemented. As part of the analysis one can take an approach toward the remaining temperature drop between the outlet of the NGNP and the maximum temperature delivered to the HPP by dividing it evenly between the IHX and PHX, and, if present, the secondary heat exchanger (SHX).

The effectiveness of a heat exchanger ε can be written as Eq. 6.6, as suggest by Krieith (1964) [36], knowing that the temperature drop between the NGNP and the production plant drives the requirements for the heat exchanger.

$$\varepsilon = \frac{(\dot{m}\,c_{\mathrm{p}})_{\mathrm{hot}}\left[(T_{\mathrm{hot}})_{\mathrm{in}} - (T_{\mathrm{hot}})_{\mathrm{out}}\right]}{(\dot{m}\,c_{\mathrm{p}})_{\mathrm{min}}\left[(T_{\mathrm{hot}})_{\mathrm{in}} - (T_{\mathrm{cold}})_{\mathrm{in}}\right]} \tag{6.6}$$

where $\dot{m} = v\rho$ (v is flow rate and ρ is flow density) is the mass flow rate, c_p is the specific heat capacity at a constant pressure and is assumed to be constant, and T_{hot}

and T_{cold} are the temperatures for the hot and cold side of the heat exchanger, $_{in}$ and $_{out}$ refer to the inlet and outlet ends of the heat exchangers, and $_{min}$ refers to the minimum value for the hot and cold sides.

Equation 6.6 applies when a counter-flow type heat exchanger is chosen, which requires less surface area (i.e., more compact shape and presumably more cost effective from a production point of view) than is required by a parallel-flow type.

If we encounter the condition where the value of $\dot{m}\,c_p$ is the same for the hot and cold flow streams, then the effectiveness temperature-wise depends on the inlet and outlet temperature. An approximation can also be made to analyze the required heat transfer area A_{ht} in order to size the heat exchanger, where this is given by:

$$A_{ht} = \frac{\varepsilon\left(\dot{m}\,c_p\right)_{min}\left[\left(T_{hot}\right)_{in} - \left(T_{cold}\right)_{in}\right]}{U\Delta T} \tag{6.7}$$

where U is the overall heat transfer coefficient and ΔT is the LMTD, which is calculated as follows:

$$\Delta T = \frac{\Delta T_a - \Delta T_b}{\ln\left(\Delta T_a/\Delta T_b\right)} \tag{6.8}$$

where ΔT_a is the temperature difference between the hot and cold fluid streams at one end of the heat exchanger and ΔT_b is the temperature difference at the other end.

Example 3.1 Calculate the heat transfer area for heat exchangers given the following values:

Flow rate = 20,000 m/h
Density of fluid = 1020 kg/m^3
Specific heat = 3.95 kJ/kg.K
Overall heat transfer coefficient = 5000 W/m^2.K
Temperature change = 30 °C
Temperature difference = 20.8 °C
Effectiveness = 1

Solution Using Eq. 3.7, we can calculate the heat transfer area as:

$$A = \frac{1 \times 20000 \times 1020 \times 3.95 \times 30}{3600 \times 20.8 \times 5000} = 6.5\text{m}^2$$

The overall heat transfer coefficient is calculated from the heat transfer coefficients on both sides of the exchanger and the thermal conductivity and thickness of the metal. The heat transfer coefficients and the thermal conductivity are assumed constant over the length of the heat exchanger. For turbulent flow, the heat transfer coefficients are calculated using the Dittus-Boelter correlation, with a leading coefficient of 0.021 for gases and 0.023 for liquids (INEEL 2003a) [34]. For laminar flow, the heat transfer coefficients are calculated from the exact solution for fully developed flow with constant heating rate [37, 30]. The thermal conductivity of the metal is calculated assuming Alloy 800 and varies between 18 and 26 W/m-K over the temperature range of interest.

Further analysis related to the subject of sizing of heat exchangers, including estimating of the pumping power and efficiency evaluation, can be found in the books by Zohuri and colleagues [5, 13, 30, 38].

The inner diameters of the hot and cold leg pipes in the heat transport loop are sized to produce a given pressure drop. The thickness of the piping is based on the results of the stress analysis. The heat loss is calculated using an overall heat transfer coefficient, which accounts for the thermal resistance of the heat transfer coefficient at the inner and outer surfaces, the pipe metal, and the insulation [39].

Pumping power Q_p is given by the approximation analysis provided by Glasstone and Sesonske (1967) [40] as follows:

$$Q_p = \frac{\dot{m}\,\Delta P}{\rho} \tag{6.9}$$

where $\dot{m}$ is the mass flow rate, ΔP is the pressure drop, and ρ is the fluid density, which is based on the temperature at the inlet to the reactor for the hot stream of the IHX and on the temperature of the cold stream entering the IHX or the SHX for the intermediate and tertiary loops.

Further research, analysis, and development are required going forward using concepts of utilizing NGNPs coupled with HPP is required since NGNPs are still at the conceptual stage of design and their thermal efficacies are constantly challenged [8, 9].

6.8 Next-Generation Nuclear Plant Intermediate Heat Exchanger Acquisition

As part of the acquisition strategy for the correct choice of IHX for NGNPs, first we need to select the right generation of nuclear power plant for this purpose that is able to deliver a quality product at the high temperature required. The DOE has selected the HTGR design for the NGNP Project to be coupled with an HPP via an IHX and its subcomponents that are most efficient and cost effective.

The NGNP will demonstrate the use of nuclear power for electricity and hydrogen production. The reactor design will be a graphite-moderated, helium-cooled, and prismatic or pebble-bed thermal neutron spectrum reactor. The NGNP will use very high burn-up, low-enriched uranium, Tri-Isotopic (TRISO)-coated fuel and have a projected plant design service life of 60 years. The HTGR concept is considered to be the nearest-term reactor design among the six choices designated as GEN-IV nuclear power plants of the future and has the capability to efficiently produce hydrogen by being coupled to such a hydrogen plant via its designated IHX set (see Fig. 6.20). The plant size, reactor thermal power, and core configuration will ensure a passive decay heat removal system without fuel damage or radioactive material releases during accidents as part of the safety net imposed on GEN-IV.

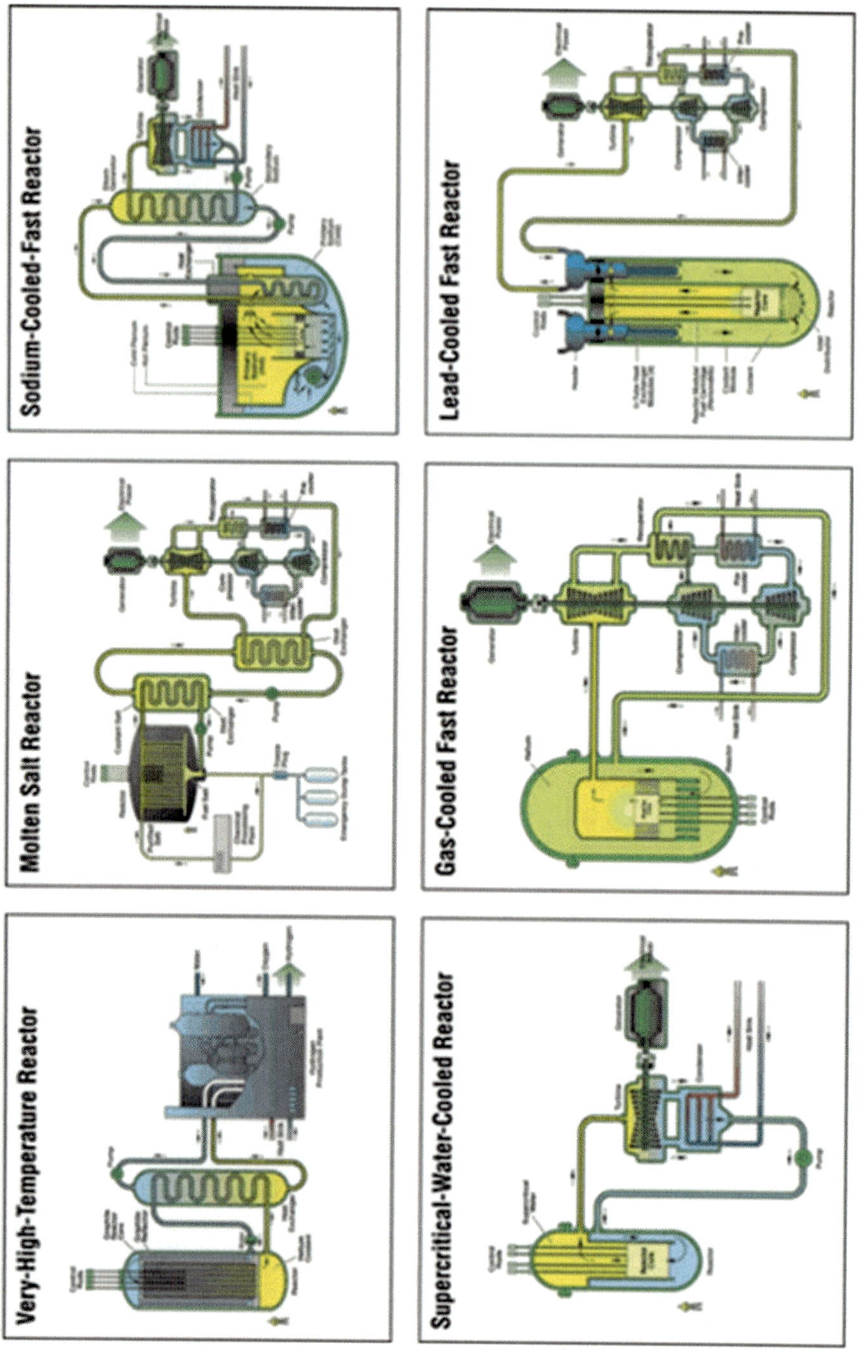

Fig. 6.20 Six proposed Generation-IV systems

Figure 6.20 shows the six proposed GEN-IV systems that are on the table for consideration as part of the efforts by the GEN-IV International Forum (GIF) initiated in 2001. As can be seen from this figure, these proposed designs and the main desired aspects behind them address inherent safety, proliferation resistance, and, last but not least, the efficiency and process heat capability that they share as a common denominator.

Figure 6.21 provides a conceptual illustration of a HTGR coupled with a HPP along with a power conversion unit. There are various iterations on this concept, most notably the VHTR, a GEN-IV concept proposed by IAEA, pebble bed reactor or the HTTR. In all cases they share a number of common characteristics and can be grouped under the HTGRs. In principle all of them consist of a graphite-moderated reactor core using an inert gas, likely helium, as a cooling fluid.

As part of the core design and fuel options for the HTGR, the fuel would be either in the form of roughly 60 mm graphite spheres inside of which are suspended a matrix of silicon carbide and uranium dioxide particles, or stacks of prismatic blocks of a similar composition. By including thorium dioxide within the fuel mix, we can also part utilize the thorium fuel cycle. However, as thorium has no naturally occurring fissile isotope, it must be used in a mixed mode with uranium (see Fig. 6.22).

One of the key advantages of HTGRs is the high operating temperatures, ranging from a minimum of around 650 °C all the way up to 1000 °C. This has several useful advantages:

1. We can now utilize the Brayton cycle, which can potentially allow for thermal efficiencies of up to 65% (although 45–55% is more typical) compared with a maximum of 47% for the more conventional Rankine (with a typical range of 33–40%) [8, 9].
2. Alternatively, if the operating temperature can be maintaining well below 800 °C, we can utilize the S–I process to manufacture hydrogen directly using heat energy with a high level of energy efficiency [8, 9].

The operating conditions for the NGNP represent a major departure from existing water-cooled reactor technologies. Few choices exist for metallic alloys for use at NGNP conditions and the design lifetime considerations for the metallic components may restrict the maximum operating temperature. Qualification of materials for successful and long-life application at the high-temperature conditions planned for the NGNP is a large portion of the effort in the NGNP Materials R&D Program [42].

Selection of the technology and design configuration for the NGNP must consider both the cost and risk profiles to ensure that the demonstration plant establishes a sound foundation for future commercial deployments. The NGNP challenge is to achieve a significant advancement in nuclear technology while at the same time setting the stage for an economically viable deployment of the new technology in the commercial sector soon after 2020 [42].

Now that we have a better understanding of the VHTR concept and how they operate, it is time to return to the original topic of this section and the strategy of selecting the right IHX for this type of reactor to be used to couple the nuclear side

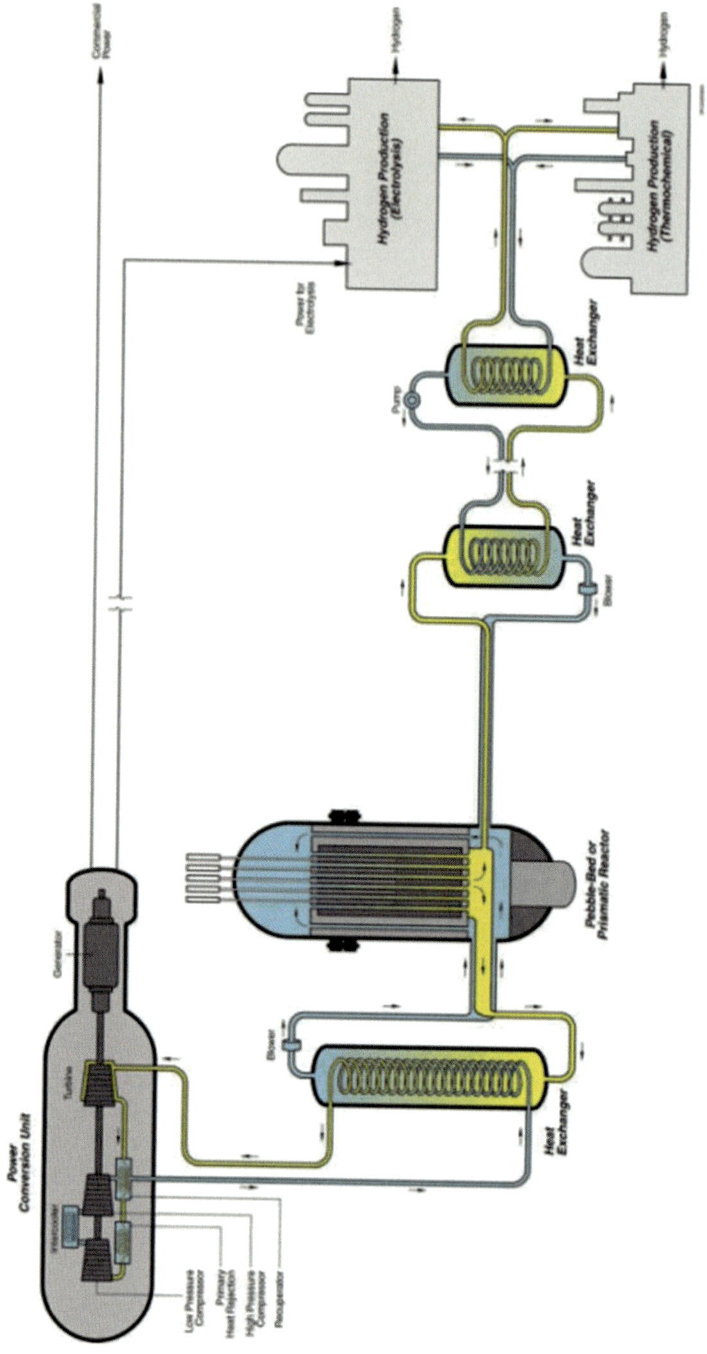

Fig. 6.21 High-temperature gas-cooled reactor (HTGR) schematic, with optional hydrogen production plant attached [41]

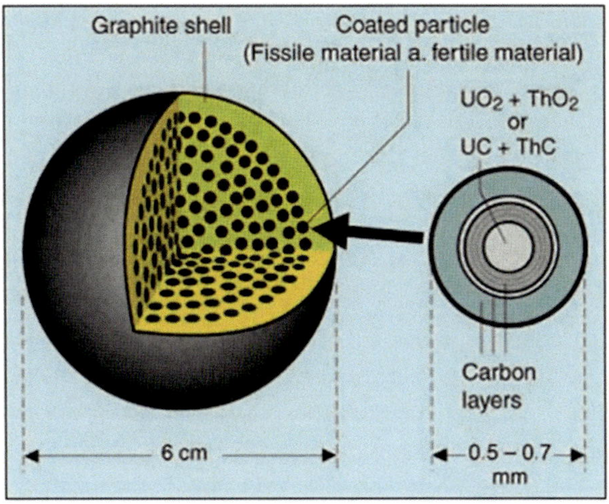

Fig. 6.22 High-temperature gas-cooled reactor (HTGR) fuel pellet

with hydrogen generating plant side. As already mentioned, the major component of the NGNP driving the HPP is the IHX.

This component will transfer heat to secondary systems that will generate electricity or hydrogen. The IHX will be operated in flowing, impure helium on the primary and secondary side at temperatures up to 950 °C. There are major high temperature design, materials availability, and fabrication issues that need to be addressed. The prospective materials are Alloys 617, 230, 800H and 800XR, with Alloy 617 being the leading candidate for use at 950 °C.

Developing an acquisition strategy policy is part of the NGNP Materials R&D Program. The objective of the NGNP Materials R&D Program is to provide the essential studies and laboratory investigations needed to support the design and licensing of the reactor and balance of the plant, excluding the hydrogen plant. The materials R&D Program was initiated prior to the design effort to ensure that materials R&D activities are started early enough to support the design process. The thermal, environmental, and service life conditions of the NGNP will make selection and qualification of the high-temperature materials a significant challenge; thus, new materials and approaches may be required. The mission of the NGNP Materials Program must support the objectives associated with the NGNP in the Energy Policy Act of 2005 and provide any materials-related support required during the development of the NGNP.

As a result of this energy policy act the selection of IHX should take place based on certain assumptions listed here [42]:

- The NGNP will be a full-sized reactor plant capable of electricity generation with a hydrogen demonstration unit of an appropriate size.
- The reactor design will be a helium-cooled, graphite-moderated core design fueled with TRISO design fuel particles in carbon-based compacts or pebbles.
- The NGNP must demonstrate the capability to obtain an NRC operating license. The design, materials, and construction will need to meet appropriate quality

assurance methods and criteria and other nationally recognized codes and standards.

- The demonstration plant will be designed to operate for a nominal 60 years.
- The NGNP Program, including the materials program, will continue to be directed by the INL based on the guidelines given in the Energy Policy Act of 2005. The scope of work will be adjusted to reflect the level of congressional appropriations.
- Application for an NRC operating license and fabrication of the NGNP will occur with direct interaction and involvement of one or more commercial organizations.

Certain issues are arising due to these assumptions that require some attention in respect to the design effort for NGNP-driven hydrogen generation plants and associated IHX to perform such a duty. These general concerns are listed here [38]:

- The last HTGR design reactor built in the United States was the Fort Saint Vrain (FSV) gas-cooled reactor [39], which was constructed in the early 1970s, generated the first power sent to the grid in 1976, and was taken out of service in 1989. The fact that there has been no HTGR construction in the United States since then, along with the long gap in construction of light water reactors (LWRs), puts the NGNP in the situation where there is a lack of current industry technical information and experience with regard to the materials of construction and fabrication practices associated with the NGNP designs currently under consideration.
- The design effort needs to be completed, which will include a final IHX design, so a material acquisition list can be developed. There needs to be new information developed regarding the primary metals producers who can produce the high-temperature alloys in the required product forms specified for use in the IHX. For compact IHX designs, joining and inspection R&D will be necessary.
- Vessel fabrication vendors with the appropriate American Society of Mechanical Engineers (ASME) certifications to perform nuclear work need to be identified. The number of these firms has declined over the last 20 years and the NGNP will be competing for these services with resurgent orders for LWRs and chemical process facility components in a world market. There is significant competition for these fabrication resources.
- To meet the NGNP start-up date of 2021, these IHXs must be delivered much earlier. The required delivery date must be identified and a schedule for material acquisition and fabrication must be developed. For a given desired delivery date the following steps need to be completed with the appropriate completion dates:

    1. Place material order with primary metal producer to obtain position in the melting schedule to secure material for fabrication.
    2. Finalize material shapes and sizes (tubing, sheet, forgings, plate) and choose the appropriate specifications for the intermediate product mill.
    3. Secure fabrication vendor services and ship material to his facility.
    4. Completion date for fabrication.
    5. Shipment to Idaho.
    6. Installation of the IHX and other major equipment to meet start-up schedule.

In accordance with the pros and cons discussed, two alternative IHX designs were developed based on the printed circuit heat exchanger (PCHE) concept with one design developed by Heatric Corporation and the second by Toshiba Corporation. These designs consist of metal plates that are diffusion bonded together with flow channels that are chemically milled into the plate. The PCHE concept allows for simultaneous high-temperature and high-pressure operation with relatively thin wall thicknesses between the primary and secondary coolants. The PCHE designs are typically four to six times smaller than conventional shell-and-tube heat exchangers of equivalent duty and designs have been developed with thermal effectiveness greater than 98%.

An alternative design using a shell-and-tube, counter-flow heat exchanger using a helically coiled tube was developed by Toshiba Corporation. For an equivalent heat duty and long mean temperature difference (LMTD), this type of heat exchanger is considerably larger than a PCHE. This design allows for in-service inspection of the heat transfer tubes and has successfully operated in the HTTR [38, 39].

6.9 Applicability of Heat Exchanger to Process Heat Applications

The strategic goal of the Advanced Reactor Concept Program is to broaden the environmental and economic benefits of nuclear energy in the United States economy from power production to meet the energy needs to also demonstrate applicability to market sectors not being served LWRs. The advanced high-temperature reactor (AHTR) offers unique advantages for a variety of markets beyond power production because of the high reactor outlet temperature (ROT) and superior heat transport characteristics of molten salt. Increased ROT would expand the AHTR's applicability to many other applications (see the report by Sabharwall et al. [43]).

The integration of AHTR technology with conventional chemical industrial processes is presented in this section. The process heats industrial applications being considered are hydrogen production via steam methane reforming of natural gas and HTSE, substitute natural gas production, oil sands recovery via steam-assisted gravity drainage, coal-to-liquid production, natural gas-to-liquids production, methanol-to-gasoline production, ammonia production, ex situ oil shale, and in situ oil shale. The temperature ranges of applications that could be coupled to the AHTR with the current ROT (green band) and others that could potentially be coupled if the ROT was raised (red band) are shown in Fig. 6.23. These are representative and should not be considered inclusive of all potential applications.

In Situ Oil Definition

This relatively new method is mainly used to access bitumen in oil sand that is buried too deep below the Earth's surface to be recovered with a truck and

(continued)

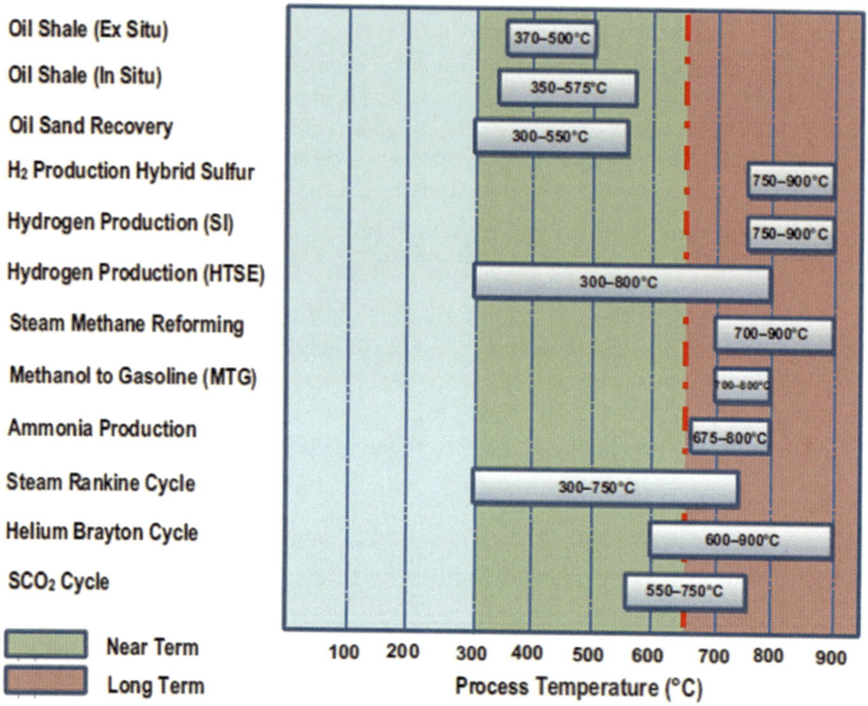

Fig. 6.23 Process applications for advanced high-temperature reactor (AHTR) versus process required temperature range [40]

shovel. In situ technology injects steam deep beneath the Earth to separate the viscous bitumen from the sand and pump it up to the surface. The bitumen then goes through the same upgrading process as it would in the mining method.

In their analyses, Sabharwall et al. (2011) [43] reported heat from an AHTR which was transferred from the reactor core by the primary liquid-salt coolant to an intermediate heat-transfer loop through an IHX. The intermediate heat-transfer loop circulates intermediate liquid-salt coolant through an SHX to move the heat to a power conversion system or for process industrial application, as shown in Fig. 6.24.

The ROT, reported by Sabharwall et al. (2011) [40], is currently 704 °C but will possibly increase to 900–1000 °C for the *n*th-of-a-kind. With its ability to provide higher ROT, the process heat application becomes an attractive option. Even though the ROT for the AHTR is 700 °C (~ 704 °C), the maximum available temperature for any process application is 650 °C, as shown by the dashed line in Fig. 6.24. Further detail on process heat applications are given in TEV-1160, "AHTR Technical Evaluation" [44].

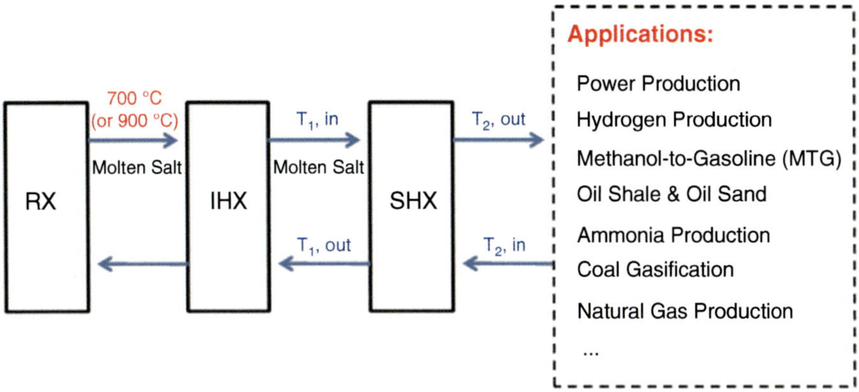

Fig. 6.24 Thermal energy transfer in an advanced high-temperature reactor (AHTR) for power or process application [40]

For their analysis, Sabharwall et al. (2011) [41] assumed the following conditions in order to carry out their calculation:

- The ROT for AHTR is assumed to be 700 °C.
- An AHTR ROT should be sufficiently larger (~ 50 °C) than the process application temperature requirement.
- Any power production/industrial application requiring greater than 650 °C is referred to as a long-term objective.
- The minimum AHTR HX temperature should be maintained high enough to avoid molten salt freezing (>500 °C), which will provide about a 50 and 65 °C temperature threshold before fluoride salts such as LiF-NaF-KF (FLiNaK) and chloride salts such as $KCl\text{-}MgCl_2$ freeze.
- Heat exchanger tube material should have sufficient mechanical integrity to sustain pressure difference across the tube wall (will depend on the application).

With the ROT of 704 °C and the maximum available temperature of 650 °C for process heat applications, the current AHTR could provide process heat for the following applications:

- Near-term integration (<650 °C):
  - Power production cycles (steam Rankine cycle, helium Brayton cycle, SCO_2 cycle)
  - Oil shale (in situ)
  - Oil shale (ex situ)
  - Oil sands.
- Long-term integration (>650 °C):
  - Hydrogen production via steam methane reforming
  - Substitute natural gas production
  - Coal-to-liquid production
  - Natural gas-to-liquid production

– Methanol-to-gasoline production
– Ammonia production.

An AHTR, when compared to gas-cooled reactors for process heat applications, benefits from a higher reactor inlet temperature because the molten salt is a more efficient heat transport medium and can transfer all of the reactor's heat with less temperature drop. This allows smaller and more efficient heat exchange equipment and produces smaller thermal stresses on those components.

References

1. C. Hongli, W. Yican, S. Konishi, J. Hayward, High temperature DEMO blanket concept for hydrogen production, 8th international symposium on fusion nuclear technology (ISFNT-8). Proceedings
2. Y. Bilge, M.S. Kazimi, Efficiency of hydrogen production systems using alternative nuclear energy technologies, International Association for Hydrogen Energy (Published by Elsevier Ltd, 2005)
3. B. Zohuri, Small modular reactors as renewable energy sources, 1st edn. (Springer Publishing Company, 2018)
4. B. Zohuri, P. McDaniel, *Combined cycle driven efficiency for next generation nuclear power plants: an innovative design approach*, 2nd edn. (Springer Publishing Company, 2018)
5. B. Zohuri, P. McDaniel, C.R.E. de Oliveira, Advanced nuclear open air-Brayton cycles for highly efficient power conversion. Nucl. Technol. **192** (2015)
6. B. Zohuri, P. McDaniel, C.R.E. de Oliveira, *A comparison of a recuperated open cycle (air) Brayton power conversion system with the traditional steam Rankine cycle for the next generation nuclear power plant* (ANS Transactions, 2014)
7. P.J. McDaniel, C.R.E. de Oliveira, B. Zohuri, J. Cole, *A Combined Cycle Power Conversion System for the Next Generation Nuclear Power Plant* (ANS Transactions, 2012)
8. P.J. McDaniel, B. Zohuri, C.R.E. de Oliveira, A combined cycle power conversion system for small modular LMFBRs, in *ANS Transactions*, (2014)
9. Technical Evaluation Study, Project No. 23843, Nuclear-Integrated Hydrogen Production. The INL is a U.S. Department of Energy National Laboratory Analysis, 05/15/10
10. http://www-pub.iaea.org/MTCD/Publications/PDF/Pub1577_web.pdf
11. http://www.world-nuclear.org/info/Non-Power-Nuclear-Applications/Transport/Transport-and-the-Hydrogen-Economy/
12. INL/EXT-09-16326 Feasibility study of hydrogen production at existing nuclear power plants, electric transportation applications
13. B. Zohuri, *Combined cycle driven efficiency for next generation nuclear power plants: an innovative design approach* (Springer Publishing Company, 2015)
14. B. Zohuri, *Innovative open-air Brayton combined cycle systems for the next generation nuclear power plants* (University of New Mexico Publications, 2014)
15. C. Forsberg, P. McDaniel, B. Zohuri, Variable electricity and steam from salt, helium, and sodium cooled base-load reactors with gas turbines and heat storage, Proceedings of ICAPP 2015 May 03–06, 2015 – Nice (France) Paper 15115
16. National Renewable Energy Laboratory (NREL) Report on "Summary of electrolytic hydrogen production", 2004
17. M. Mintz, J. Gillette, H^2A delivery scenario model and analysis, Argonne National Laboratory, February 8, 2005
18. M. W. Melaina, O. Antonia, and M. Penev, Blending Hydrogen into Natural Gas Pipeline Networks: A Review of Key Issues. Technical Report NREL/TP-5600-51995 March 2013

19. J.M. Ohi, Hydrogen codes and standards: an overview of U.S. DOE Activities, WHEC 16/13–16 June 2006 – Lyon, France
20. G.E. Besenbruch, L.C. Brown, J.F. Funk, S.K. Showalter, High efficiency generation of hydrogen fuels using nuclear power, GA Project 30047 September 2000
21. NUREG-0800, 2.2.1–2.2.2, Rev. 3, March 2007
22. International Energy Outlook 2000: .DOE/EIA-0484 (2000)]. The Energy Information Administration of the Department of Energy (www.eia.doe.gov)
23. Annual Energy Outlook 2000 With Projections to 2020: DOE/EIA-0383 (2000). The Energy Information Administration of the Department of Energy (www.eia.doe.gov)
24. Analysis of the impacts of an early start for compliance with the Kyoto Protocol: SR/OIAF/99–02. The Energy Information Administration of the Department of Energy (www.eia.doe.gov)
25. Impacts of the Kyoto Protocol on U.S. energy markets and economic activity: SR/OIAF/98–03. The Energy Information Administration of the Department of Energy (www.eia.doe.gov)
26. C. B Davis, R. B Barner, S. R Sherman, D. F Wilson, Thermal-hydraulic analyses of heat transfer fluid requirements and characteristics for coupling a hydrogen product plant to a high-temperature nuclear reactor, The INL is a U.S. Department of Energy National Laboratory operated by Battelle Energy Alliance, June 2005
27. International Atomic Energy Agency (IAEA) Nuclear Energy Series No. NP-T-42 Report, Hydrogen production using nuclear energy
28. PF. Peterson, Intermediate heat exchanger dynamic thermal response model, Eugenio Urquiza Fernández U.C. Berkeley Report UCBTH-07-006 August 31, 2007
29. F.P. Per, H. Zhao, G. Fukuda U.C. Berkeley Report UCBTH-03-004 December 5, 2003
30. B. Zohuri, *Thermal-hydraulic analysis of nuclear reactors*, 2nd edn. (Springer Publishing Company, 2017)
31. www.nuscalepower.com
32. K. Vernondern, T. Nishihara, Valuation of the safety concept of the combined nuclear/chemical complex for hydrogen production with HTTR, JUEL-4135. 2004
33. S.H. Crandall, N.C. Dahl, T.J. Lardner, *An introduction to the mechanics of solids*, 2nd edn. (McGraw-Hill Book Company, New York, 1972)
34. S.J. Dewson, B. Thonon, The development of high efficiency heat exchangers for helium gas cooled reactors, in *Paper 3213, ICAPP03*, (2003)
35. Dostal, V., M. J. Driscoll, P. Hejzlar, A supercritical carbon dioxide cycle for next generation nuclear reactors, MIT-ANP-TR-100, 2004
36. Krieth, F., Principles of heat transfer, International Textbook Company, Scranton. General Atomics, 1996, Gas Turbine-Modular Helium Reactor (GT-MHR) Conceptual Design Description Report, GA Project No. 7658, 910720 Revision 1, July 1996 (1964)
37. W.M. Kayes, M.E. Crawford, *Convective heat and mass transfer*, 2nd edn. (McGraw-Hill Book Company, New York, 1980)
38. B. Zohuri, P.J. McDaniel, Thermodynamics in nuclear power plant systems, Published by Springer Publishing Company, 2014
39. R.B. Bird, W.E. Stewart, E.N. Lightfoot, *Transport phenomena* (Wiley, New York, 1960)
40. S. Glasstone and A. Sesonske, Nuclear reactor engineering, Published by D. Van Nostrand Company, Inc. (1967)
41. https://daryanenergyblog.wordpress.com/ca/part-6_htgr/
42. R. E. Mizia, Next generation nuclear plant intermediate heat exchanger acquisition strategy, INL/EXT-08-14054, 2008
43. Sabharwall P., E. S. Kim, M. McKellar, N. Anderson, M. Patterson, Process heat exchanger options for the advanced high temperature reactor INL/EXT-11-21584 Revision 1, Idaho National Laboratory June 2011
44. Sabharwall, P., and E. S. Kim, Fluoride High Temperature Reactor Integration with Industrial Process Applications , Idaho National Laboratory, TEV-1160, 2011
45. Diehl, H. and E. Bodman, Alloy 800 Specifications in Compliance with Component Requirements, Journal of Nuclear Materials, **171**, pp. 63–70, 1990

Chapter 7
Large-Scale Hydrogen Production

The fast-paced growth of the need and demand for—and dependency on—raw materials such as hydrogen in today's industries is high on the list of the political economy of most industrial countries around the globe. In particular, besides classical applications of hydrogen in industry, we have come to realize that it is a good source of renewable energy during the on- and off-peak demand for electricity imposed on the grid by fast-growing industrial countries and their populations. However, the question of where the hydrogen comes from and how we can produce it remains. The "sustainable" routes are still too expensive. Steam reforming of hydrocarbons is considered to be the most feasible route today.

Hydrogen from bio-fuels, wind energy, solar energy, or, recently, nuclear energy is still expensive, leaving fossil fuels as the most feasible feedstock for hydrogen generation in the near term. while still researchers are thinking to use the mentioned source of energy in long term production of hydrogen. If carbon dioxide sequestration is accepted, then fossil fuels may play an important role in a future "hydrogen economy." This will happen by use of the reforming technologies. In this chapter we discuss large-scale hydrogen production in stationary plants using a steam reforming approach from a top-level point of view. For a more granular overview, readers are referred to other sources in the literature [6–8].

To understand this aspect of hydrogen production using the steam reforming process, one needs to understand combustion as well; thus, most of this chapter is dedicated to combustion and its introduction.

7.1 Introduction

Society is more progressive today than it has ever been in terms of innovation. Ranging from Moore's Law on the exponential growth of technology to the life-changing advancements in the medical field, scientists constantly push the question of how can this be better? One paramount example is the evolutionary need for more

© Springer International Publishing AG, part of Springer Nature 2019
B. Zohuri, *Hydrogen Energy*, https://doi.org/10.1007/978-3-319-93461-7_7

efficient vehicles. Indisputably, our greenhouse gas emissions have skyrocketed in recent years, making this endeavor imperative for the future. Consequently, the energy economy has been desperately looking for alternate solutions with hydrogen (H_2) being the most notable candidate as a clean gas. The only caveat with the H_2 gas alternative is that it cannot be found naturally, forcing researchers to explore methods of separating it from water (H_2O) into its constituent elements. Thus, to maximize the pollutant-free aspect of H_2, separation through fossil fuels is overlooked and rather directed towards the effects of nuclear power plants with either electrolysis or a thermodynamic process [13].

Couple options that can be considered for producing hydrogen as part of the future involving this precious, yet clean element for a green and decarbonized environment, are listed here.

- *Option 1: Production* via *Electrolysis*

Essentially, H_2 can be produced when a system generates enough electricity to heat water from liquid to steam, which is achieved through nuclear fission—the separation of certain atoms for electrical energy. Because nuclear fission coincides perfectly with the needs of electrolysis, investigation on the economic efficiency is currently underway for the three current processes: polymer electrode membrane (PEM) electrolysis, low-temperature alkaline electrolysis, and high-temperature steam electrolysis (HTSE). Both PEM and alkaline electrolysis effectively work as the inverse of each other by utilizing the attractive properties of ions and anions in relation to their respective electrodes. For example, alkaline electrolysis will take in the electricity generated from nuclear fission to break down highly ionized water into H_2 gas for extraction and usage. However, both of these methods have proven to operate with an equally lower efficiency than HTSE. As a compensation factor, the electricity demand from nuclear fissions can be significantly reduced by increasing the temperature and reaching the heat of vaporization through thermal processes, which are much more cost effective than an electrical one. By increasing the temperature, scientists are able to optimize both heat and electrical energy which maxes out at about 33% with our current nuclear power [14]. As depicted in Fig. 7.1, a feedback mechanism occurs with the feed-gas stream pumping a combination of 10% H_2 and steam into the HTSE cell (to prevent the oxidation of nickel at such a high temperature) which passes through a separator extracting about 90% of H_2 for use.

In Fig. 7.1, the gas tight electrolyte helps dispense the high heat advantageously at incremental states.

- *Option 2: Production* via *Thermochemical Cycles*

Noting the efficiency of a system at high temperatures has also shed light on the idea of a thermochemical cycle: a series of thermally driven chemical processes which decompose water into oxygen and hydrogen at rather moderate temperatures as the reverse is a direct one-step reaction requiring temperatures greater than 2800 °K for a substantial hydrogen conversion yield [14]. Supporting chemical compounds are infused internally to help recycle the system through its products

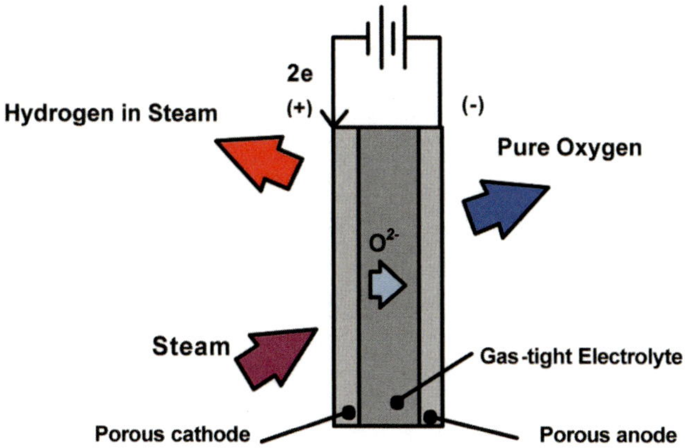

Fig. 7.1 Typical high-temperature steam electrolysis cell. (Courtesy of the US Department of Energy)

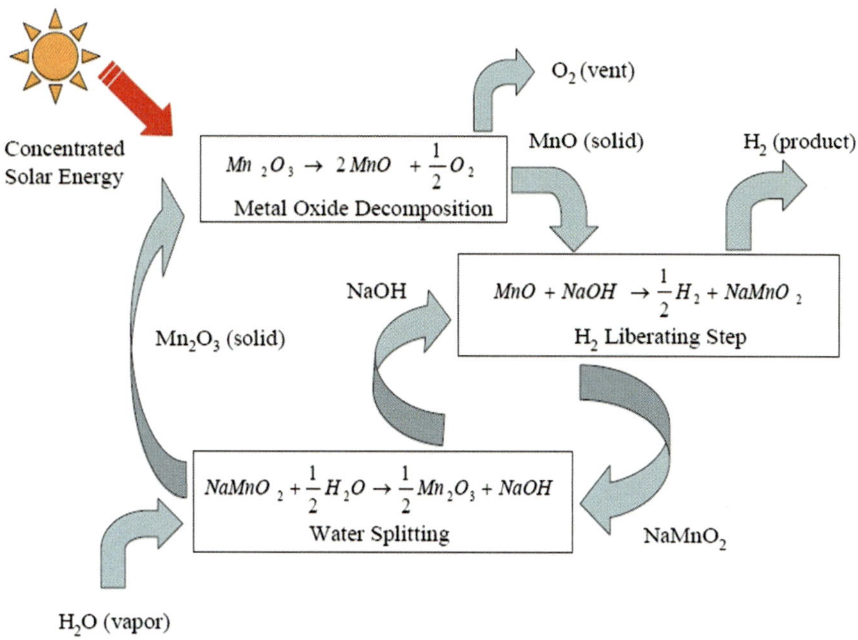

Fig. 7.2 The solar power driving process via thermochemical cycles. (Courtesy of the US Department of Energy)

with the only inputs being water and high temperature heat to help catalyze these chemical reactions. Due to the variation of usable chemical reactions to run this system, efficiency yields are still under heavy consideration. As seen in Fig. 7.2, one example that has been investigated is the MnO cycle—this cycle risks the complexity of a more than 2+ reaction cycle but is compensated with the trade-off of lower temperature requirements. Scientists are working to determine which cycle (the particular chemical substances it will consist of) will provide the maximum amount of H_2 without as many excess reagents.

In Fig. 7.2, the solar power can be used as part of heat source to drive the process, which is using more than two steps. Mn_2O_3 is reduced to MnO, which in turn goes on to react with NaOH, eventually creating $NaMnO_2$ and releasing the favored hydrogen element that we are looking to produce. This cycle is controversial in its potential complexity and is thus being investigated with other cycles for significant H_2 production. Note that *nuclear reactor power* is used as the heat source in place of *Sun power*.

A nuclear reactor operates inherently by taking as much heat as possible from nuclear fission and supplying that to generate electricity for various American activities. On average, a substantial amount of H_2 production will take about 1600 megawatt thermal (MWt) from a nuclear fission process that produces about 3000 MWt. Directly taking the thermal energy needed to power electricity for other events will have considerable transition issues as the nuclear power plant takes on new roles and intentions for society [15].

We see that, currently, the future of hydrogen is far from here—there are various factors which must be investigated thoroughly considering the implications of such an endeavor. The efficiencies are still not where they need to be in some cases for the following to be practiced immediately; however, pure running fuel is inevitably on the horizon with these evolving mechanisms.

7.2 Hydrogen Production by Steam Reforming of Hydrocarbons

Environmental movements for green reform are pushing for low sulfur gasoline and diesel fuels as a mandatory action to reduce harmful emissions drastically. Oil refiners are also encountering similar green environment conditions and they also need hydrogen in order to produce cleaner products during refining. The growing fuel cell market creates another demand for the production of hydrogen and these fuel cells will also be dependent on hydrogen as a primary fuel source.

As touched upon in Chapter 1, cells for transportation purposes, which are typically units of 50 kW to 1 MW capacity, also represent another usage for hydrogen. However, the applications in mind for such fuel cells have not grown as fast as intended and predicted, because of the return on investment (ROI) and total cost of ownership (TCO). The high investment costs and challenging technology needed for advanced gas turbine machinery is not a winning combination for fuel cells technology. The latest approach to hydrogen fuel cells with driving cells of 30 kW are more likely to attract increasing interest by manufacturers of mobile vehicles. The question is where and how to produce hydrogen that is more cost effective from an ROI point of view. The possible answers are:

1. Build large-scale hydrogen production plants that are driven by next-generation nuclear plants, solar, or wind as the source of heat energy required for production of hydrogen. Very efficient and compact intermediate heat exchangers (IHXs) will also be required, which will impact the ROI and TCO.
2. Production of hydrogen in a medium-sized plant co-existing at gas stations.
3. Smaller-scale production within mobile applications, i.e., cars.

Each of these answers imposes their own pros and cons as well as technological challenges that need to be addressed [6].

One of the common difficulties in terms of mobile applications is how to store sufficient hydrogen to have more efficient mobility in, for example, commercial vehicles, which raises the idea of having a self-sustaining hydrogen generation plant onboard with a small footprint that will make such applications feasible. With this idea in mind, having a mini-plant onboard to produce hydrogen from hydrocarbons or methanol draws our attention [9].

7.2.1 Steam Reforming Technologies

In this section we explore steam reforming technology, how it typically works, and what a typical hydrogen plant layout looks like based on this technology. As part of the hydrogen demand based on industry requirement variations, we have to understand that this element can be produced in very cost-effective manner for its consumption.

For small capacities below 0.1 million metric (MM) standard cubic feet per day (SCFD) hydrogen supplied in cylinders or production by electrolysis may be the preferred method. Hydrogen production from methanol or ammonia cracking is suitable for small, constant, or intermittent requirements as used in the food, electronics, and pharmaceutical industries. For larger capacities, hydrocarbons are used primarily as feedstock in the steam reforming process for production of hydrogen and synthesis gas.

At least in North America, the annual growth for hydrogen production is about 4% above the present production rate of 6000 MM SCFD. Most of this growth is demand by refineries for their production of ultra-low sulfur diesel and gasoline fuels. In spite of efforts to produce hydrogen by processes involving solar energy, wind energy, nuclear energy, and biofuels, fossil fuels remain the most feasible feedstock in the near term, and for commercial-scale production of pure hydrogen, steam reforming remains the most economic and efficient technology for a wide range of hydrocarbon feedstock. In order to reduce the ROI and TCO of hydrogen production for near-term needs, some refineries have installed gasification units for power production and co-generation of hydrogen as part of several existing routes for hydrogen production (see Table 7.1), and one of the important processes is their catalytic conversion method that is also known as the *steam reforming* method of hydrocarbons followed by gasification of coal, tar sands, and so on [10].

Table 7.1 Hydrogen production routes [10]

Natural gas Refinery off-gases LPG Naphtha Kerosene, gas oil	Steam reforming
Methanol, DME, NH_3	Cracking
Coal Biomass	Gasification
Water	Electrolysis

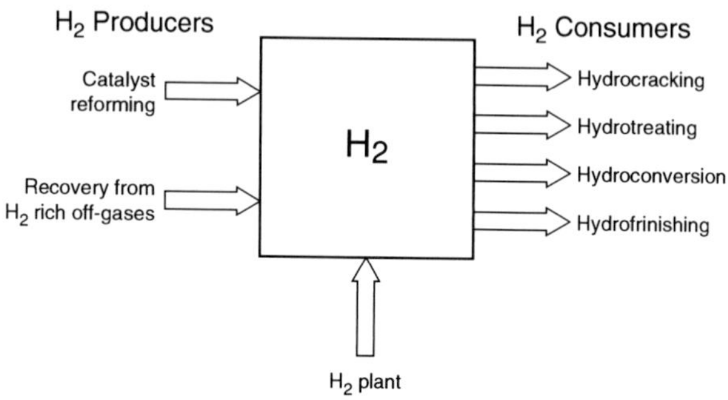

Fig. 7.3 Schematic of refinery hydrogen balance [10]

The steam reforming method for larger-scale hydrogen production plants in the short-term becomes the preferred solution cost-wise over more expensive hydrogen production plants driven by solar, wind, and nuclear energy until these plants become more efficient and their ROI and TCO are justified [11–12].

Traditionally, a major a part of the hydrogen consumption in refineries was met by hydrogen produced as a by-product from other refinery processes (110×10^9 Nm ^{3}/year), mainly catalytic reforming ("plat-forming"). A main reaction in catalytic reforming (not to be confused with catalytic steam reforming) is the conversion of paraffin into aromatics and hydrogen. As aromatics are not wanted in reformulated fuels, this means that less hydrogen will become available from catalytic reforming. Similarly, the gasoline and diesel fractions from catalytic crackers are highly unsaturated. The refinery hydrogen balance is illustrated in Fig. 7.3 [10].

In summary, there is a fast-growing need for increased hydrogen production capacity in refineries. This need is being met mainly by installation of steam reforming-based hydrogen plants.

For small-scale production, investment is a dominant factor, and simple equipment may be preferred over high-energy efficiency. Electrolysis of water accounts for less than 5% of production. For large-scale production, steam reforming of natural gas (or refinery off-gases) is the preferred solution. Gasification of heavy oil fractions may play an increasing role as these fractions are becoming more available because of falling demand.

The main application of hydrogen at the present time is as a carbon-free fuel due to green environmental and clean air aims, besides it being seen as a new source of

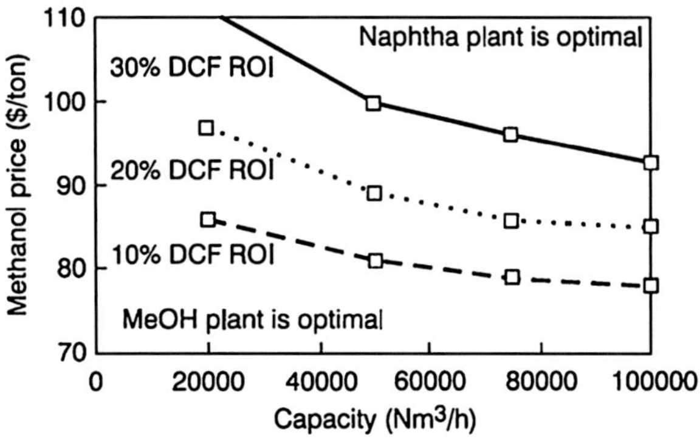

Fig. 7.4 Hydrogen production from naphtha or methanol. Naphtha price: US\$140/t. Steam credit: US\$8.3/t. When competing against natural gas at US\$13/Gcal, the methanol prices have to be about US\$10/t less than indicated. ROI means rate of return based on discounted cashflow. *DFC* direct fixed capital, *ROI* return on investment

renewable energy to meet the increased demand on electricity caused by urban growth and industry among newly industrialized countries.

As indicated at the beginning of this chapter, many technologies for production of hydrogen that do not co-produce carbon dioxide are being considered. Hydrogen production using non-fossil energy for electrolysis of water is one example. These schemes have not been introduced primarily due to reluctance concerning nuclear power and low efficiency of the electrolysis process. Hydrogen from bio-fuels, wind energy, or solar energy is still expensive, leaving fossil fuels as the most feasible feedstock for hydrogen generation in the near term, and for commercial-scale production of pure hydrogen, steam reforming remains the most economic and efficient technology for a wide range of hydrocarbon feedstocks.

In areas with a high cost of hydrocarbon feedstocks, methanol may be considered as an alternative. One possible scheme involves production of methanol in an area with very inexpensive natural gas, with subsequent transportation of the methanol to the hydrogen plant location. A methanol-based hydrogen plant is a simple unit and less costly than a natural gas and naphtha-based plant with a steam reformer. Figure 7.4 shows the conditions in which a methanol-based hydrogen plant is more economical than a naphtha-based plant [11].

A typical layout of a hydrogen plant based on steam reforming includes the following steps (reproduced from Udengaard [8]):

- Natural gas feed is preheated in coils in the waste heat section of the reformer, and sulfur is removed over a zinc oxide catalyst. Process steam is added, and the mixture of natural gas and steam is further preheated before entering the tubular reformer. Here, conversion to equilibrium of hydrocarbons to hydrogen, carbon monoxide, and carbon dioxide takes place over a nickel-based reforming catalyst.
- The gas exits the reformer and is cooled by steam production before entering the shift converter, typically a medium temperature shift. Over the shift catalyst more

hydrogen is produced by converting carbon monoxide and steam to carbon dioxide and hydrogen. The shifted gas is cooled further to ambient temperature before entering the pressure swing adsorption (PSA) unit. High-purity hydrogen product is obtained, and the off-gas from the PSA unit is used in the reformer as fuel supplemented with natural gas fuel.
- Combustion air for the tubular reformer burners can be preheated in coils in the reformer waste heat section. Part of the steam produced in the hydrogen plant is used as process steam, and the excess steam is exported.
- In many situations when natural gas is not available, higher hydrocarbons become the preferred feedstock for the reforming process. Many refineries can also benefit from flexibility in feedstock, taking advantage of the surplus of various hydrocarbon streams in the refinery.

Fired tubular reforming is generally the most competitive technology for capacities of up to more than 100 MM SCFD hydrogen.

The latest development in steam reforming process technology is the advanced steam reforming. The characteristics of this process are as follows:

- High reformer outlet temperature
- Low steam:carbon ratio
- High combustion air preheats (optional)
- Adiabatic pre-reforming (optional)
- High heat flux reformer.

Low steam:carbon ratios, typically 2.5, in hydrogen plants reduces the mass flow through the plant and thus the size of equipment. The lowest investment is therefore generally obtained for plants designed for low steam:carbon ratios. However, a low steam:carbon ratio also increases the methane leakage from the reformer. This can be compensated for by increasing the reformer outlet temperature to typically 1690 °F in hydrogen plants. Furthermore, operating at a low ratio requires the use of a non-iron-containing catalyst, that is a copper-based medium temperature shift catalyst, in order to eliminate production of by-products in the shift section.

The installation of an adiabatic pre-reformer upstream of a tubular reformer has been found to be very advantageous in naphtha-based plants and plant operating on fuel gases with higher concentrations of higher hydrocarbons. Since all higher hydrocarbons are converted over the pre-reformer catalyst, the inlet temperature of the gas inlet in the reformer can be increased to 1200 °F and the reformer can be designed for higher heat fluxes. This reduces the size of the tubular reformer, resulting in direct capital cost reduction.

High combustion air preheats temperatures and results in reduced fuel consumption and reduced steam production. The combustion air temperature can be used to adjust the steam export to a desired level. Temperatures of up to 1020 °F are industrially proven in a radiant wall reformer.

As part of the cost analysis in term of efficiency and production costs, with no steam export the theoretical energy consumption is 300 British thermal units per standard cubic foot (Btu/scf) H_2 on lower heating value (LHV) (see Sect. 7.2.2). The

Table 7.2 Heat of combustion of several fuels (MJ/kg)

	HHV	LHV	HHV:LHV ratio
Hydrogen	142	121	1.17
Methane	56	50	1.12
Gasoline	47	44	1.07
Coal (anthracite)	27	27	1.00
Wood	15	15	1.00

industrial value for natural gas-based plants is about 320 BTU/scf H_2, corresponding to 94% of the theoretical efficiency. At locations with high natural gas prices, the energy efficiency becomes critical. For a natural gas price of US$4/MM BTU, the feedstock and utility costs make about 65% of total operating costs. For further information, please refer to Udengaard [8].

7.2.2 *Heat of Combustion*

The energy that can be extracted from a fuel is often measured as the energy released as heat when the fuel undergoes complete combustion with oxygen. Table 7.2 shows this heat of combustion for several common fuels, including hydrogen.

The heat of combustion can be specified in terms of the higher heating value (HHV) or LHV. The HHV represents the entire produced heat while the LHV excludes any energy that is used to vaporize water during combustion. The difference between the two is higher for lower-carbon fuels (Table 7.2). The extent to which energy is lost to water vaporization depends on the technology; the LHV is most appropriate where large amounts of water vapor are produced at a temperature below 150 °C or where condensation of the combustion products is impractical. Both HHV and LHV are used in different sources in the literature so it is necessary to use a consistent approach.

Some countries, such as United Kingdom under their MARKAL (an acronym for MARKet ALlocation) energy efficiencies program, have theoretically calculated their data on the basis of the HHV; HHV values are used throughout this report. However, the existing data in UK MARKAL has not always been consistently calculated using HHV data; for example, the vehicle technology data uses the LHV (which inflates the costs of hydrogen storage tanks, because it effectively reduces the assumed usable energy in hydrogen). The high hydrogen HHV:LHV ratio as per Table 7.2 accentuates the discrepancies caused by using different approaches for different hydrogen technologies relative to other fuels [7].

UK MARKAL

The UK MARKAL model was originally developed to provide insights for the Energy White Paper in 2003. It was adopted and completely revised by the

(continued)

University College London (UCL) Energy Systems team in 2005 and was under constant development until 2012. The development of the original model was funded by the UK Government but the current academic version was developed and supported primarily by the UK Energy Research Centre.

UK MARKAL is a multi-time period linear optimization model. Its simplest formulation is to minimize discounted energy systems cost, under a wide variety of physical and policy constraints. This minimization takes into account evolving costs and characteristics of resources, infrastructures, technologies, taxes, and conservation measures to meet energy service demands under a range of physical and policy constraints.

MARKAL is a very large model, with 1500 technology types, 250 energy carriers plus constraints, taxes, emissions, and other model parameters. The model has well over half a million data elements.

For further information, refer to the following link: https://www.ucl.ac.uk/energy-models/models/uk-markal

For solid fuels with higher carbon content (e.g., coal and biomass), the HHV and LHV can vary substantially depending on the exact fuel composition.

For example, the efficiency of coal gasification will be substantially influenced by the composition of the coal being used. Great care must be taken to match the fuel type to the process efficiency; if necessary, the process technology should be defined several times for different fuel compositions with different process efficiencies.

7.2.3 *Reforming Reactions*

The principal process for converting hydrocarbons into hydrogen is steam reforming [6, 7], which involves the following reactions:

$$CH_4 + H_2O = CO + 3H_2 \qquad (-\Delta H_{0238} = -206\ \text{kJ/mol})$$

$$CO + H_2O = CO_2 + H_2 \qquad (-\Delta H_{0239} = \ 41\ \text{kJ/mol})$$

$$C_nH_m + nH_2O = nCO + \left(\frac{m+2n}{2}\right)H_2 \quad (-\Delta H_{0239} = -1109\ \text{kJ/mol for } nC_7H_{16})$$

The first reaction above is the steam reforming of methane. It is reversible and strongly endothermic, and according to the principle of le Chatelier it must be carried out at high temperature, high steam:methane ratio, and low pressure to achieve maximum conversion. The design of the steam reforming process is in part dictated by these constraints. The equilibrium composition out of the steam reformer is

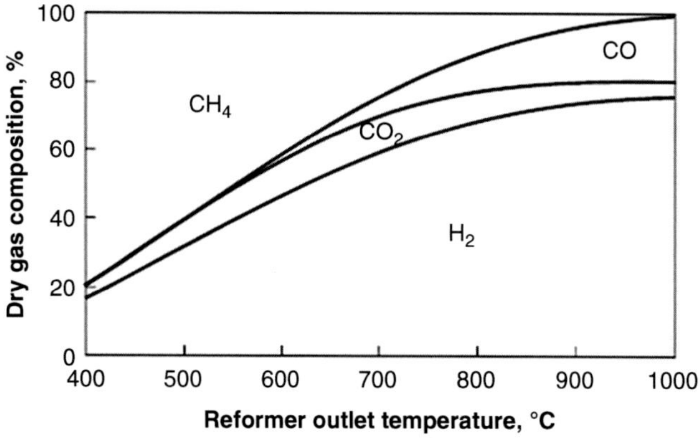

Fig. 7.5 Equilibrium composition out of a steam reformer at 26 bars with a feed steam:methane ratio of 2.5

shown in Fig. 7.5 as a function of steam reformer outlet temperature under typical industrial conditions (26 bar with a feed steam:methane ratio of 2.5).

The reader is referred to Rostrup-Nielsen and Rostrup-Nielsen [9] for further information.

7.3 Introduction to Combustion

Chemical combustion is the major source of energy used for transportation and the production of electricity. In this section, thermodynamic concepts important to the study of combustion are examined. A review of the concepts of basic property relations for ideal gases and ideal-gas mixtures and first law of thermodynamics are covered here as they are integral to the study of combustion.

Combustion is a rapid exothermic reaction that liberates substantial energy as heat and has the ability to propagate through a suitable medium. This propagation results from the strong coupling of the reaction with the molecular transport process. The chemistry and physics of combustion involves the destruction and rearrangement of certain molecules and a rapid energy release within a few millionths of a second. Currently, the study of combustion is a mature discipline and an integral element of diverse research and development programs from fundamental studies of the physics of flames and high-temperature molecular chemistry to applied engineering projects involved with developments such as advanced coal-burning equipment and improved combustion furnaces, boilers, and engines. These developments are important in optimizing fuel use and controlling the emission of pollutants.

The study of combustion starts with the *mass* and *energy balances* that bound the combustion process. Then, the energy characteristics of various important fuel resources and their physical and chemical properties are considered. Finally, the practical stoichiometry and thermochemical requirements that apply during combustion processes, including chemical reactions, equilibrium compositions and temperatures, are discussed.

Combustion is the conversion of a substance called a fuel into products of combustion by combination with an oxidizer. The combustion process is an exothermic chemical reaction—a reaction that releases energy as it occurs. Thus, combustion may be represented symbolically by:

$$\text{Fuel} + \text{Oxidizer} \rightarrow \text{Products of combustion} + \text{Energy}$$

Here the fuel and the oxidizer are reactants, that is, the substances present before the reaction takes place. This relation indicates that the reactants produce combustion products and energy. Either the chemical energy released is transferred to the surroundings as it is produced, or it remains in the combustion products in the form of elevated internal energy (temperature), or some combination thereof.

Fuels are evaluated, in part, based on the amount of energy or heat that they release per unit mass or per mole during combustion of the fuel. Such a quantity is known as the fuel's heat of reaction or heating value.

Heats of reaction may be measured in a calorimeter, a device in which chemical energy release is determined by transferring the released heat to a surrounding fluid. The amount of heat transferred to the fluid in returning the products of combustion to their initial temperature yields the heat of reaction. In combustion processes, the oxidizer is usually air but could be pure oxygen, an oxygen mixture, or a substance involving some other oxidizing element such as fluorine. Only oxygen-based oxidizers are considered here. Chemical fuels exist in gaseous, liquid, or solid form. Natural gas, gasoline, and coal are the most widely used examples of these three forms. Each is a complex mixture of reacting and inert compounds. The analysis process proceeds in three steps:

- Concepts and definitions related to element conservation;
- A definition of enthalpy that accounts for chemical bonds; and
- First-law concepts defining heat of reaction, heating values, and so on, and adiabatic flame temperature.

Combustion is actually a result of dynamic, or time-dependent, events that occur on a molecular level among atoms, molecules, radicals, and solid boundaries. The rapid reactions produce gradients that transport processes convert into heat and species fluxes that speed-up the reactions.

At the heart of fossil-fueled power plant operation is the combustion process, through which modern power plant burns fuel to release the energy that generates steam—energy that ultimately is transformed into electricity. Yet, while the combustion process is one of a power plant's most fundamental processes, it is also one of the most complex.

Combustion, or the conversion of fuel to useable energy, must be carefully controlled and managed. Only the heat released that is successfully captured by the steam is useful for generating power. Hence, the ability of the steam generator to successfully transfer energy from the fuel to steam is driven by the combustion process, or, more precisely, the characteristics of the combustion process.

A chemical reaction may be defined as the rearrangement of atoms due to redistribution of electrons. In a chemical reaction, the terms "reactants" and "products" are frequently used. *Reactants* comprise the initial constituents which start the reaction while *products* are the final constituents which are formed by the chemical reaction. Although the basic principles, which are discussed in this chapter, apply to any chemical reaction, the focus is on combustion.

7.4 Chemical Combustion

Combustion is a chemical process in which a substance reacts rapidly with oxygen and gives off heat. The original substance is called the fuel, and the source of oxygen is called the oxidizer. The fuel can be a solid, liquid, or gas. For most forms of transportation propulsion, the fuel is usually a liquid. The oxidizer, likewise, could be a solid, liquid, or gas, but is usually a gas (air). Rockets, on the other hand, usually carry their own oxidizer in addition to their fuel.

During combustion, new chemical substances are created from the fuel and the oxidizer. These substances will be called the exhaust. Most of the exhaust comes from chemical combinations of the fuel and oxygen. When a hydrogen–carbon-based fuel (such as gasoline) burns, the exhaust includes water (hydrogen + oxygen) and carbon dioxide (carbon + oxygen). However, the exhaust can also include chemical combinations from the oxidizer alone. If the gasoline is burned in air, which contains 21% oxygen and 78% nitrogen, the exhaust can also include nitrous oxides (NOX; nitrogen + oxygen). The temperature of the exhaust is high because of the heat that is transferred to the exhaust during combustion. Because of the high temperatures, exhaust usually occurs as a gas, but there can be liquid or solid exhaust products as well. Soot (i.e., a black powdery or flaky substance consisting largely of amorphous carbon, produced by the incomplete burning of organic matter), for example, is a form of solid exhaust that occurs in some combustion processes (see Fig. 7.6).

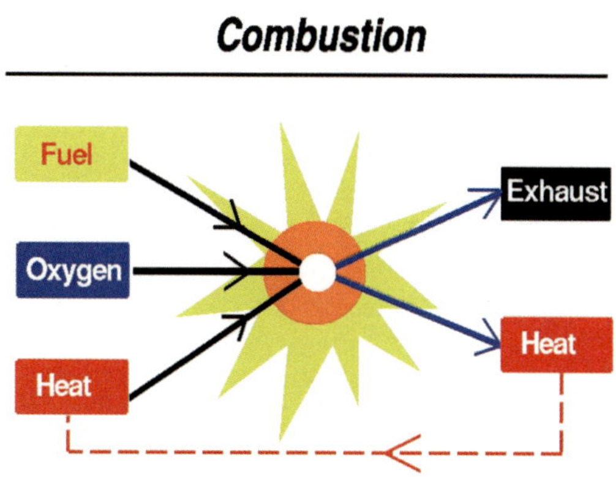

Fig. 7.6 Simple chemical combustion. (Courtesy of the National Aeronautics and Space Administration [NASA])

During the combustion process, as the fuel and oxidizer are turned into exhaust products, heat is generated. Interestingly, a source of heat is also necessary to start combustion. Gasoline and air are both present in your automobile fuel tank, but combustion does not occur because there is no source of heat. Since heat is both required to start combustion and is itself a product of combustion, we can see why combustion takes place very rapidly. Once combustion gets started, we do not have to provide the heat source because the heat of combustion will keep things going. We do not have to keep lighting a campfire, it just keeps burning.

To summarize, for combustion to occur three things must be present:

1. A fuel to be burned;
2. A source of oxygen; and
3. A source of heat.

Because of combustion, exhaust products are created, and heat is released. You can control or stop the combustion process by controlling the amount of fuel available, the amount of oxygen available, or the source of heat.

In fact, combustion is a result of dynamic, or time-dependent, events that occur on a molecular level among atoms, molecules, radicals, and solid boundaries. Therefore, this chapter presents chemical kinetics that includes kinetic theory of gases, elementary reactions, and reaction rate theory. Furthermore, the rapid reactions produce gradients that transport processes convert into heat and species fluxes that speed up the reactions.

7.5 Combustion Equations

A simple chemical-reaction equation is the combustion of propane in a pure oxygen environment. The chemical reaction is represented by:

$$C_3H_8 + 5O_2 \rightarrow 3CO_2 + 4H_2O \tag{7.1}$$

Note that the number of moles on the left-hand side may not equal the number of moles on the right-hand side. However, the number of atoms of an element must remain the same before, after, and during a chemical reaction; this demands that the mass of each element be conserved during combustion.

In writing the equation some knowledge of the products of the reaction was assumed. *Complete combustion* was assumed. The products of *complete combustion* of a hydrocarbon fuel will be H_2O and CO_2. *Incomplete combustion* results in products that contain H_2, CO, C, and/or OH.

For a simple chemical reaction, such as Eq.7.1, writing down a balanced chemical equation is straightforward. For reactions that are more complex the following systematic method proves useful [5]:

1. Set the number of moles of fuel equal to 1.
2. Balance CO_2 with number of C from the fuel.

3. Balance H_2O with H from the fuel.
4. Balance O_2 from CO_2 and H_2O.

For the combustion of propane, it was assumed that the process occurred in a pure oxygen environment. Such a combustion process normally occurs in air. Nominally, air consists of 21% O_2 and 79% N_2 by volume so that for each mole of O_2 in a reaction there are 3.76 moles of N_2.

$$\frac{79}{21} = 3.76\,\frac{\text{mol N}_2}{\text{mol O}_2} \tag{7.2}$$

Thus, on the (simplistic) assumption that N_2 will not undergo any chemical reaction, Eq. 7.1 is replaced by:

$$C_3H_8 + 5(O_2 + 3.76N_2) \rightarrow 3CO_2 + 4H_2O + 18.8N_2 \tag{7.3}$$

The minimum amount of air that supplies sufficient O_2 for the complete combustion of the fuel is called *theoretical air* or *stoichiometric air*. When complete combustion is achieved with theoretical air, the products contain no O_2, as in the reaction of Eq. 7.3. In practice, it is often found that if complete combustion is to occur, air must be supplied in an amount greater than theoretical air. This is due to the chemical kinetics and molecular activity of the reactants and products. The term *percent theoretical air* is used to compare the actual air provided to the combustion process compared to *stochiometric* air (Eq. 7.4a).

$$\%\text{theoretical air} = 100\% + \%\text{excess air} \tag{7.4a}$$

Slightly insufficient air results in CO being formed; some hydrocarbons may result from larger deficiencies [5]. So, in summary, a mixture of air and fuel is called *stoichiometric* if it contains just sufficient oxygen for the complete combustion of the fuel. Moreover, the percentage of excess air is given by Eq. 7.4b:

$$\text{Percentage excess air} = \frac{\text{Actual }(A/F)\text{ ratio} - \text{Stoichiometric }(A/F)\text{ ratio}}{\text{Stoichiometric }(A/F)\text{ ratio}} \tag{7.4b}$$

where A denotes air and F denotes Fuel.

The parameter that relates the amount of air used in a combustion process is the *air:fuel ratio* (*AF*), which is the ratio of the mass of air to the mass of fuel. The reciprocal is the *fuel:air ratio* (*FA*) (Eq. 7.5). Thus:

$$AF = \frac{m_{\text{air}}}{m_{\text{fuel}}} \qquad FA = \frac{m_{\text{fuel}}}{m_{\text{air}}} \tag{7.5}$$

Considering propane combustion with theoretical air as in (Eq. 7.3) [13], the *air: fuel ratio* is as follows (Eq. 7.6).

$$AF = \frac{m_{\text{air}}}{m_{\text{fuel}}} = \frac{(5)(4.76)(29)}{(1)(44)} = 15.69\,\frac{\text{kg air}}{\text{kg fuel}} \tag{7.6}$$

where the molecular weight of air is taken as 29 kg/kmol and that of propane as 44 kg/kmol. If for the combustion of propane, $AF > 15.69$, a *lean* or *weak mixture* occurs; if $AF < 15.69$, a *rich mixture* results.

For solid and liquid fuels, the ratio is expressed by mass, while for gaseous fuels the ratios are normally expressed by volume. For boiler plant, the mixture is usually greater than 20% lean; for gas turbines, it can be as much as 300% lean. Petrol engines have to meet various conditions of load and speed and operate over a wide range of mixture strengths. The following definition is then used (Eq. 7.7):

$$\text{Mixture strength} = \frac{\text{Stoichiometric } (A/F) \text{ ratio}}{\text{Actual } (A/F) \text{ ratio}} \tag{7.7}$$

In this situation, the working values range between 80% (lean) and 120% (rich). Where fuels contain some oxygen (e.g., ethyl alcohol C_2H_6O), this oxygen is available for the combustion process, and so the fuel requires a smaller supply of air [6].

The combustion of hydrocarbon fuels involves H_2O in the products of combustion. The calculation of the dew point of the products is often of interest; it is the saturation temperature at the partial pressure of the water vapor. If the temperature drops below the dew point, the water vapor begins to condense. The condensate usually contains corrosive elements, and thus it is often important to ensure that the temperature of the products does not fall below the dew point.

Example 7.1 Butane is burned with dry air at an air:fuel ratio of 20. Calculate (a) the percent excess air; (b) the volume percentage of CO_2 in the products; and (c) the dew-point temperature of the products.

Solution The reaction equation for theoretical air is:

$$C_4H_{10} + 6.5(0_2 + 3.76N_2) \rightarrow 4CO_2 + 5H_2O + 24.44N_2$$

(a) The air:fuel ratio for theoretical air is:

$$AF = \frac{m_{\text{air}}}{m_{\text{fuel}}} = \frac{(6.5)(4.76)(29)}{(1)(58)} = 15.47 \frac{\text{kg air}}{\text{kg fuel}}$$

This represents 100% theoretical air. The actual air:fuel ratio is 20. The excess air is then:

$$\%\text{excess air} = \left(\frac{AF_{\text{act}} - AF_{\text{th}}}{AF_{\text{th}}}\right)(100\%) = \frac{20 - 15.47}{15.47}(100\%) = 29.28\%$$

(b) The reaction equation with 129.28% theoretical air is:

$$C_4H_{10} + (6.5)(1.2928)(O_2 + 3.76N_2) - +4CO_2 + 5H_2O + 1.9030_2 + 31.6N_2$$

The volume percentage is obtained using the total moles in the products of combustion. For CO_2 we have:

$$\%CO_2 = \left(\frac{4}{42.5}\right)(100\%) = 9.41\%$$

(c) To find the dew-point temperature of the products we need the partial pressure of the water vapor. It is found using the mole fraction to be:

$$p_{H_2O} = y_{H_2O} p_{atm} = \left(\frac{5}{42.5}\right) * 101.325 \text{ kPa} = 11.76 \text{ kPa}$$

where we have assumed an atmospheric pressure of 101.325 kPa. Using Appendix A-14.2 of Zohuri and McDaniel. [13] we find the dew-point temperature to be $T_{d \cdot p \cdot}$ = 49 °C.

When a chemical reaction occurs, there may be considerable change in the chemical composition of a system. The problem this creates is that for a control volume the mixture that exits is different from the mixture that enters.

7.6 Mass and Mole Fractions

The amount of a substance present in a sample may be indicated by its mass or by the number of moles of the substance. A *mole* is defined as the mass of a substance equal to its molecular mass or molecular weight. Molecular weights for substances of interest are given in the Appendix of Zohuri and McDaniel [13]. Compound molecular weights can be obtained by adding up the atomic weights of the constituents.

The composition of a mixture may be given as a list of the fractions of each of the substances present. Thus, we define the mass fraction, of a component i, mf_i, as the ratio of the mass of the component, m_i, to the mass of the mixture (Eq. 7.8), m:

$$mf_i = \frac{m_i}{m} \tag{7.8}$$

It is evident that the sum of the mass fractions of all the components must be 1. Thus (Eq. 7.9):

$$mf_1 + mf_2 + \cdots = 1 \tag{7.9}$$

The mole fraction of component i, x_i, is the ratio of the number of moles of component i, n_i, to the total number of moles in the mixture (Eq. 7.10), n:

$$x_i = \frac{n_i}{n} \tag{7.10}$$

The total number of moles, n, is the sum of the number of moles of all the components of the mixture (Eq. 7.11):

$$n = n_1 + n_2 + n_3 \ldots \quad (7.11)$$

It follows that the sum of all the mole fractions of the mixture (Eq. 7.12) must equal 1.

$$x_1 + x_2 + \cdots = 1 \quad (7.12)$$

The mass of component i in a mixture is the product of the number of moles of i and its molecular weight, M_i. The mass of the mixture is therefore the sum, $m = n_1M_1 + n_2M_2 + \cdots$ over all components of the mixture. Substituting x_in for n_i, the total mass becomes (Eq. 7.13):

$$m = (x_1M_1 + x_2M_2 + \cdots)n \quad (7.13)$$

But the average molecular weight of the mixture is the ratio of the total mass to the total number of moles. Thus, the average molecular weight is as follows (Eq. 7.14):

$$M = m/n = x_1M_1 + x_2M_2 + \cdots \quad (7.14)$$

Example 7.2 Express the mass fraction of component 1 of a mixture in terms of: (a) the number of moles of the three components of the mixture, n_1, n_2, and n_3, and (b) the mole fractions of the three components. (c) If the mole fractions of carbon dioxide and nitrogen in a three-component gas containing water vapor are 0.07 and 0.38, respectively, what are the mass fractions of the three components?

Solution (a) Because the mass of i can be written as $m_i = n_iM_i$, the mass fraction of component i can be written as:

$$mf_i = \frac{n_iM_i}{n_1M_1 + n_2M_2 + n_3M_3 + \ldots}$$

For the first of the three components, $i = 1$, this becomes:

$$mf_1 = \frac{n_1M_1}{n_1M_1 + n_2M_2 + n_3M_3}$$

Similarly, for $i = 2$ and $i = 3$:

$$mf_2 = \frac{n_2M_2}{n_1M_1 + n_2M_2 + n_3M_3}$$

$$mf_3 = \frac{n_3M_3}{n_1M_1 + n_2M_2 + n_3M_3}$$

(b) Substituting$n_1 = x_1 n_1$, $n_2 = x_2 n_2$, etc. in the earlier equations and simplifying, we obtain for the mass fractions:

$$mf_1 = x_1 M_1/(x_1 M_1 + x_2 M_2 + x_3 M_3)$$
$$mf_2 = x_2 M_2/(x_1 M_1 + x_2 M_2 + x_3 M_3)$$
$$mf_3 = x_3 M_3/(x_1 M_1 + x_2 M_2 + x_3 M_3)$$

(c) Identifying the subscripts 1, 2, and 3 with carbon dioxide, nitrogen, and water vapor, respectively, we have $x_1 = 0.07$, $x_2 = 0.38$, and $x_3 = 1 - 0.07 - 0.038 = 0.55$. Then:

$$\begin{aligned} mf_1 &= (0.07)(44)/[(0.07)(44) + (0.38)(28) + (0.55)(18)] \\ &= (0.07)(44)/(23.62) = 0.1304 \\ mf_2 &= (0.38)(28)/(23.62) = 0.4505 \\ mf_3 &= (0.55)(18)/(23.62) = 0.4191 \end{aligned}$$

As a check we sum the mass fractions: 0.1304 + 0.4505 + 0.4191 = 1.0000.

For a mixture of gases at a given temperature and pressure, the ideal gas law shows that $pV_i = n_i \mathfrak{R} T$ holds for any component, and $pV = n\mathfrak{R}T$ for the mixture as a whole. Forming the ratio of the two equations, we observe that the mole fractions have the same values as the volume fraction (Eq. 7.7):

$$x_i = V_i/V = n_i/n \tag{7.15}$$

Similarly, for a given volume of a mixture of gases at a given temperature, $pV_i = n_i \mathfrak{R} T$ for each component and $pV = n\mathfrak{R}T$ for the mixture. The ratio of the two equations shows that the *partial pressure* of any component i is the product of the mole fraction of i and the pressure of the mixture (Eq. 7.16):

$$p_i = pn_i/n = px_i \tag{7.16}$$

Example 7.3 What is the partial pressure of water vapor in Example 7.2 if the mixture pressure is two atmospheres?

Solution The mole fraction of water vapor in the mixture of Example 7.2 is 0.55. The partial pressure of the water vapor is therefore (0.55) (2) = 1.1 atm.

7.7 Enthalpy of Formation

In order to deal with the heat produced in a chemical reaction, a reference point is required so that changes in enthalpy can be computed. The standard reference conditions are 25 °C (77 °F, 298 K, 537 °R) and 1 atmosphere pressure. The enthalpy of a substance at the reference state is usually identified as h^0. At these conditions many of the elements in their normal form are defined to have a 0.0 enthalpy of formation or *heat of formation*, h_f^0. Typically, this includes gases such as oxygen, nitrogen, and hydrogen as well as the solid form of carbon. Other gases such as carbon dioxide and water vapor have a negative heat of formation at standard conditions. This means that when they are formed by burning carbon with oxygen, or hydrogen with oxygen, a certain amount of energy will be given off. The reaction is exothermic. When oxygen, nitrogen, or hydrogen is decomposed to oxygen ions, nitrogen ions, or hydrogen ions, energy is required. The heat of formation for these reactions is positive because energy must be added to the molecule to break it down into ions. Therefore, they are endothermic reactions.

$$\begin{aligned}
&\mathrm{H_2} + \tfrac{1}{2}\mathrm{O_2} \rightarrow 2\mathrm{H_2O} && -241{,}820\ \mathrm{kJ/kg\text{-}mol}\\
&\mathrm{C} + \mathrm{O_2} \rightarrow \mathrm{CO_2} && -393{,}520\ \mathrm{kJ/kg\text{-}mol}\\
&\mathrm{O_2} \rightarrow 2\mathrm{O} && 249{,}170\ \mathrm{kJ/kg\text{-}mol}\\
&\mathrm{N_2} \rightarrow 2\mathrm{N} && 472{,}680\ \mathrm{kJ/kg\text{-}mol}\\
&\mathrm{H_2} \rightarrow 2\mathrm{H} && 218{,}000\ \mathrm{kJ/kg\text{-}mol}
\end{aligned}$$

The negative sign for the heats of formation means that when the reaction occurred, energy was given up by the reactants.

The First Law for a chemical reaction can be written as Eq. 7.17:

$$Q = H_P - H_R \tag{7.17}$$

H_P is the enthalpy of the products of combustion that leave the combustion chamber and H_R is the enthalpy of the reactants that enter the combustion chamber. If the reactants are stable elements and the reaction occurs at constant temperature and pressure at the reference state (77 °F and 1 atm) then each H represents the heat of formation for the substances involved. If the temperature deviates from the reference state, each of the enthalpies must be corrected for the temperature changes.

The general equation for a flowing system is Eq. 7.18:

$$Q - W_s = \sum_{\text{prod}} N_p\left[h_{\mathrm{f}}^o + (h(T) - h^o)\right]_p - \sum_{\text{react}} N_r\left[h_{\mathrm{f}}^o + (h(T) - h^o)\right]_\gamma$$

$$N_p = \text{moles of products} \quad N_r = \text{moles of reactants} \tag{7.18}$$

$$h(T) - h^o = \text{Change in enthalpy from the reference state}$$

The general equation for a rigid chamber is Eq. 7.19 [13]:

$$
\begin{aligned}
Q - W_s = U_p - U_r &= \sum_{\text{prod}} N_p\left[h_f^o + (h(T) - h^o) - Pv\right]_p \\
&\quad - \sum_{\text{react}} N_r\left[h_f^o + (h(T) - h^o) - Pv\right]_r \\
Q - W_s = U_p - U_r &= \sum_{\text{prod}} N_p\left[h_f^o + (h(T) - h^o) - \Re\mathrm{T}\right]_p \\
&\quad - \sum_{\text{react}} N_r\left[h_f^o + (h(T) - h^o) - \Re\mathrm{T}\right]_r
\end{aligned}
\tag{7.19}
$$

The changes in enthalpy from the reference state can be calculated by the following techniques.

For a solid or liquid:

$\Delta h = C\Delta T$

For gases:

$$\Delta h = C_p \Delta T$$

1. Use tabulated values for Δh.

2. Use generalized charts for a real gas.

3. Use tables for vapors such as the steam tables.

Example 7.4 Volumetric analysis of the products of combustion of an unknown hydrocarbon measured on a dry basis gives the following mole percent:

$$
\begin{aligned}
CO_2 &= 10.4\% \\
CO &= 1.2\% \\
O_2 &= 2.8\% \\
N_2 &= 85.6\%
\end{aligned}
$$

Determine the composition of the hydrocarbon and the percent theoretical air.

Solution The combustion equation is:

$$C_aH_b + c(O_2 + 3.76N_2) \rightarrow 10.4CO_2 + 1.2CO + 2.8O_2 + 85.6N_2 + dH_2O$$

Writing equations to balance each of the species gives:

C :	a = 10.4 + 1.2	a = 11.6
N :	3.76c = 85.6	c = 22.8
O :	2c = 10.4(2) + 1.2 + 2.8(2) + d	d = 2(22.8) − 20.8 − 1.2 − 5.6 = 18
h :	b = 2d = 36	

The actual equation for 100% theoretical air must be:

$$C_{11.6}H_{36} + 20.6(O_2 + 3.76N_2) \rightarrow 11.6CO_2 + 77.5N_2 + 18H_2O$$

22.8 moles of air were used and only 20.6 were needed, so percent theoretical air = 110.7%.

Example 7.5 Methane is burned with dry air at an air:fuel ratio of 5. Calculate the percent excess air and the percentage water vapor in the exhaust. Estimate the dew point temperature of the products.

Solution The combustion equation is:

$$CH_4 + 2O_2 + 2(3.76)N_2 \rightarrow 2H_2O + CO_2 + 2(3.76)N_2$$

The stochiometric air-fuel mixture is: $AF_{st} = \frac{2(28.9669)}{16.043} = 3.611$.
The percent excess air is 100 × (5–3.611)/3.611 = 38.47%. So the balance equation is:

$$CH_4 + 1.385(2O_2 + 2(3.76)N_2) \rightarrow 2H_2O + CO_2 + 0.77O_2 + 10.4N_2$$

There are 2 + 1 + 0.77 + 10.4 = 14.17 moles of products.
The mole percent of H_2O in the exhaust is 2/14.17 = 14.1%.
The partial pressure of H_2O in the exhaust is (2/14.17) × 101,325 = 14.3 kPa.
The saturation temperature (dew point) at 14.3 kPa is 326 K = 53 °C.

Example 7.6 Gaseous methyl alcohol and air enter a combustion chamber at 25 °C and 1 atm and leave at 550 K and 1 atm. Assume 150% theoretical air. Estimate the heat transfer to the chamber.

Solution The reaction equation is:

$$CH_3OH + 1.5\{1.5[O_2 + (3.76)N_2]\} \rightarrow 2H_2O + CO_2 + 0.75O_2 + 2.25(3.76)N_2$$

The First Law gives:

$$Q = \sum_{\text{prod}} N_p\left(h_f^o + \Delta h\right)_p - \sum_{\text{react}} N_r\left(h_f^o + \Delta h\right)_r$$

Products:

$$\begin{array}{ll} H_2O & h_f = -241,820 \Delta h = h(550) - h(298) = 17,489.1 - 8,853.3 = 8635.8 \\ CO_2 & h_f = -393,520 \Delta h = h(550) - h(298) = 18,878.8 - 8,378.4 = 10, \\ 500.4O_2 & h_f = 0 \Delta h = h(550) - h(298) = 15,363.4 - 7,766.2 = 7,597.2 \\ N_2 & h_f = 0 \Delta h = h(550) - h(298) = 15,095.3 - 7,754.3 = 7,431.0 \end{array}$$

Reactants:

$$CH_3OH \quad h_f = -200,890$$

The balance equation is:

$$\begin{aligned} Q = &\left(2^*(-241,820 + 8635.8) + (-393,520 + 10,500.4)\right. \\ &+ 8.46^*7431.0 + 0.75^*7597.2 - (-200,890)Q \\ = &-579,930 \frac{\text{kJ}}{\text{kmol}} \textit{ of Methyl Alcohol} \end{aligned}$$

Table 7.2-1 Selected enthalpies of combustion and enthalpies of vaporization

Substance	Formula	Higher heating value (kJ/kmol)	h_{fg}
Hydrogen	$H_2(g)$	−285,840	
Carbon	C(s)	−393,520	
Carbon monoxide	CO(g)	−282,990	
Methane	$CH_4(g)$	−890,360	
Acetylene	$C_2H_2(g)$	−1,299,600	
Ethylene	$C_2H_4(g)$	−1,410,970	
Ethane	$C_2H_6(g)$	−1,559,900	
Propylene	$C_3H_6(g)$	−2,058,500	
Propane	$C_3H_8(g)$	−2,220,000	15,060
n-Butane	$C_4H_{10}(g)$	−2,877,100	21,060
n-Pentane	$C_5H_{12}(g)$	−3,536,100	26,410
n-Hexane	$C_6H_{14}(g)$	−4,194,800	31,530
n-Heptane	$C_7H_{16}(g)$	−4,853,500	36,520
n-Octane	$C_8H_{18}(g)$	−5,512,200	41,460

The above calculation assumes the water in the exhaust is liquid, which at the temperature of the exhaust products is not very likely. So an additional enthalpy must be added to the Δh for H_2O to account for the vaporization of water.

$$\Delta h_{\text{fg}} = 2256.6 \ \text{kJ/kg} = 40,619.0 \ \text{kJ/Kmol}$$

Adding this to the negative Q gives $Q = -539{,}311$ kJ/kmol *of Methyl Alcohol*

Note that the heat of formation for nitrogen and oxygen is zero on both sides of the equation. If the methyl alcohol had entered as a liquid, we would have had to add its heat of vaporization to the reactant side of the equation, further reducing the heat available per mole of the fuel.

7.8 Enthalpy of Combustion

With most hydrocarbons, normal combustion occurs with oxygen in the air. Therefore, the enthalpy change for the complete combustion of a substance with oxygen is called the *heat of combustion*. Several heats of combustion are given in Table 7.2. If the products of combustion contain water in the vapor state, an allowance for the heat required to vaporize the water must be included in the change in enthalpy from the reference state. If the products of combustion include water in the liquid state, the heat of vaporization is not subtracted, and this gives the HHV for the heat of combustion for this fuel.

7.9 Adiabatic Flame Temperature

In many cases, the heat released in the combustion reaction will determine the final temperature of the products and the gases present in the combustion chamber. The temperature achieved assuming no heat transfer to the surroundings is called the adiabatic flame temperature. Since the change in enthalpy of the products will depend on this temperature, the adiabatic flame temperature must be found by iteration. Note also that when a fuel is burned in air, the heat required to bring the nitrogen in the air up to the adiabatic flame temperature must be included in the analysis. The adiabatic flame temperature can be lowered by adding excess air above that required for complete combustion. Typically, this will be required due to peak temperature restrictions on the materials used for construction, for example the strength of turbine blades downstream of the combustion process.

It is worth pointing out at this point that combustion in itself is an irreversible process and to achieve complete combustion typically requires a pressure loss in the combustion chamber. Since the major part of the fluid flowing through an air combustion chamber does not participate in the chemical reactions (primarily the nitrogen), the combustion process can be thought of as simply heating the working fluid. A nuclear heated heat exchanger can often accomplish the same heating with a lower pressure loss. However, in the nuclear heated system the temperature drops are in the opposite direction requiring heat exchanger walls to operate at a higher temperature than the combustion chamber walls.

Example 7.7 Calculate the enthalpy of combustion of gaseous octane and liquid octane assuming the reactants and products to be at the reference state of 25 °C and 1 atmosphere. Assume liquid water in the products exiting the steady-flow combustion chamber.

Solution The reaction is:

$$C_8H_{18} + 12.5(O_2 + 3.76N_2) \rightarrow 8CO_2 + 9H_2O + 47N_2$$

Products:

$$\begin{aligned} &H_2O \quad h_f = -285,830 \\ &CO_2 \quad h_f - 393,520 \\ &N_2 \quad\;\; h_f = 0 \end{aligned}$$

Reactants:

$$\begin{aligned} &C_8H_{18}(l) \quad h_f = -208,450 \quad h_{fg} = 41,460 \\ &C_8H_{18}(g) \quad h_f = -208,450 \end{aligned}$$

Liquid octane:

$$\begin{aligned} Q &= 8(-393,520) + 9(-285,830) - (-208,450 - 41,460) \\ &= 5.4707 \times 10^6 \ \text{kJ/kmol} \end{aligned}$$

Gaseous octane:

$$Q = 8(-393,520) + 9(-285,830) - (208,450) = 5.5122 \times 10^6 \text{ kJ/kmol}$$

Example 7.8 Kerosene is burned with theoretical air in a jet engine. Estimate the adiabatic flame temperature. Kerosene can be treated as n-Dodecane—$C_{12}H_{26}$—and has a heat of formation of −291,010 kJ/kmol.

Solution The balance equation is:

$$C_{12}H_{26} + 18.5(O_2 + 3.76N_2) \rightarrow 12CO_2 + 13H_2O + 69.5N_2$$

For the adiabatic flame temperature $Q = 0$, so.
Products:

$$\begin{array}{lll} H_2O & h_f = -241,820 & \Delta h(H2O) = ? \\ CO_2 & h_f = -393,520 & \Delta h(CO2) = ? \\ N_2 & & \Delta h(N2) = ? \end{array}$$

Reactants:

$$C_{12}H_{26} \; h_f = -291,010$$

$$\begin{aligned} Q = 0 = {} & 12(-393,520 + \Delta h(CO_{21})) + 13(-241820 + \Delta h(H_2O)) \\ & + 69.5\Delta h(N_2) - (-291,010) \end{aligned}$$

At 25 °C we would have:

$$Q = 12(-393,520) + 13(-241,820) + 291,010 = 7.5749 \times 10^6 \text{ kJ/kmol}$$

Treating all of the products as nitrogen the dominant product, we would have 94.5 kg-moles of product. Thus, the change in enthalpy per mole is 80,158 kJ/kmol. At 25 °C nitrogen has an enthalpy of 7754.3 kJ/kmol. So the gas tables are entered looking for an enthalpy of 87,912 kJ/kmol. This corresponds to a temperature of 2660 K. Then evaluating the Δh values for H_2O and CO_2 at 2660 K gives:

$$\Delta h(H_20) = 116547.5 - 8853.3 = 107690.0$$

$$\Delta h(CO_2) = 140041.3 - 8378.4 = 131660.0$$

$$\Delta h(N_2) = 87846.0 - 7754.3 = 80092.0$$

At 2660 K the net heat balance is:

$$\begin{aligned} Q &= -7.5749 \times 10^6 + 13(107690.0) + 12(131660.0) + 69.5(80092.0) \\ &= 97,138 \text{ kJ/kmol} \end{aligned}$$

Since it is positive, the temperature does not quite reach 2660 K. Now noting that both H_2O and CO_2 had larger enthalpy changes than nitrogen, the number of moles can be adjusted to represent them with the all nitrogen model.

$$\Delta h(\mathrm{H_2O})/\Delta h(\mathrm{N_2}) = 107690.0/80092 = 1.35$$

$$\Delta h(\mathrm{CO_2})/\Delta h(\mathrm{N_2}) = 131660/80092 = 1.64$$

So, the new number of moles will be $N = 69.5 + 1.35(13) + 1.64(12) = 106.7$.

Dividing the 25 °C enthalpy excess by 106.7 gives $7.5749 \times 10^6/106.7 = 71{,}017$ kJ/kmol.

Add this to the nitrogen enthalpy at 25 °C to get 78,771 kJ/kmol. The gas table gives a temperature of about 2410 K.

$$\Delta h(\mathrm{H_2O}) = 103045.9 - 8853.3 = 94,193$$

$$\Delta h(\mathrm{CO_2}) = 124671.2 - 8378.4 = 116,290.0$$

$$\Delta h(\mathrm{N_2}) = 78,688.1 - 7754.3 = 70,934.0$$

$$\begin{aligned} Q &= -7.5749 \times 10^6 + 13(94,193) + 12(116,290) + 69.5(70934) \\ &= -24,998 \ \ \mathrm{kJ/kmol} \end{aligned}$$

So now the temperature is bracketed. The new mole effectiveness ratios for H_2O and CO_2 are 1.33 and 1.64. Therefore, 2410 °K is very close and in fact the best answer obtained by another iteration is $T = 2416.3$ K, which is probably 2 or 3 too many digits for the round-offs that have been made in the solution process.

It is obvious that this would be a very stressing temperature, and so most jet engines are running at a mixture ratio quite a bit lower than stochiometric.

References

1. F. El-Mahallawy, S. El-Din Habik, *Fundamentals and Technology of Combustion* (Elsevier, Amsterdam, 2002)
2. V.A. Kirillin, V.V. Sychev, A.E. Sheindlin, *Engineering Thermodynamics* (Mir Publishers, Moscow, 1976)
3. World Energy Outlook, WEO, Chapter. 3, International Energy Agency, 1998 37. Last Edited by Jonathan Cobb at 28/01/2011
4. M.C. Potter, C.W. Somerton, *Thermodynamics for Engineers*, 2nd edn. (McGraw-Hill Schaum's Outlines Series, Boston, 2006)
5. T.D. Eastop, A. McConkey, *Applied Thermodynamics for Engineering Technologists*, 5th edn. (Pearson Education Ltd, London, 1993)
6. http://www.topsoe.com/sites/default/files/topsoe_large_scale_hydrogen_produc.pdf
7. P.E. Dodds, W. McDowall, *A review of hydrogen production technologies for energy system models*, UKSHEC Working Paper No. 6, 2012, UCL Energy Institute, (University College London Central House, 14 Upper Woburn Place London WC1H 0NN, UK)

8. N.R. Udengaard, *Hydrogen Production by Steam Reforming of Hydrocarbons* (Haldor Topsoe Inc, Houston, 1997)
9. J.R. Rostrup-Nielsen, T. Rostrup-Nielsen, Phys Chem Phys **3**, 283 (2001)
10. R. Pitt, World Refining, Jan/Feb. (2001), p. 6
11. B. Zohuri, *Combined Cycle Driven Efficiency for Next Generation Nuclear Power Plants: an Innovative Design Approach* (Springer Publishing Company, Cham, 2015)
12. B. Zohuri, *Innovative Open-air Brayton Combined Cycle Systems for the Next Generation Nuclear Power Plants* (University of New Mexico Publications, Albuquerque, 2014)
13. B. Zohuri, P. McDaniel, *Thermodynamics in Nuclear Power Plant Systems*, 1st edn. (Springer Publishing Company, New York, 2015)
14. S. Abanades et al., Screening of water-splitting thermochemical cycles. Energy **31**, 2805 (2006)
15. G. Rothwell, E. Bertel, and K. Williams, Can nuclear power compete in the hydrogen economy? in *Nuclear production of hydrogen: Third Information Exchange Meeting, Oarai, Japan, 5–7 October 2005* (OECD, 2006)

Chapter 8
Hydrogen Storage Processes and Technologies

Hydrogen storage is a significant challenge for the development and viability of hydrogen-powered vehicles. Onboard hydrogen storage in the range of approximately 5–13 kg is required to enable a driving range of greater than 300 miles for the full platform of light-duty automotive vehicles using fuel cell power plants. In addition to production and distribution, costs are associated with hydrogen storage. Little public information is available on the cost of bulk gas storage, meaning a significant error margin exists for the assumptions on storage costs given in this chapter. Hydrogen storage is a key enabling technology for the advancement of hydrogen and fuel cell technologies in applications such as stationary power, portable power, and transportation. Hydrogen has the highest energy per mass of any fuel; however, its low ambient temperature density results in a low energy per unit volume, requiring the development of advanced storage methods that have the potential for higher energy density.

8.1 Introduction

Increased energy demand worldwide while simultaneously sources of fossil fuels are becoming depleted will eventually lead to an increased share of renewable energies within the global energy portfolio. Long-term national energy strategies are reflecting this trend already today.

Hydrogen is becoming increasingly acknowledged as the energy carrier of choice for the twenty-first century. Clean and inexhaustible, substitution of hydrogen for petroleum for use as an automotive fuel would largely eliminate smog in inner cities and health concerns related to airborne particulates, and would reduce the dependence on oil reserves. Coupled with the high efficiency of proton exchange membrane (PEM) fuel cells as an automotive power plant, simultaneous significant

© Springer International Publishing AG, part of Springer Nature 2019

B. Zohuri, *Hydrogen Energy*, https://doi.org/10.1007/978-3-319-93461-7_8

increases in vehicle fuel economy can be made. Indeed, the recent flurry of strategic alliances in the automotive fuel cell world and the automotive companies that are involved such as Ford, Daimler-Benz, Ballard, General Motors, and Toyota attests to the seriousness with which the automotive industry views fuel cell propulsion, and since fuel cells fundamentally depend on hydrogen fuel, attests to the increasing importance and prominence of hydrogen production, storage, and distribution for the future.

Matching of supply and demand of wind and solar energy as the renewable energy resources having the largest technical potential requires the use of storage technologies capable of storing electric energy. Both short-term storage and long-term seasonal-type storage need to be considered. While short-term storage can be technically handled using established storage technologies such as pumped hydro and batteries as well as advanced technologies such as adiabatic compressed air storage, seasonal energy storage requires different solutions.

Clearly the hydrogen star is rising. What is less certain is whether the hydrogen bulk supplier community will be able to accommodate the increased consumer demand for hydrogen in the near-, mid-, and far-term timeframes. The use of liquid hydrocarbons (gasoline, methanol, DME) in onboard reformer fuel cell vehicles is under active development as an interim step to a full hydrogen economy. Such onboard chemical reformation plants capable of converting widely available fuels such as gasoline into a hydrogen-rich reformate stream for use by the fuel cell have the major advantage of not requiring major fuel infrastructure alternations. However, creating a load following, highly efficient, compact, low-cost reformer is technically challenging and necessarily compromises vehicle system performance compared to a pure hydrogen system. For this reason, onboard hydrocarbon reformation is viewed as an interim link between today's gasoline internal combustion engine automobiles and tomorrow's pure hydrogen fuel cell vehicles.

Whether for vehicular onboard storage or stationary bulk storage, the storage of hydrogen has been problematic due to hydrogen's low volumetric density and resulting high cost. This chapter outlines current hydrogen storage techniques for both vehicular and stationary storage and discusses future hydrogen research trends.

The Fuel Cell Technology Office (FCTO) under the United States Department of Energy (DOE) is developing onboard automotive hydrogen storage systems that allow for a driving range of more than 300 miles, while meeting cost, safety, and performance requirements.

As we have learned in previous chapters of this book, in addition to production and distribution, costs are associated with hydrogen storage. Little public information on the cost of bulk gas storage is available. Therefore, a significant error margin exists for the following assumptions on storage cost.

The initial investment cost for hydrogen storage can be as low as €13/kWh based on low heating value (LHV) when using compressed gas storage in steel cylinders assuming a cylinder cost of €350 only.

What Is Low Heating Value and High Heating Value?
Low heating value (LHV) calculations assume that the water component of a combustion process is in vapor state at the end of combustion, as opposed to the higher heating value (*HHV*) (also known as gross calorific value or gross CV), which assumes that all of the water in a combustion process is in a liquid state after a combustion process.

Heating value is the amount of heat produced by a complete combustion of *fuel* and is measured as a unit of energy per unit mass or volume of substance (e.g., kcal/kg, kJ/kg, J/mole, and Btu/m^3). The heat of *combustion of fuels* is expressed by the HHV and *LHV*.

The cost expected in bulk storage facilities such as caverns can be estimated from the development of salt caverns in natural gas storage of €250–750 per cubic meter of working gas capacity [1]. From the capital and operating cost reported in for natural gas cavern storage, storage costs of approximately €0.75/kg can be extrapolated for hydrogen when assuming the same volume-specific cost as for natural gas.

The cost of hydrogen is heavily influenced by the primary energy used for production. At the present time, fossil primary energies are the least expensive. In the future renewable energies will become more and more cost competitive.

Besides cost of production, lack of infrastructure is currently one of the most serious technical obstacles for a broad direct introduction of hydrogen. Options such as the substitution of fossil-based hydrocarbon fuels by synthetic fuels using hydrogen as an intermediate might become a pathway into a hydrogen-based future energy world.

8.2 Hydrogen Storage Technologies

As already stated, demand for hydrogen for consumption as a clean source of energy comes with demand for the storage of this clean source of energy, whether for vehicular onboard storage or stationary bulk storage, such as refueling stations for the vehicles or side-by-side hydrogen production plants for use in industry.

The storage of hydrogen has been problematic due to its low volumetric density and resulting high cost, and in this section, I evaluate the overview of hydrogen storage technologies, both from mobile and stationary point of views.

The following methods of hydrogen storage are of interest [2]:

- *Liquid hydrogen (LH_2)*: Liquid hydrogen storage is currently the bulk hydrogen storage medium of choice and has a very impressive safety record. The hydrogen is typically liquefied at the production site in large quantities (10–30 tons per day) and then trucked cross-country in 11,000 gal LH_2 tankers with no boil-off losses. Unfortunately, the energy requirements of liquefaction are high, typically 30% of the hydrogen's heating value, leading to relatively high hydrogen cost compared with gaseous hydrogen. LH_2 will likely remain the main technique of bulk, stationary hydrogen storage for the foreseeable future.

Vehicular LH_2 systems have the highest hydrogen (H_2) mass fractions and one of the lowest system volumes, along with near-zero development risk, good fast fill capability, and acceptable safety characteristics. They seem to be an excellent choice except for two adverse factors: dormancy and infrastructure impact. Dormancy concerns arise due to boil-off losses that will inevitably concern the average car owner, although daily use or proper planning for route or fleet applications can remove most if not all dormancy concerns: (i) the liquefaction process is costly; (ii) small-scale LH_2 production is impractical; and (iii) low-volume distribution/dispensing of LH_2 is expensive. Consequently, LH_2 systems will not easily support a transition from weak start-up to a robust H_2 economy. Overall, LH_2 storage is a most appropriate for a mature H_2 economy where the inherent difficulties (and high cost) of large-scale remote LH_2 production and very small-scale LH_2 dispensing are least encountered.

- *Compressed Gaseous Hydrogen (GH_2)*: Vehicular compressed hydrogen systems consisting of 34.5 MPa (5000 psi) gaseous hydrogen in metal- or plastic-lined, carbon fiber wound pressure vessels offer simplicity of design and use, high H_2 fraction, rapid refueling capability, excellent dormancy characteristics, minimal infrastructure impact, high safety due to the inherent strength of the pressure vessel, and little to no development risk. The disadvantages are system volume and use of high pressure. Integrating the moderate-to-large system volume will clearly challenge the automotive designer, but such a tank volume can be packaged into a "clean sheet" vehicle. In this author's opinion, the many advantageous features of compressed gas storage outweigh its larger volume. Compressed gas storage is supportable by small-scale H_2 production facilities (onsite natural gas reforming plants, partial oxidation burners, and electrolysis stations) as well as larger-scale LH_2 production facilities. Thus, a plausible H_2 infrastructure transition pathway exists. For these reasons, room temperature compressed gas storage is viewed as the most appropriate fuel storage system for PEM fuel cell vehicles.

  For stationary hydrogen storage, GH_2 also offers the advantages of simplicity and stable storage (no boil-off losses) but at a considerably greater volume than LH_2. Even accounting for compression costs, high pressure gaseous hydrogen is cheaper than LH_2. However, except for pipeline transmission, GH_2 lacks the bulk transportability of LH_2. Consequently, GH_2 will mostly be employed for storage of limited hydrogen quantities, for long-term storage, or when the cost of liquefaction is prohibitive. Remaining issues for GH_2 include its safety perception, and the current high cost of the pressure vessels and hydrogen compressors.
- *Metal Hydrides:* Metal hydrides can be subdivided into two categories: low dissociation-temperature hydrides and high dissociation-temperature hydrides. The low-temperature hydrides suffer from low H_2 fraction (−2%). The high-temperature hydrides require a heat source to generate the high temperature of dissociation (−300 °C). Both systems offer fairly dense H_2 storage and good

safety characteristics. Indeed, it is the bad characteristics of dissociation (high temperature, high energy input) that create the good safety characteristics (no or slow H_2 release in a crash). Overall for vehicular hydrogen storage, metal hydrides are either much too heavy or their operating requirements are poorly matched to PEM vehicle systems. Without a dramatic breakthrough achieving high weight fraction, low temperature, low dissociation energy, and fast charge time, metal hydrides will not be an effective storage medium for PEM fuel cell vehicles. For stationary storage, the high weight of metal hydride system is not an adverse factor. Consequently, their attributes of high volumetric storage density and stability make them quite attractive. Improving resistance to gaseous contaminants and increasing system cycle life remain as obstacles to overcome.

- *Carbon Adsorption:* Gaseous hydrogen can be adsorbed onto the surface of carbon to attain storage volumetric densities greater than liquid hydrogen. Adhesion capacity is greatly increased by low temperature (particularly cryogenic temperatures) and by high pressure. Indeed, significant fractions of the hydrogen contained in carbon adsorbent systems are actually held in gaseous form within the interstitial volume of the carbon adsorbent. Carbon nanofibers are a special type of carbon adsorbent systems which may exploit a fundamentally different mechanism of hydrogen storage and thereby achieve dramatically improved storage capability. However, development and evaluation of nanofibers is at an early stage of development and system characterization is speculative.
- *Microspheres:* Microsphere hydrogen storage systems consist of hollow glass spheres that are "charged" with hydrogen (300–500 °C, 27–62 MPa for an hour), and discharged by heating (200–250 °C) and reducing pressure. The microspheres can be pumped or poured from one tank to another, making them viable for vehicular hydrogen storage. Overall, system characterization is immature. Microsphere shelf life remains a concern.

In summary, multiple techniques of hydrogen storage are viable for both vehicular storage and bulk stationary storage. However, no one storage mechanism is ideal. As demand for hydrogen grows, industry must respond by supplying (and storing) hydrogen is ways suitable for the new class of consumers and must educate the public in its safe use [2].

8.3 How Does Hydrogen Storage Work?

Hydrogen storage is a key enabling technology for the advancement of hydrogen and fuel cell technologies in applications including stationary power, portable power, and transportation.

Hydrogen has the highest energy per mass of any fuel; however, its low ambient temperature density results in a low energy per unit volume, requiring the development of advanced storage methods that have potential for higher energy density.

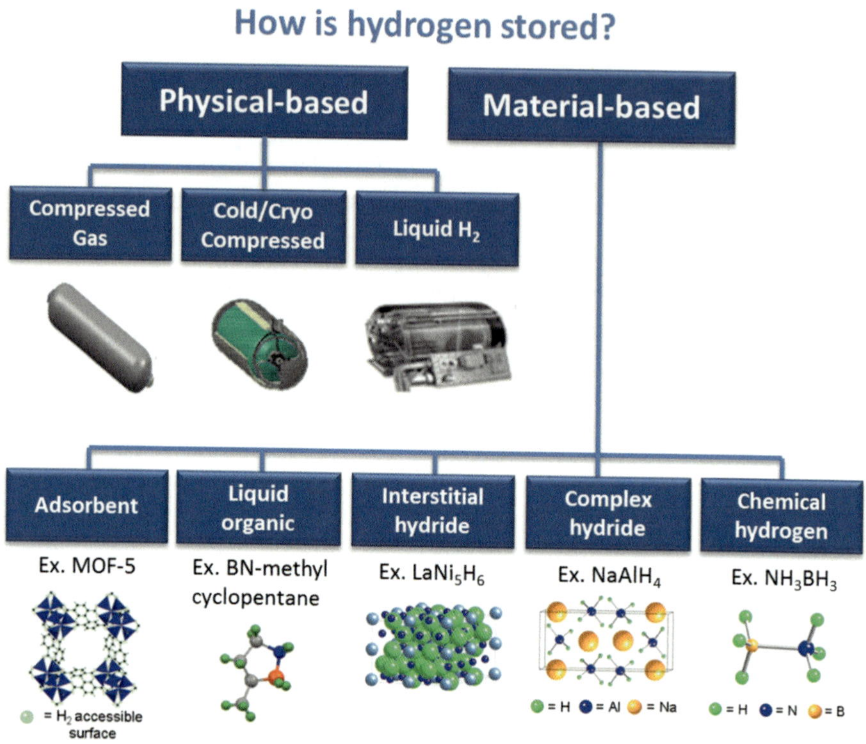

Fig. 8.1 Hydrogen storage technologies overview. (Courtesy of the US Department of Energy)

Hydrogen can be stored physically as either a gas or a liquid. Storage of hydrogen as a gas typically requires high-pressure tanks (350–700 bar [5000–10,000 psi] tank pressure). Storage of hydrogen as a liquid requires cryogenic temperatures because the boiling point of hydrogen at 1 atmosphere pressure is −252.8 °C. Hydrogen can also be stored on the surfaces of solids (by adsorption) or within solids by absorption.

How to store hydrogen efficiently, economically, and safely is one of the challenges to be overcome to make hydrogen an economic source of energy. An overview of present hydrogen storage technologies, namely high-pressure gas compression, liquefaction, metal hydride storage, and carbon nanotube adsorption is illustrated in Fig. 8.1.

The energy efficiency, economic aspects, and environmental and safety issues of various hydrogen storage technologies have been compared. Presently, high-pressure gas compression is favorable due to its high energy efficiency as well as low capital and operation costs. Liquefaction is mainly employed in space applications because of its high volumetric and gravimetric efficiency. The disadvantages are low energy efficiency and high cost. Due to their high volumetric efficiency, metal hydride storage and carbon nanotube adsorption are promising hydrogen storage technologies and are expected to play a key role in hydrogen economy in the future.

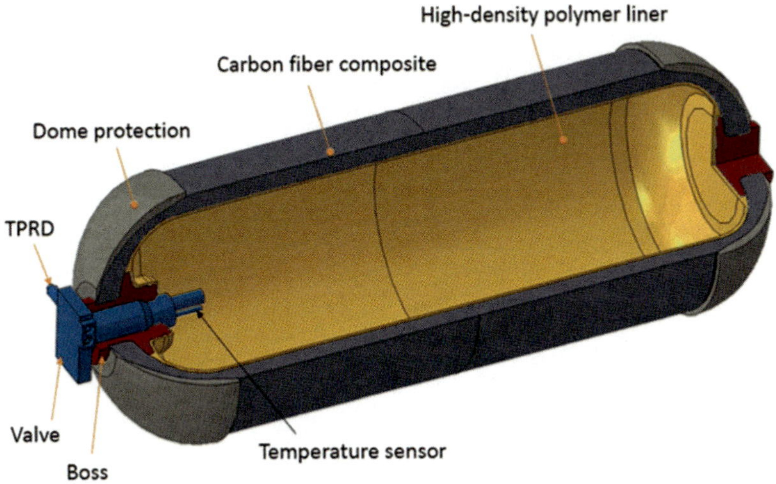

Fig. 8.2 Thermally activated pressure relief device (TPRD). (Courtesy of Process Modeling Group, Nuclear Engineering Division of Argonne National Laboratory)

8.4 Physical Hydrogen Storage

Physical storage is the most mature hydrogen storage technology. The current near-term technology for onboard automotive physical hydrogen storage is 350 and 700 bar (5000 and 10,000 psi) nominal working-pressure compressed gas vessels—that is, "tanks."

While low-pressure liquid hydrogen, near the normal boiling point of 20 K, is routinely used for bulk hydrogen storage and transport, there is currently little activity in developing it for onboard automotive use. Although compressed hydrogen typically is stored at near-ambient temperatures, "cold" (sub-ambient but greater than 150 K) and "cryogenic" (150 K and below) compressed hydrogen storage are being investigated due to the higher hydrogen densities that can be achieved at reduced temperatures (Fig. 8.2).

8.5 Technical Targets and Status

Compressed hydrogen storage systems have been demonstrated in hundreds of prototype fuel cell vehicles and are available commercially at low production volumes. While physical storage has not yet met all of the DOE targets for onboard automotive storage, many targets have been achieved with only a few key areas requiring further improvement, including gravimetric density, volumetric density, and cost.

Table 8.1 Project performance table[a]

Storage system targets	Gravimetric density [kWh/kg system ($kg\ H_2/kg$ system)]	Volumetric density [kWh/L system ($kg\ H_2/L$ system)]	Cost [US$/kWh (US$/kg H_2)]
2020	1.5 (0.045)	1.0 (0.030)	10 (333)
Ultimate	2.2 (0.065)	1.7 (0.050)	$8 (266)
Current status (from Argonne National Laboratory)			
700 bar compressed (type IV, single tank)	1.4 (0.042)	0.8 (0.024)	$15[c] ($500)

Courtesy of the US Department of Energy
[a]Assumes a storage capacity of 5.6 kg of usable hydrogen
[b]Cost projections are estimated at 500,000 units per year and are reported in 2007$
[c]Cost projection from Strategic Analysis (November 2015)

8.5.1 Status and Targets for Gravimetric Density, Volumetric Density, and Cost

Table 8.1 provides the projected performance and cost of compressed automotive hydrogen storage systems compared to 2020 and ultimate DOE targets.

8.5.2 Performance Against All Department of Energy 2010 Onboard Vehicle Storage Targets

The system projections graph in Fig. 8.3 shows how a conventional 700 bar Type IV compressed hydrogen storage system at 300 K compares against all of the DOE's 2020 onboard vehicle storage targets. The blue space indicates current performance and the white space indicates the areas in which 700 bar compressed systems currently fall short of the DOE's 2020 targets.

8.5.3 Distribution of 350 Bar and 700 Bar Compressed Hydrogen System Costs

The cost of current compressed gas systems for automotive applications is dominated by the carbon fiber composite with a significant impact from balance of plant components. Current physical storage research and development efforts focus on reducing the cost of the fiber-reinforced composite portion of the pressure vessel to help meet DOE targets. Figure 8.4 shows cost distributions for 350 bar and 700 bar Type IV single-tank compressed hydrogen systems at 500,000 units [3].

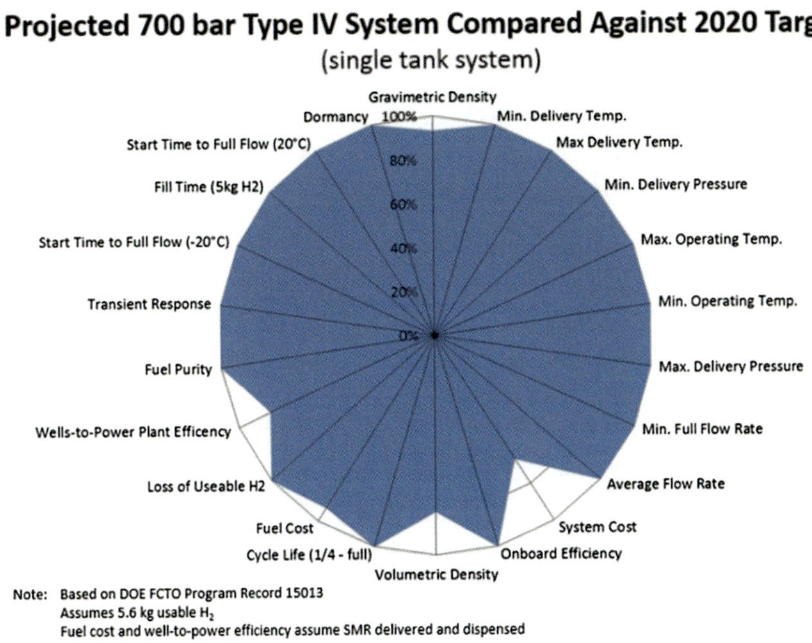

Fig. 8.3 Performance against all US Department of Energy (DOE) 2020 targets. (Courtesy of the US Department of Energy)

Fig. 8.4 700 bar compressed hydrogen storage system cost breakout based on single tank system, from 2015 Department of Energy (DOE) Fuel Cell Technologies Office (FCTO) Record #15013 [3]. BOP balance of plant. (Courtesy of the US Department of Energy)

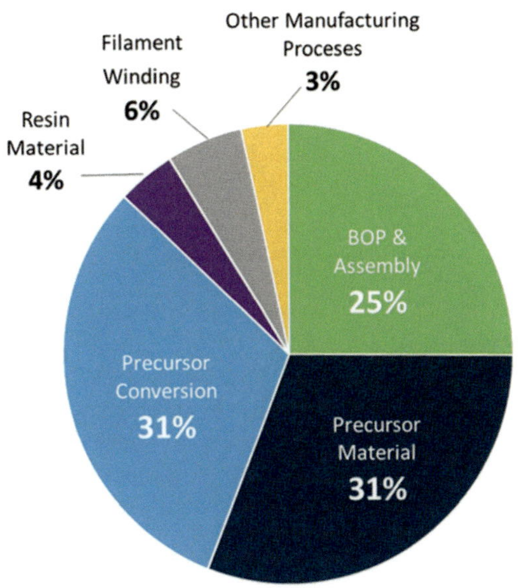

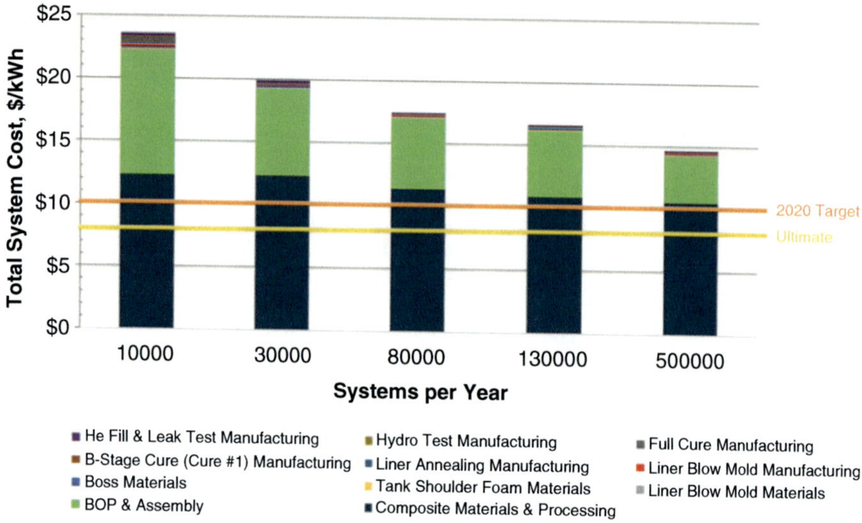

Fig. 8.5 700 bar compressed hydrogen storage system cost breakout by volume from 2015 Department of Energy (DOE) Fuel Cell Technologies Office (FCTO) Record #15013 [3]. BOP balance of plant. (Courtesy of the US Department of Energy)

8.5.4 System Cost Based on Production Volume

The Fuel Cell Technologies Office (FCTO) has also conducted analysis to determine the cost for the low volumes that are expected during the initial ramp up of fuel cell electric vehicles (FCEVs). Figure 8.5 shows how the estimated system costs vary based on production volume for 350 bar (top) and 700 bar (bottom) compressed hydrogen storage systems [3].

8.5.5 Pathways to Reduced Cost

Finally, the DOE has determined potential pathways to reduce the cost of compressed tanks. This includes reducing the cost of carbon fiber composites and/or developing lower-cost alternative fiber-reinforced composites, better utilizing and therefore reducing the amount the fiber reinforcement included in the tank, and reducing the amount and cost of balance of plant components. Figure 8.6 shows one potential cost reduction strategy for 700 bar compressed systems.

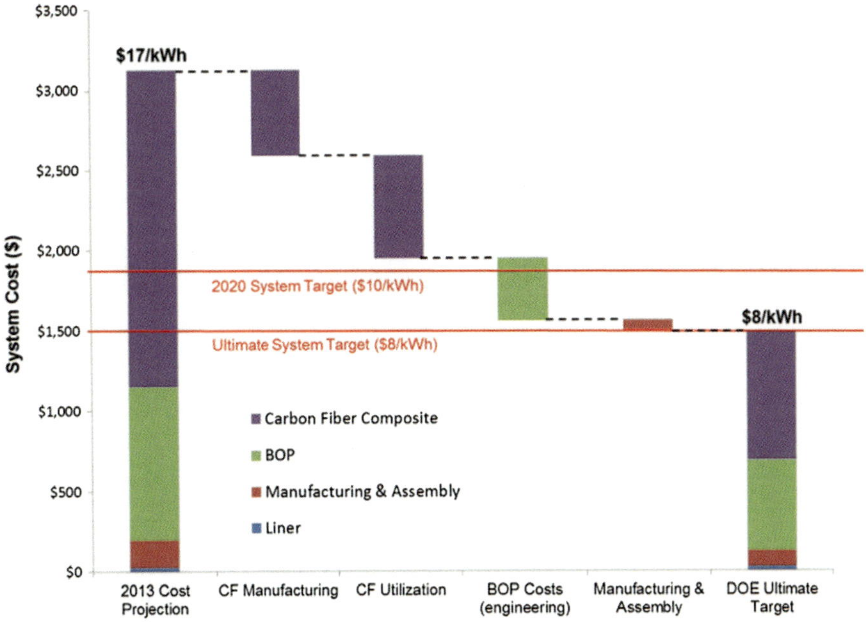

Fig. 8.6 Potential pathways to cost reduction. BOP balance of plant, CF carbon fiber, DOE Department of Energy. (Courtesy of the US Department of Energy)

8.6 Research and Development Goals

The FCTO conducts research and development activities to advance hydrogen storage systems technology and develop novel hydrogen storage materials (see Sect. 8.7). The goal is to provide adequate hydrogen storage to meet the DOE hydrogen storage targets for onboard light-duty vehicle (see Sect. 8.8), material-handling equipment (see Sect. 8.9), and portable power (see Sect. 8.10) applications. By 2020, the FCTO aims to develop and verify onboard automotive hydrogen storage systems achieving targets that will allow hydrogen-fueled vehicle platforms to meet customer performance expectations for range, passenger and cargo space, refueling time, and overall vehicle performance.

Specific system targets include the following:

- 1.5 kWh/kg system (4.5 wt.% hydrogen)
- 1.0 kWh/L system (0.030 kg hydrogen/L)
- US$10/kWh (US$333/kg stored hydrogen capacity).

Details on DOE-funded hydrogen storage activities can be found elsewhere [4].

8.7 Materials-Based Hydrogen Storage

The FCTO's applied materials-based hydrogen storage technology research, development, and demonstration (RD&D) activities focus on developing materials and systems that have the potential to meet DOE 2020 light-duty vehicle system targets with an overarching goal of meeting ultimate full-fleet, light-duty vehicle system targets.

Materials-based research is currently being pursued on metal hydride, chemical hydrogen storage, and sorbent materials.

- *Metal hydride materials* research focuses on improving the volumetric and gravimetric capacities, hydrogen adsorption/desorption kinetics, cycle life, and reaction thermodynamics of potential material candidates.
- *Chemical hydrogen storage materials* research focuses on improving volumetric and gravimetric capacity, improving transient performance, reducing release of volatile impurities, and developing efficient regeneration processes for the spent storage material.
- *Sorbent materials* research focuses on increasing effective adsorption temperature through increase of the dihydrogen binding energies and improving volumetric and gravimetric storage capacities through optimizing the material's pore size, increasing pore volume and surface area, and investigating effects of material densification.

A key component for advancing storage materials is the use of reliable material property measurement techniques. It is imperative to understand how the hydrogen storage properties of a material can be significantly influenced by not only individual sample characteristics—including chemical composition and distribution and microscopic and macroscopic material structure—but also pressure, temperature, and sample size. To help researchers better understand the proper measurement techniques, the FCTO commissioned a best practices manual [4] that gives a detailed overview of the recommended best practices in measuring the hydrogen storage properties of materials.

In 2015, the FCTO launched the Hydrogen Materials Advanced Research Consortium (HyMARC) [5], a collaborative research effort comprising a core of national laboratories and competitively selected individual projects. HyMARC will focus on conducting foundational research to understand the interaction of hydrogen materials in relation to the formation and release of hydrogen from hydrogen storage materials.

8.7.1 Technical Targets and Status

Materials-based research offers a long-term solution to the challenge of onboard automotive storage, as well as opportunities for stationary and portable power applications, with the potential to significantly reduce the required storage pressure,

Table 8.2 Projected performance and cost of materials-based automotive hydrogen storage systems compared to the 2020 and ultimate US Department of Energy (DOE) targets[a]

Storage system targets	Gravimetric density [kWh/kg system (kg H_2/kg system)]	Volumetric density [kWh/L system (kg H_2/L system)]	Cost [US $/kWh (US $/kg H_2)]
2020	1.5 (0.045)	1.0 (0.030)	10 (333)
Ultimate	2.2 (0.065)	1.7 (0.050)	8 (266)
Current status (from HSECoE)			
Metal hydride: $NaAlH_4$	0.4 (0.012)	0.4 (0.012)	43 (1430)
Sorbent: MOF-5, 100 bar, 80 K	1.3 (0.038)	0.7 (0.021)	15 (490)
Chemical hydrogen storage: Off-board regenerable[a, b]	1.5 (0.046)	1.3 (0.040)	17 (550)

HSECoE Hydrogen Storage Engineering Center of Excellence
[a]Assumes a storage capacity of 5.6 kg of usable hydrogen
[b]Metal hydride reflects status at the end of phase I; chemical hydrogen and sorbent reflect status at the end of phase II

increase gravimetric and volumetric capacity, and reduce cost. From 2005 through 2010, the DOE Hydrogen Storage Program supported three collaborative efforts—the Metal Hydride Center of Excellence [6], the Hydrogen Sorption Center of Excellence [7], and the Chemical Hydrogen Storage Center of Excellence [8]—as well as independent projects that investigated more than 400 materials for potential use in hydrogen storage applications. Analysis activities in the collaborative Hydrogen Storage Engineering Center of Excellence (HSECoE) [9], which conducts analysis activities to determine the current status of materials-based storage system technologies, have determined the current status of systems using these materials. HSECoE has also developed system projection graphs [10] showing three modeled systems for each material class and how they compare against all of the DOE's 2020 targets.

Table 8.2 is presenting the DOE, project performance and cost of materials-based automotive hydrogen storage systems compared to the 2020 and ultimate DOE target. See the comments at the bottom of Table 8.2.

Figure 8.7 shows hydrogen gravimetric capacity as a function of hydrogen release temperature for many of the unique hydrogen storage materials investigated by the FCTO.

8.8 Onboard Hydrogen Storage for Light-Duty Vehicles

The DOE technical targets for onboard hydrogen storage for light-duty vehicles are summarized in the Table 8.3. These targets were established through the US DRIVE Partnership, a partnership between the DOE, the United States Council for

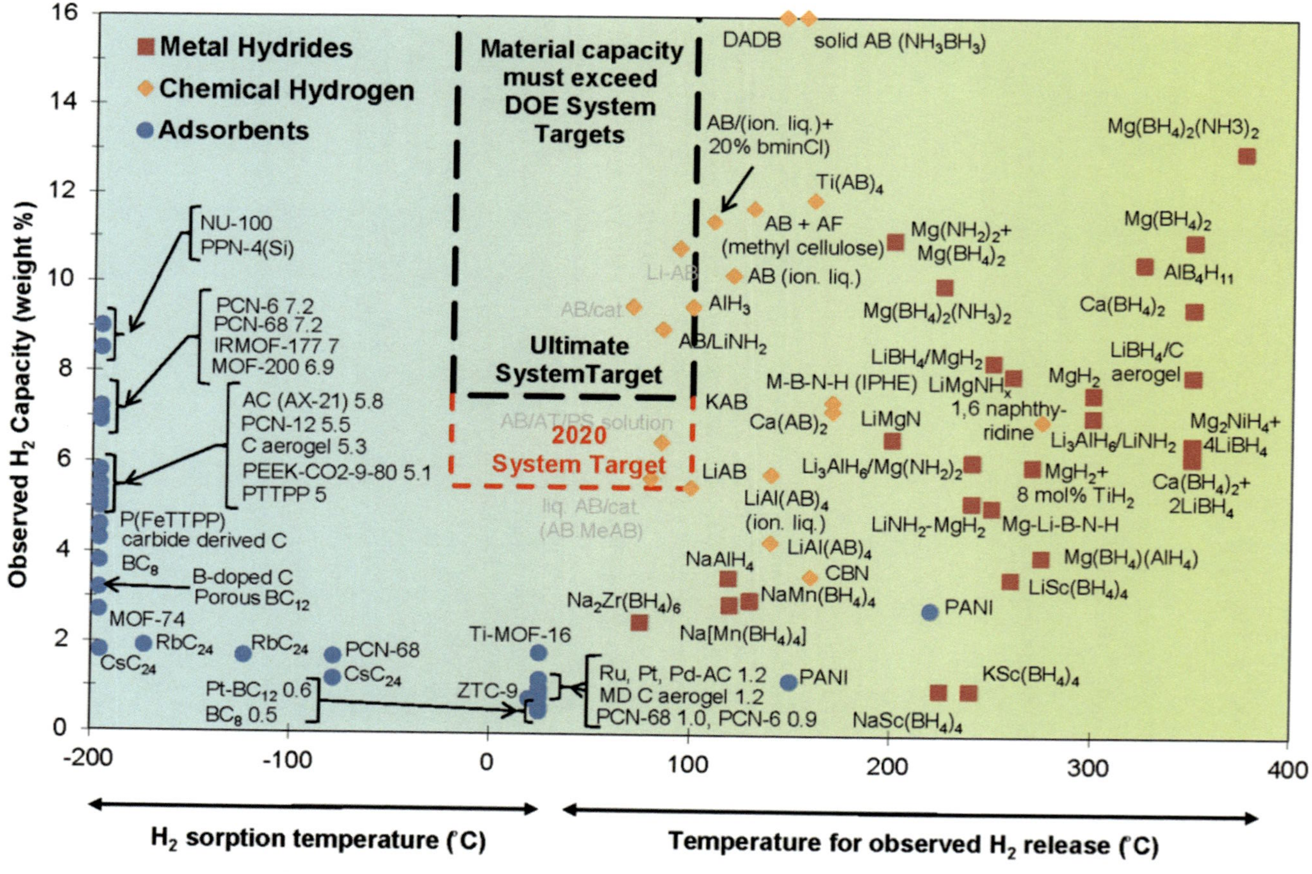

Fig. 8.7 Hydrogen gravimetric capacity as a function of hydrogen release temperature for hydrogen storage

Table 8.3 Technical system targets: onboard hydrogen storage for light-duty fuel cell vehicles[a]

Storage parameter	Units	2020	2025	Ultimate
System gravimetric capacity				
Usable, specific-energy from H_2 (net useful energy/ max system mass)[b]	kWh/kg (kg H_2/kg system)	1.5 (0.045)	1.8 (0.055)	2.2 (0.065)
System volumetric capacity				
Usable energy density from H_2 (net useful energy/ maximum system volume)[b]	kWh/L (kg H_2/L system)	1.0 (0.030)	1.3 (0.040)	1.7 (0.050)
Storage system cost				
Fuel cost[c]	US$/kWh net (US$/kg H_2) US$/gge at pump	10 333 4	9 300 4	8 266 4
Durability/operability				
Operating ambient temperature[d]	°C	−40/ 60 (sun)	−40/ 60 (sun)	−40/60 (sun)
Minimum/maximum delivery temperature	°C	−40/ 85	−40/ 85	−40/85
Operational cycle life (1/4 tank to full)	Cycles	1500	1500	1500
Minimum delivery pressure from storage system	Bar (abs)	5	5	5
Maximum delivery pressure from storage system	Bar (abs)	12	12	12
Onboard efficiency[e]	%	90	90	90
"Well" to power plant efficiency[f]	%	60	60	60
Charging/discharging rates				
System fill time[g]	Min	3–5	3–5	3–5
Minimum full flow rate (e.g., 1.6 g/s target for 80 kW rated fuel cell power)	(g/s)/kW	0.02	0.02	0.02
Average flow rate	(g/s)/kW	0.004	0.004	0.004
Start time to full flow (20 °C)	s	5	5	5
Start time to full flow (−20 °C)	s	15	15	15
Transient response at operating temperature 10–90% and 90–100% (based on full flow rate)	s	0.75	0.75	0.75
Fuel quality				
Fuel quality (H_2 from storage)[h]	% H_2	Meet or exceed SAE J2719		
Dormancy				
Dormancy time target (minimum until first release from initial 95% usable capacity)	Days	7	10	14
Boil-off loss target (maximum reduction from initial 95% usable capacity after 30 days)	%	10	10	10
Environmental health and safety				
Permeation and leakage[i]	–	Meet or exceed SAE J2579 for system safety		

(continued)

Table 8.3 (continued)

Storage parameter	Units	2020	2025	Ultimate
Toxicity	–	Meet or exceed applicable standards		
Safety	–	Conduct and evaluate failure analysis		

SAE Society of Automotive Engineers

Useful constants: 0.2778 kWh/MJ; Lower heating value for H_2 is 33.3 kWh/kg H_2; 1 kg $H_2 \approx$ 1 gal gasoline equivalent (gge) on energy basis.

[a]For a normalized comparison of system performance to the targets, a usable H_2 storage capacity of 5.6 kg H_2 should be used at the lower heating value of hydrogen (33.3 kWh/kg H_2). Targets are for a complete system, including tank, material, valves, regulators, piping, mounting brackets, insulation, added cooling capacity, and all other balance-of-plant components. All capacities are defined as usable capacities that could be delivered to the fuel cell system. All targets must be met at the end of service life

[b]Capacities are defined as the usable quantity of hydrogen deliverable to the fuel cell system divided by the total mass/volume of the complete storage system, including all stored hydrogen, media, reactants (e.g., water for hydrolysis-based systems), and system components. Capacities must be met at end of service life. Tank designs that are conformable and have the ability to be efficiently packaged on board vehicles may be beneficial even if they do not meet the full volumetric capacity targets

[c]Hydrogen threshold fuel cost is calculated to be competitive with a gasoline hybrid vehicle, and thus is independent of pathway. It is defined as the untaxed cost of hydrogen produced, delivered, and dispensed to the vehicle [11]. For material-based storage technologies, the impact of the technology on the hydrogen threshold fuel cost (e.g., off-board cooling, off-board regeneration of chemical hydrogen storage materials, etc.) must be taken into account

[d]Stated ambient temperature plus full solar load (i.e., full exposure to direct sunlight). No allowable performance degradation from −20 to 40 °C. Allowable degradation outside these limits is to be determined

[e]Onboard efficiency is the energy efficiency for delivering hydrogen from the storage system to the fuel cell power plant, i.e., accounting for any energy required operating pumps, blowers, compressors, heating, etc., required for hydrogen release

[f]Well-to-power-plant efficiency includes onboard efficiency plus off-board efficiency, i.e., accounting for the energy efficiency of hydrogen production, delivery, liquefaction, compression, dispensing, regeneration of chemical hydrogen storage materials, etc., as appropriate. The Hydrogen Analysis (H2A) Project within DOE and Hydrogen Delivery Scenario Analysis Model (HDSAM) analyses should be used for projecting off-board efficiencies. Efficiencies less than the target may be acceptable if evidence can be given that well-to-power-plant carbon intensity (including delivery and dispensing of H_2) can achieve less than 5 kg CO_{2e}/kg H_2. Argonne National Laboratory's GREET model [12] should be used to calculate the carbon intensity of well-to-power-plant energy use

[g]When applicable, the fill time should comply with SAE J2601 [13], the Fueling Protocol for Light-Duty Gaseous Hydrogen Surface Vehicles

[h]Hydrogen storage systems must be able to deliver hydrogen that meets acceptable hydrogen quality standards for fuel cell vehicles (see SAE J2719 [14] and ISO/PDTS 14687–2 [15]). Note that some storage technologies may produce contaminants for which effects are unknown and not addressed by the published standards; these will be addressed by system engineering design on a case-by-case basis as more information becomes available

[i]Total hydrogen lost into the environment as H_2; relates to hydrogen accumulation in enclosed spaces. Storage systems must comply with applicable standards for vehicular fuel systems including but not limited to SAE J2579 [16] and the United Nations Global Technical Regulation No. 13 (hydrogen and fuel cell vehicles). This includes any coating or enclosure that incorporates the envelope of the storage system

[j]Dormancy targets assume vehicle is parked in 35 °C ambient temperature and dormancy performance is maintained over the 15-year life of the vehicle

Automotive Research (USCAR), energy companies, and utility companies and organizations. A detailed explanation of these targets and the process used in deriving them can be found in the Hydrogen Storage section of the Fuel Cell Technologies Office's Multi-Year Research, Development, and Demonstration Plan [17–19].

8.9 Material Handling Equipment

Table 8.4 summarizes hydrogen storage technical performance targets for material handling equipment. These targets were developed with input to the DOE through extensive communication with various stakeholders, industry developers, and end users, including through a 2012 request for information and workshops, as well as additional national laboratory assessments.

Table 8.4 Technical system targets[a], material handling equipment

Storage parameter	Units	2015	2020
Hydrogen capacity			
Hydrogen capacity	Kg	2	2
System volumetric capacity			
Usable energy density from H_2 (net useful energy/maximum system volume)[b]	kWh/L (kg H_2/L system)	1.0 (0.03)	1.7 (0.05)
Storage system cost			
System cost	US$/kWh net (US$/kg $H_{2\ stored}$)	20 (667)	15 (500)
Durability/operability			
External operating temperature range[c]	°C	−40/60	−40/60
Minimum/maximum delivery temperature[d]	°C	−40/85	−40/85
Operational cycle life (1/10 tank to full)	Cycles	5000 (5 years)	10,000 (10 years)
Minimum delivery pressure from storage system	Bar (abs)	3	3
Maximum delivery pressure from storage system	Bar (abs)	12	12
Shock and vibration			
Shock	g	40	40
Vibration	g	5@10 Hz–0.75@200 Hz	10@10 Hz–1@200 Hz
Charging/discharging rates			
System fill time (2 kg)	Min (kg H_2/min)	4.0 (0.5)	2.8 (0.7)
Minimum full flow rate	(g/s)/kW	0.02	0.02
Start time to full flow (20 °C)	s	5	5
Start time to full flow (−20 °C)	s	15	15
Transient response 10–90% and 90%–0%	s	0.75	0.75

(continued)

Table 8.4 (continued)

Storage parameter	Units	2015	2020
Fuel purity			
Fuel purity (H_2 from storage)[e]	% H_2	SAE J2719 and ISO/PDTS 14687–2 (99.97% dry basis)	
Environmental health and safety			
Permeation and leakage[f]	–	Meets or exceeds applicable standards (e.g., CSA HPIT 1)	
Toxicity			
Safety			
Loss of usable H_2[g]	(g/h)/kg $H_{2\ stored}$	0.1	0.05

SAE Society of Automotive Engineers; *ISO* The International Organization for Standardization; *PDTS* Preliminary Draft Technical Specification; *HPIT* Hydrogen Powered Industrial Trucks; *CSA* Canadian Standards Association

[a]The targets are based on the lower heating value of hydrogen, without consideration of the conversion efficiency of the fuel cell power plant. Targets are for the complete hydrogen storage and delivery system, including tank, material, valves, regulators, piping, mounting brackets, insulation, added cooling or heating capacity, and/or other balance-of-plant components. All capacities are defined as usable capacities that could be delivered to the fuel cell power plant during normal use. All targets must be met at the end of service life. Since most applications of material handling equipment (MHE) require extra mass as a counterbalance, the system gravimetric capacity is not specified as it can vary widely among types of MHE. However, system gravimetric capacity should be considered when developing hydrogen storage systems for MHE applications. All targets must be met at the end of service life

[b]"Net useful energy" or "net" excludes unusable energy (i.e., hydrogen left in a tank below minimum fuel cell power plant pressure, flow, and temperature requirements) and hydrogen-derived energy used to extract the hydrogen from the storage medium (e.g., fuel used to heat a material to initiate or sustain hydrogen release)

[c]Stated ambient temperature. No allowable performance degradation from −20 °C to 40 °C. Allowable degradation outside these limits is to be determined

[d]Delivery temperature refers to the inlet temperature of the hydrogen to the fuel cell

[e]Hydrogen storage systems must be able to deliver hydrogen meeting acceptable hydrogen quality standards, such as CSA HPIT 1: Compressed Hydrogen Powered Industrial Trucks (forklifts) On-Board Fuel Storage and Handling Components. Note that some storage technologies may produce contaminants for which effects are unknown and not addressed by the published standards; these will be addressed by system engineering design on a case-by-case basis as more information becomes available

[f]Total hydrogen lost into the environment as H_2; relates to hydrogen accumulation in enclosed spaces. Storage system must comply with appropriate standards, for example CSA HPIT 1: Compressed Hydrogen Powered Industrial Trucks (forklifts) On-Board Fuel Storage and Handling Components. This includes any coating or enclosure that incorporates the envelope of the storage system

[g]Total hydrogen lost from the storage system, including leaked or vented hydrogen; relates to loss of operational time

8.10 Portable Power Equipment

Tables 8.5, 8.6, and 8.7 summarize the DOE hydrogen storage technical performance targets for portable power applications. These targets were developed with input to DOE through extensive communications with various stakeholders, industry developers, and end users, including through a 2012 request for information and workshops, as well as additional national laboratory assessments.

Table 8.5 Technical performance targets[a]: hydrogen storage systems for low-power (≤2.5 W) portable equipment

Storage parameter	Units	2015		2020	
		Single-use	Rechargeable	Single-use	Rechargeable
Hydrogen capacity					
Hydrogen capacity	g H_2	≤1	≤1	≤1	≤1
System gravimetric capacity[b]					
Usable, specific-energy from H_2 (net useful energy/maximum system mass)[c]	kWh/kg (kg H_2/kg system)	0.7 (0.02)	0.5 (0.015)	1.3 (0.04)	1.0 (0.03)
System volumetric capacity					
Usable energy density from H_2 (net useful energy/maximum system volume) [c]	kWh/L (kg H_2/L system)	1.0 (0.03)	0.7 (0.02)	1.7 (0.05)	1.3 (0.04)
Storage system cost					
System cost	US$/Wh net (US$/g $H_{2\ stored}$)				

Table 8.6 Technical performance targets[a]: hydrogen storage systems for medium-power (>2.5–150 W) portable equipment

Storage parameter	Units	2015		2020	
		Single-use	Rechargeable	Single-use	Rechargeable
Hydrogen capacity					
Hydrogen capacity	g H_2	>1–50	>1–50	>1–50	>1–50
System gravimetric capacity[b]					
Usable, specific-energy from H_2 (net useful energy/maximum system mass)[c]	kWh/kg (kg H_2/kg system)	0.7 (0.02)	0.5 (0.015)	1.3 (0.04)	1.0 (0.03)
System volumetric capacity					
Usable energy density from H_2 (net useful energy/maximum system volume)[c]	kWh/L (kg H_2/L system)	1.0 (0.03)	0.7 (0.02)	1.7 (0.05)	1.3 (0.04)
Storage system cost					
System cost	US$/Wh net (US$/g $H_{2\ stored}$)				

Table 8.7 Portable power durability and operational targets[a]

Storage parameter	Units	2015 Single-use and rechargeable	2020 Single-use and rechargeable
Durability/operability			
External operating temperature range[b]	°C	−40/60	−40/60
Minimum/maximum delivery temperature[c]	°C	10/85	10/85
Minimum delivery pressure from storage system	Bar (abs)	1.5	1.5
Maximum delivery pressure from storage system	Bar (abs)	3	3
External temperature[d]	°C	≤40	≤40
Discharging rates			
Minimum full flow rate	(g/s)/kW	0.02	0.02
Start time to full flow (20 °C)	s	5	5
Start time to full flow (−20 °C)	s	10	10
Transient response 10–90% and 90–0%	s	5	2
Fuel purity			
Fuel purity (H_2 from storage)[e]	% H_2	Meets applicable standards	
Environmental health and safety			
Toxicity		Meets ISO-16111:2008; IEC 62282 part 6; or other applicable standards as appropriate or required for the application and targeted usage	
Safety			
Loss of usable H_2[f]			

[a]The targets are based on the lower heating value of hydrogen, without consideration of the conversion efficiency of the fuel cell power plant. Targets are for the complete hydrogen storage and delivery system, including tank, material, valves, regulators, piping, mounting brackets, insulation, added cooling or heating capacity, and/or other balance-of-plant components. All capacities are defined as usable capacities that could be delivered to the fuel cell power plant during normal use. All targets must be met at the end of service life

[b]Stated ambient temperature plus full solar load (i.e., if exposed to direct sunlight or stored within a container exposed to direct sunlight for extended periods of time). No allowable performance degradation from −20 °C to 40 °C. Allowable degradation outside these limits is to be determined

[c]Delivery temperature refers to the inlet temperature of the hydrogen to the fuel cell

[d]The external device temperature is the maximum temperature generated at the external surface of the hydrogen storage container during operation

[e]Hydrogen storage systems must be able to deliver hydrogen meeting acceptable hydrogen quality standards, such as ISO-16111:2008 and IEC 62282 Part 6. Note that some storage technologies may produce contaminants for which effects are unknown and not addressed by the published standards; these will be addressed by system engineering design on a case-by-case basis as more information becomes available

[f]Total hydrogen lost into the environment as H_2; relates to hydrogen accumulation in enclosed spaces. Storage system must comply with appropriate standards, such as ISO-16111:2008 and IEC 62282 Part 6. This includes any coating or enclosure that incorporates the envelope of the storage system

[g]Generally the "full" mass (including hydrogen) is used; for systems that gain weight on hydrogen release, the highest mass during discharge is used (e.g., hydrogen release through hydrolysis reaction resulting in the formation of oxides/hydroxides). All capacities are net usable capacity able to be delivered to the fuel cell power plant. Capacities must be met at end of service life

[h]"Net useful energy" or "net" excludes unusable energy (i.e., hydrogen left in a tank below minimum fuel cell power plant pressure, flow, and temperature requirements) and hydrogen-derived energy used to extract the hydrogen from the storage medium (e.g., fuel used to heat a material to initiate or sustain hydrogen release)

8.11 High-Density Hydrogen Storage Challenges

High-density hydrogen storage is a challenge for stationary and portable applications and remains a significant challenge for transportation applications. Presently, available storage options typically require large-volume systems that store hydrogen in gaseous form. This is less of an issue for stationary applications, where the footprint of compressed gas tanks may be less critical.

However, as Fig. 8.8 indicates, fuel cell-powered vehicles require enough hydrogen to provide a driving range of more than 300 miles with the ability to quickly and easily refuel the vehicle. While some light-duty hydrogen FCEVs that are capable of this range have emerged onto the market, these vehicles will rely on compressed gas onboard storage using large-volume, high-pressure composite vessels. The required large storage volumes may have less impact for larger vehicles but providing sufficient hydrogen storage across all light-duty platforms remains a challenge. The importance of the 300-mile range goal can be appreciated by looking at the sales distribution by range shown in Fig. 8.8, which shows that most vehicles sold today are capable of exceeding this minimum.

On a mass basis, hydrogen has nearly three times the energy content of gasoline—120 MJ/kg for hydrogen versus 44 MJ/kg for gasoline. On a volume basis, however, the situation is reversed; liquid hydrogen has a density of 8 MJ/L whereas gasoline has a density of 32 MJ/L, as shown in Fig. 8.9 comparing energy densities of fuels based on lower heating values. Onboard hydrogen storage capacities of 5–13 kg hydrogen will be required to meet the driving range for the full range of light-duty vehicle platforms.

To overcome these challenges, the FCTO is pursuing two strategic pathways, targeting both near-term and long-term solutions. The near-term pathway focuses on compressed gas storage, using advanced pressure vessels made of fiber-reinforced composites that are capable of reaching 700 bar pressure, with a major emphasis on system cost reduction.

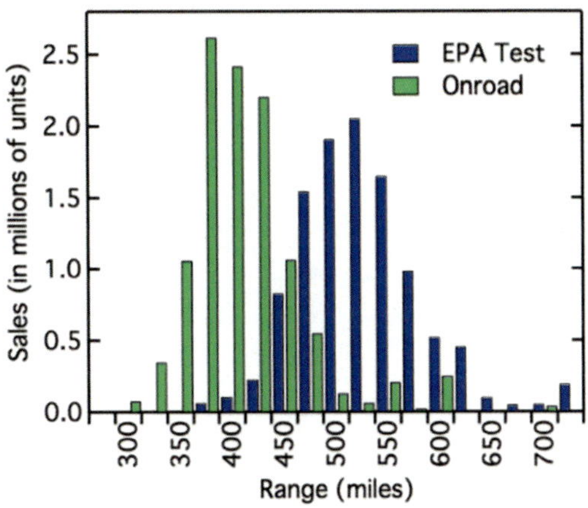

Fig. 8.8 2010 United States light-duty vehicle sales distribution by driving range

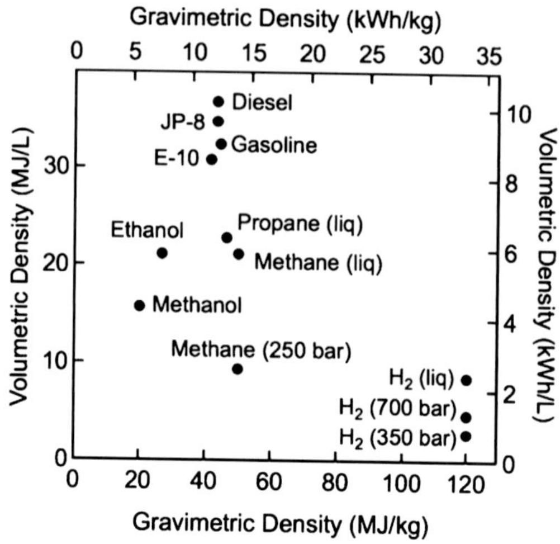

Fig. 8.9 Volumetric density of different fuel types

The long-term pathway focuses on both:

1. Cold or cryo-compressed hydrogen storage, where increased hydrogen density and insulated pressure vessels may allow for DOE targets to be met; and
2. Materials-based hydrogen storage technologies, including sorbents, chemical hydrogen storage materials, and metal hydrides (see Sect 8.7), with properties having potential to meet DOE hydrogen storage targets.

References

1. Federal Energy Regulatory Commission. Current state of and issues concerning underground natural gas storage (2004)
2. B.D. James, *Overview of Hydrogen Storage Technologies* (Directed Technologies, Inc., Arlington)
3. Onboard type IV compressed hydrogen storage system-cost and performance status 2015, DOE Hydrogen and Fuel Cells Program Record #: 15013, Dated: 9/30/2015, Update to: 13010, Originated by Grace Odaz, Cassidy Hochins, and Thanh Hua
4. https://www.energy.gov/eere/fuelcells/downloads/recommended-best-practices-characterization-storage-properties-hydrogen-0
5. https://www.energy.gov/eere/fuelcells/hymarc-hydrogen-materials-advanced-research-consortium
6. L. Klebanoff, J. Keller, Final report for the DOE Metal Hydride Center of Excellence, Sandia Report, SAND2012–0786, Unlimited Release, Printed February 2012
7. L. Simpson, *Hydrogen Sorption Center of Excellence (HSCoE)*, U.S. Department of Energy Office of Energy Efficiency and Renewable Energy Fuel Cell Technologies Program
8. K.C. Ott, *Final report for the DOE Chemical Hydrogen Storage Center of Excellence*, LA-UR-2274, Los Alamos National Laboratory
9. https://www.energy.gov/eere/fuelcells/hydrogen-storage-engineering-center-excellence

10. https://www.energy.gov/eere/fuelcells/hydrogen-storage-engineering-center-excellence#graphs
11. Hydrogen threshold cost calculation. Program Record (Offices of Fuel Cell Technologies) Record#: 11007, Dated: March 25, 2011, Originator: Mark Ruth and Fred Joseck
12. https://greet.es.anl.gov/
13. https://www.sae.org/standards/content/j2601_201003/
14. https://saemobilus.sae.org/content/J2719_201511
15. https://www.iso.org/standard/43994.html
16. https://saemobilus.sae.org/content/J2579_201303
17. Fuel Cell Technologies Office. Multi-year research, development, and demonstration plan. www.vehicles.energy.gov/about/partnerships/usdrive.html or www.uscar.org
18. https://www.energy.gov/eere/fuelcells/downloads/fuel-cell-technologies-office-multi-year-research-development-and-22
19. https://www.energy.gov/eere/fuelcells/hydrogen-storage-related-links

Index

© Springer International Publishing AG, part of Springer Nature 2019

B. Zohuri, *Hydrogen Energy*, https://doi.org/10.1007/978-3-319-93461-7

Made in the USA
Middletown, DE
06 August 2021

45489030R00170